Algebra II

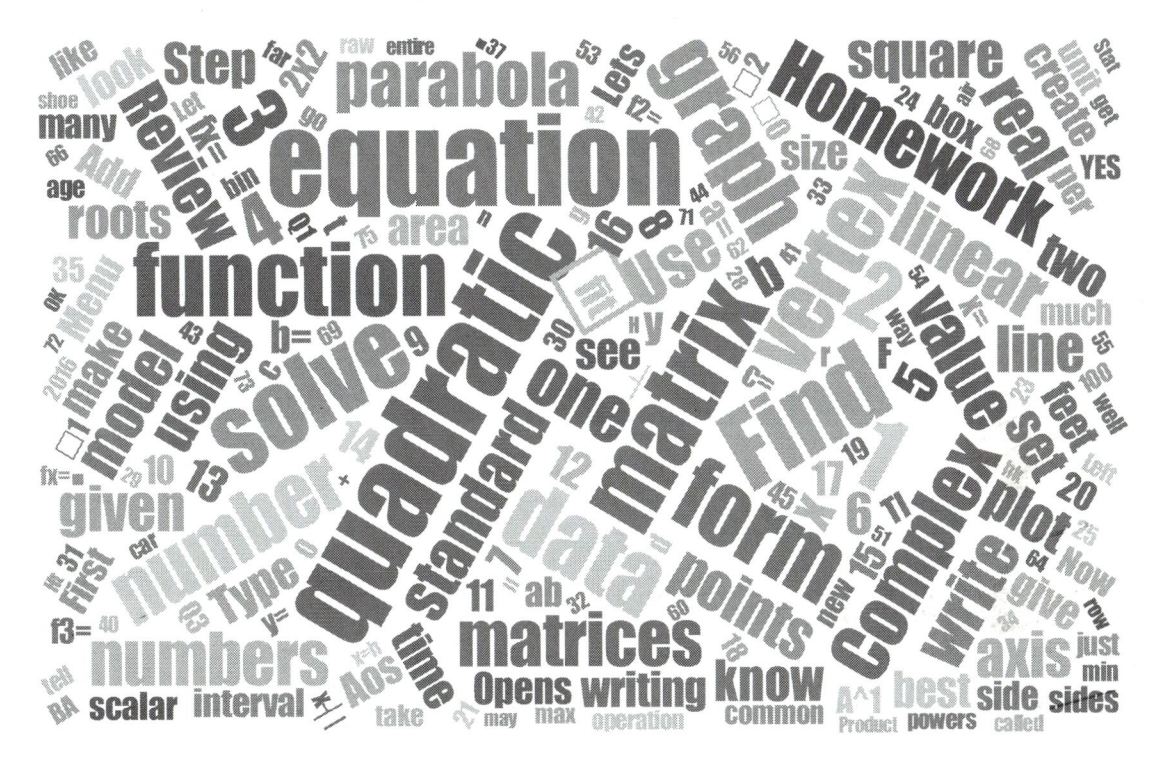

Woodward Academy Fall 2019

Robert Daugherty
Ken Kirschner

WOODWARD ACADEMY
ESTABLISHED 1900

Copyright © 2019 Robert Daugherty, Ken Kirschner

All rights reserved. No part of this publication may be reproduced, distributed, or transmitted in any form or by any means, including photocopying, recording, or other electronic or mechanical methods, without the prior written permission of the publisher, except in the case of brief quotations embodied in critical reviews and certain other noncommercial uses permitted by copyright law. For permission requests, write to the publisher, addressed "Attention: Permissions Coordinator," at the address below.

ISBN-13:978-1074496753

Any references to historical events, real people, or real places are used fictitiously. Names, characters, and places are products of the author's imagination.

Publisher
PO Box 87392
College Park, GA, 30337

Algebra II

TABLE OF CONTENTS

1. *Matrices* .... *1.1—6*
    1.1. Introduction, Addition and Subtraction, Scalar Multiplication .... 1.1—6
    1.2. Matrix Multiplication .... 1.2—14
    1.3. Multiplicative Properties, Determinants and Inverses .... 1.3—22
    1.4. Matrix Equations .... 1.4—34
    1.5. Solving Systems of Linear Equations with Matrices .... 1.5—40
    1.6. Solving Systems of Linear Equations with the TI-Nspire .... 1.6—48
    1.7. Review .... 1.7—57
2. *Linear Programming* .... *1.7—61*
    2.1. Systems of Inequalities, Feasible Regions, Objective Functions .... 2.1—61
    2.2. Linear Programming Application: Maximization .... 2.2—72
    2.3. Linear Programming Application: Minimization .... 2.3—80
    2.4. Review .... 2.4—90
    2.5. Review Additional Linear Programming Problems .... 2.5—96
3. *Data Analysis* .... *2.5—102*
    3.1. One variable statistics, Dot Plots, & Box Plots .... 3.1—103
    3.2. Histograms, Interpreting Two-variable data, and Scatter Plots .... 3.2—118
    3.3. Lines of Best Fit & Linear Regression Models .... 3.3—130
    3.4. Monopoly Line of Best Fit .... 3.4—142
    3.5. Review .... 3.5—143
4. *Piecewise Linear Functions* .... *3.5—148*
    4.1. Relations/Function/Domain/Range .... 4.1—148
    4.2. Introduction to piecewise functions .... 4.2—158
    4.3. Graphing Piecewise Functions .... 4.3—168
    4.4. Writing Equations of Piecewise Functions .... 4.4—178
    4.5. Review .... 4.5—186

5. Quadratic Functions .... 4.5—192

- 5.1. Complex Numbers, Addition, Subtraction, Multiplication & Powers ... 5.1—192
- 5.2. Complex Numbers, Division .... 5.2—201
- 5.3. Graphing Vertex Form of Quadratics .... 5.3—207
- 5.4. Graphing Intercept Form of Quadratics .... 5.4—217
- 5.5. Graphing Standard Form of Quadratics, Quadratic Inequalities .... 5.5—225
- 5.6. Review .... 5.6—235

6. Quadratic Equations .... 5.6—240

- 6.1. Solving Quadratic Equations by Square Roots .... 6.1—241
- 6.2. Solve by factoring .... 6.2—249
- 6.3. Solve by Completing the Square .... 6.3—256
- 6.4. Solve by quadratic formula .... 6.4—265
- 6.5. The Discriminant .... 6.5—272
- 6.6. Solve by any method .... 6.6—281
- 6.7. Review .... 6.7—288
- 6.8. Writing Quadratic equations .... 6.8—293
- 6.9. Converting between forms for Quadratic equations .... 6.9—301
- 6.10. Review .... 6.10—310

7. Quadratic Modeling .... 6.10—317

- 7.1. Working with specified quadratic models .... 7.1—318
- 7.2. Finding quadratic models from data with the TI-Nspire .... 7.2—329
- 7.3. Projectile Motion .... 7.3—335
- 7.4. Quadratic Modeling with Borders .... 7.4—344
- 7.5. Quadratic Modeling with fencing .... 7.5—349
- 7.6. Review .... 7.6—355
- 7.7. Algebra 2 Grades .... 7.7—360

ALGEBRA II

Day One

The height of a basketball shot, $H(t)$, is given by the equation
$$H(t) = -7t^2 + 16t + 6$$
where t = time in seconds that the shot is in the air, and H = the height in feet after t seconds. Assume the rim of the basket is 10 feet tall.

What is the initial height of the basketball?	What is the approximate maximum height of the basketball?	How much time elapses from the moment the shot is taken until it enters the basket?
a. 5 feet b. 10 feet (c.) 6 feet d. Not enough information *plug in 0 for t*	a. ~~10 feet~~ b. ~~6 feet~~ (c.) 15 feet d. ~~50 feet~~	a. ~~.5 seconds~~ b. ~~5 seconds~~ c. ~~10 seconds~~ (d.) 2 seconds
What shoes should the shooter wear? a. Nike Air Jordan, circa 1998 b. Nike Lebron VIII, circa 2010 c. Under Armour Curry 6 (d.) Woodward Regulation Sperry Boat Shoes		

Over the course of the semester we'll fill in a lot of the details that allow you to interpret much more about this problem.

1. MATRICES

1.1. Introduction, Addition and Subtraction, Scalar Multiplication

A **matrix is a rectangular array of numbers**. By the end of this unit we will use matrices to solve systems of linear equations like the one below:

$$x + 2y + z = 5$$
$$2x + 3y + z = 9$$
$$5x + 6y - z = 11$$

With a good deal of algebraic manipulation, we can solve the system. But much of the work is with the coefficients, not the actual variables, $x, y,$ and z.

We will learn to write this system as a **matrix equation** and solve it.

$$\begin{bmatrix} 1 & 2 & 1 \\ 2 & 3 & 1 \\ 5 & 6 & -1 \end{bmatrix} \begin{bmatrix} x \\ y \\ z \end{bmatrix} = \begin{bmatrix} 5 \\ 9 \\ 11 \end{bmatrix}$$

By using matrices, we will work only with the numbers, and turn much of that algebra into simple arithmetic. But before we can work a matrix equation like the one above, we must cover some basic properties of matrices.

Definition: A **matrix** is a rectangular array of numbers.

Matrices are a way to represent data in array form.

$$\begin{array}{c} \downarrow \text{Column 1} \\ \text{Row 1} \to \begin{bmatrix} 2 & -1 \\ \pi & 7 \end{bmatrix} \\ \text{Row 2} \to \uparrow \text{Column 2} \end{array}$$

The horizontal components of a matrix are called: __row__

The vertical components of a matrix are called: __column__

1.1—7

The dimensions of a matrix, or its size, is described by the number of rows and columns.

Consider matrix A below.

$$A = \begin{bmatrix} a_{11} & a_{12} \\ a_{21} & a_{22} \\ a_{31} & a_{32} \end{bmatrix}$$

Number of rows in matrix A	Number of columns in matrix A	Dimension of A
3	2	3 × 2

We would describe A as being a 3x2, or "three by two" matrix.

How many numbers or elements are in a 3x2 matrix? __6__

The element a_{11} is in the __1__ row and __1__ column.

The element a_{32} is in the __3__ row and __2__ column.

Consider $B = \begin{bmatrix} 1 & 2 & 5 \\ x & 23 & -7 \end{bmatrix}$

Number of rows in matrix B	Number of columns in matrix B	Dimension of B
2	3	2 × 3

Square matrices have the same number of rows and columns. The dimensions are n x n.

Row matrices have only one row. Their dimensions are 1 x n.

Column matrices have only one column. Their dimensions are n x 1.

Classify the matrices below as either square, column, or row matrices and give the dimensions.

$\begin{bmatrix} 1 & -2 \\ 4 & x \end{bmatrix}$	$\begin{bmatrix} 1 & 3 & 0 \end{bmatrix}$	$\begin{bmatrix} 0 \\ 0 \\ 0 \\ 1 \\ 2 \end{bmatrix}$	$\begin{bmatrix} 1 & 0 & 0 \\ 0 & 1 & 0 \\ 0 & 0 & 1 \end{bmatrix}$
2 x 2	1 x 3	5 x 1	3 x 3
Square	Row	Column	Square
$\begin{bmatrix} 1 & -1 & -2 \\ 4 & 0 & x \end{bmatrix}$	$\begin{bmatrix} 1 & 0 & 3 & 0 \end{bmatrix}$	$[0]$	$\begin{bmatrix} 1 & 4 & 0 & 0 \\ 0 & 5 & 1 & 0 \\ 0 & 8 & 0 & 1 \\ 4 & 91 & 2 & 3 \end{bmatrix}$
2 x 3	1 x 4	1 x 1	4 x 4
n/a	Row	Square Row Column	Square

Equality of Matrices

Matrices are equal if they have the same dimension, and every element of the matrices is equal.

Under what conditions are the two matrices below equal?

$$\begin{bmatrix} 1 & x \\ y & 4 \end{bmatrix} = \begin{bmatrix} 1 & 5 \\ 7 & 4 \end{bmatrix}$$
$$2 \times 2 \qquad 2 \times 2$$

$x = 5$
$y = 7$

Addition and Subtraction of Matrices

Matrices may be combined through subtraction and addition if the dimensions are identical. For addition or subtraction, you add/subtract each corresponding element.

$$\begin{bmatrix} 1 & x \\ y & 4 \end{bmatrix} + \begin{bmatrix} 1 & 5 \\ 7 & 4 \end{bmatrix} = \begin{bmatrix} 2 & x+5 \\ y+7 & 8 \end{bmatrix}$$
$$2 \times 2 \qquad 2 \times 2 \qquad\qquad 2 \times 2$$

Consider the matrices defined below. Perform the operations where defined. Write "undefined" if an operation cannot be performed.

$$A = \begin{bmatrix} 2 & 4 \\ -1 & 7 \end{bmatrix} \quad B = \begin{bmatrix} 6 & 0 \\ 1 & 2 \\ -1 & -3 \end{bmatrix} \quad C = \begin{bmatrix} -2 & 0 \\ 1 & -4 \end{bmatrix} \quad D = \begin{bmatrix} x & 0 \\ -1 & 2 \\ 1 & \pi \end{bmatrix}$$

$A + C$	$\begin{bmatrix} 2 & 4 \\ -1 & 7 \end{bmatrix} + \begin{bmatrix} -2 & 0 \\ 1 & -4 \end{bmatrix} = \begin{bmatrix} 0 & 4 \\ 0 & 3 \end{bmatrix}$ 2×2 2×2 2×2
$D - B$	$\begin{bmatrix} x & 0 \\ -1 & 2 \\ 1 & \pi \end{bmatrix} - \begin{bmatrix} 6 & 0 \\ 1 & 2 \\ -1 & -3 \end{bmatrix} = \begin{bmatrix} x-6 & 0 \\ -2 & 0 \\ 2 & \pi+3 \end{bmatrix}$ 3×2 3×2 3×2
$B + C$ 3×2 + 2×2	undefined
$A + D$ 2×2 + 3×2	undefined
$D + B$	$\begin{bmatrix} x & 0 \\ -1 & 2 \\ 1 & \pi \end{bmatrix} + \begin{bmatrix} 6 & 0 \\ 1 & 2 \\ -1 & -3 \end{bmatrix} = \begin{bmatrix} x+6 & 0 \\ 0 & 4 \\ 0 & \pi-3 \end{bmatrix}$ 3×2 3×2 3×2

1.1—10

Scalar Multiplication

To wrap up this lesson, scalar multiplication refers to the multiplication of a matrix by a single value.

How do you think you would perform the operation below?

$$2\begin{bmatrix} 2 & 4 \\ -1 & 7 \end{bmatrix} = \begin{bmatrix} 4 & 8 \\ -2 & 14 \end{bmatrix}$$

With scalar multiplication, each element of the matrix is multiplied by the scalar value.

$$5\begin{bmatrix} 6 & x \\ -3 & 0 \end{bmatrix} = \begin{bmatrix} 30 & 5x \\ -15 & 0 \end{bmatrix}$$

So, putting it all together, how do we solve the equations below?

Solve for x and y.	Solve for matrix A.
$2\left(\begin{bmatrix} 2x & -3 \\ 5 & -y \end{bmatrix} + \begin{bmatrix} -1 & 4 \\ 3 & 5 \end{bmatrix}\right) = \begin{bmatrix} 10 & 2 \\ 16 & 14 \end{bmatrix}$	$\begin{bmatrix} 3 \\ 7 \\ -2 \end{bmatrix} + A = \begin{bmatrix} 5 \\ 2 \\ -3 \end{bmatrix}$
$2\left(\begin{bmatrix} 2x-1 & 1 \\ 8 & -y+5 \end{bmatrix}\right) = \begin{bmatrix} 10 & 2 \\ 16 & 14 \end{bmatrix}$	$3 \times 1 \qquad 3 \times 1$
$\begin{bmatrix} 4x-2 & 2 \\ 16 & -2y+10 \end{bmatrix} = \begin{bmatrix} 10 & 2 \\ 16 & 14 \end{bmatrix}$	$A = \begin{bmatrix} 5 \\ 2 \\ -3 \end{bmatrix} - \begin{bmatrix} 3 \\ 7 \\ -2 \end{bmatrix}$
$4x-2 = 10 \qquad -2y+10 = 14$	$A = \begin{bmatrix} 2 \\ -5 \\ -1 \end{bmatrix}$
$4x = 12 \qquad -2y = 4$	
$\boxed{x = 3} \qquad \boxed{y = -2}$	

1.1—11

Homework 1.1 Matrices Introduction, Addition, Subtraction, Scalar Multiplication

Perform the following matrix operations. Write the dimension under each matrix. Write "undefined" for any operation that is not possible.

1. $\begin{bmatrix} -6 & -6 & 3 \\ -5 & 4 & -4 \end{bmatrix} + \begin{bmatrix} 2 & 4 & 3 \\ 1 & -4 & -2 \end{bmatrix} = \begin{bmatrix} -4 & -2 & 6 \\ -4 & 0 & -6 \end{bmatrix}$

   2×3   2×3

2. $[1 \quad -4 \quad 1] + \begin{bmatrix} -2 & -2 \\ 1 & -2 \end{bmatrix} =$ undefined

   1×3   2×2

3. $\begin{bmatrix} 5 & -6 \\ 0 & -4 \end{bmatrix} - \begin{bmatrix} 6 & -2 \\ 0 & -1 \end{bmatrix} = \begin{bmatrix} -1 & -4 \\ 0 & -3 \end{bmatrix}$

   2×2   2×2

4. $[-5 \quad 0 \quad 5 \quad -3] + [-1 \quad -1 \quad 3 \quad -3] = [-6 \quad -1 \quad 8 \quad -6]$

   1×4   1×4

5. $\begin{bmatrix} 3 \\ 2 \\ 1 \end{bmatrix} - 2\begin{bmatrix} 0 \\ -1 \\ 2 \end{bmatrix} = \begin{bmatrix} 3 \\ 2 \\ 1 \end{bmatrix} - \begin{bmatrix} 0 \\ -2 \\ -4 \end{bmatrix} = \begin{bmatrix} 3 \\ 0 \\ 5 \end{bmatrix}$

   1×3   1×3

6. $[-1 \quad -1 \quad 3 \quad -3] - ([2 \quad 0 \quad 1 \quad -3] + [1 \quad -1 \quad -2 \quad -5]) =$

   $[-1 \quad -1 \quad 3 \quad -3] - [3 \quad -1 \quad -1 \quad -8]$

   $[-4 \quad 0 \quad 4 \quad 5]$

7. $\begin{bmatrix} 1 & 2 \\ 0 & -1 \end{bmatrix} + 7[-1 \quad -1] =$ undefined

   2×2   1×2

Homework 1.1 Matrices Introduction, Addition, Subtraction, Scalar Multiplication

8. $\begin{bmatrix} -4 & 0 & 1 \\ -2 & 4 & -4 \end{bmatrix} - 3\begin{bmatrix} 2 & 4 & 3 \\ 1 & -4 & -2 \end{bmatrix} =$ $3\begin{bmatrix} 2 & 4 & 3 \\ 1 & -4 & -2 \end{bmatrix} = \begin{bmatrix} 6 & 12 & 9 \\ 3 & -12 & -6 \end{bmatrix}$

   2×3   2×3

   $\begin{bmatrix} -4 & 0 & 1 \\ -2 & 4 & -4 \end{bmatrix} - \begin{bmatrix} 6 & 12 & 9 \\ 3 & -12 & 6 \end{bmatrix} = \begin{bmatrix} -10 & -12 & -8 \\ -5 & 16 & -10 \end{bmatrix}$

9. $4\left(\begin{bmatrix} 1 \\ -1 \end{bmatrix} - 2\begin{bmatrix} 3 \\ -3 \end{bmatrix}\right) =$ $\left(\begin{bmatrix} 1 \\ -1 \end{bmatrix} - \begin{bmatrix} 6 \\ -6 \end{bmatrix}\right) 4$

   $\left(\begin{bmatrix} -5 \\ 5 \end{bmatrix}\right) 4 = \begin{bmatrix} -20 \\ 20 \end{bmatrix}$

Solve each equation

10. $\begin{bmatrix} 1 \\ 2 \end{bmatrix} + A = \begin{bmatrix} 3 \\ 3 \end{bmatrix}$ $A = \begin{bmatrix} 3 \\ 3 \end{bmatrix} - \begin{bmatrix} 1 \\ 2 \end{bmatrix}$

    $-\begin{bmatrix} 1 \\ 2 \end{bmatrix} \quad -\begin{bmatrix} 1 \\ 2 \end{bmatrix}$ $A = \begin{bmatrix} 2 \\ 1 \end{bmatrix}$

11. $\begin{bmatrix} 3x+1 \\ 8y \\ 1 \\ 3z-8 \end{bmatrix} = \begin{bmatrix} -17 \\ -32 \\ 1 \\ -25 \end{bmatrix}$

    $3x+1 = -17 \qquad 8y = -32$
    $3x = -18 \qquad\quad y = 4$
    $x = -6$

    $3z - 8 = -25$
    $3z = -17$
    $z = -17/3$

12. $\begin{bmatrix} 1 & -3 \end{bmatrix} - 3Z = \begin{bmatrix} 4 & -18 \end{bmatrix}$

    $\begin{bmatrix} 1 & -3 \end{bmatrix} - 3\begin{bmatrix} z_1 & z_2 \end{bmatrix} = \begin{bmatrix} 4 & -18 \end{bmatrix}$

    $1 - 3z_1 = 4 \qquad -3z_2 = -15$
    $-3z_1 = 3 \qquad\quad \boxed{z_2 = 5}$
    $\boxed{z_1 = -3}$

Homework 1.1 Matrices Introduction, Addition, Subtraction, Scalar Multiplication

CHALLENGE (Challenge questions in homework are problems that are perhaps a bit deeper or increased level of difficulty than covered in class. They are good practice for the bonus questions on assessments, and good practice for those wanting to solidify the concepts.)

Solve for $x, y,$ and z.

$$\begin{bmatrix} x & 2 \\ -1 & y \end{bmatrix} + \begin{bmatrix} 3y & z \\ 4 & -4x \end{bmatrix} = \begin{bmatrix} -7 & 10 \\ 3 & -11 \end{bmatrix}$$

$$\begin{bmatrix} x+3y & 2z \\ 3 & y-4x \end{bmatrix} = \begin{bmatrix} -7 & 10 \\ 3 & -11 \end{bmatrix}$$

$x + 3y = -7$ $2z = 10$
$x = -7$ $3y = -7$ $z = 5$
         $y = \frac{-7}{3}$

$y - 4x = -11$
$y = -11$ $-4x = -11$
              $x = \frac{11}{4}$

1.2. Matrix Multiplication

Up to this point we've discussed:
    Matrices, definition
    Dimension
    Addition
    Subtraction
    Scalar multiplication

Today we will look at what it means to multiply two matrices.

Consider $A = \begin{bmatrix} 7 & 13 \\ 11 & 14 \end{bmatrix}$ and $B = \begin{bmatrix} 1.25 \\ 2.25 \end{bmatrix}$

         2×2   2×1

What would it mean to multiply A and B?

$$AB = \begin{bmatrix} 7 & 13 \\ 11 & 14 \end{bmatrix} \cdot \begin{bmatrix} 1.25 \\ 2.25 \end{bmatrix}$$

Before we look at the rules and mechanics of matrix multiplication, consider the following scenario of sales of ice cream cones.

Flavor	Small Cones	Large Cones
Vanilla	7	13
Chocolate	11	14

Sales Price	
Small Cone	$1.25
Large Cone	$2.25

What is the revenue from all sales of vanilla?

    7 × 1.25 + 13 × 2.25 = $38

What is the revenue from all sales of chocolate?

    11 × 1.25 + 14 × 2.25 = $45.25

Mathematically, we just did this calculation, multiplying matrices A and B.

$$\begin{bmatrix} 7 & 13 \\ 11 & 14 \end{bmatrix} * \begin{bmatrix} 1.25 \\ 2.25 \end{bmatrix} = \begin{bmatrix} 7*1.25 + 13*2.25 \\ 11*1.25 + 14*2.25 \end{bmatrix} = \begin{bmatrix} 38 \\ 45.25 \end{bmatrix}$$

   2×②   ②×1   2×1
           → must match

Just like with addition and subtraction, there are restrictions on the matrix dimensions that tell us when we can perform matrix multiplication.

$$A \cdot B = AB$$
$$(m \times n) \quad n \times p \quad m \times p$$

*Inside # must match
*Outside = final matrix

In order to perform the multiplication of matrices, the number of columns in the first matrix must match the number of rows in the second matrix.

In other words, when you multiply AB, the inner dimensions of A and B must match, and the outer dimensions give the dimension of the final product of the multiplication.

Consider the matrices below:

$$A = \begin{bmatrix} 1 \\ 4 \\ -7 \end{bmatrix} \quad B = \begin{bmatrix} 3 & \pi \\ 1 & x \end{bmatrix} \quad C = \begin{bmatrix} 2 \\ 2 \end{bmatrix} \quad D = \begin{bmatrix} 1 & 1 & 2 \\ 2 & 7 & -1 \\ 0 & 0 & 3 \end{bmatrix}$$

3×1 2×2 2×1 3×3

In each case below, give the dimensions of each matrix, and the dimension of the product. If the product is undefined, write "Undefined"

A	B	AB
3 x 1	2 x 2	Undefined
B 2×2	C 2×1	BC 2×1
A 3×1	D 3×3	AD undefined
D 3×3	A 3×1	DA 3×1
C 2×1	B 2×2	CB undefined
C 2×1	D 3×3	CD undefined
D 3×3	D 3×3	DD 3×3

1.2—16

Let's look at how to multiply $A = \begin{bmatrix} 2 & 1 \\ -1 & 0 \end{bmatrix}$ times $B = \begin{bmatrix} -1 & 4 \\ 5 & 6 \end{bmatrix}$

Dimensions of A?	Dimensions of B?	Dimensions of AB?
2×2	2×2	2×2

$$A \cdot B = AB$$

$$\begin{bmatrix} 2 & 1 \\ -1 & 0 \end{bmatrix} \begin{bmatrix} -1 & 4 \\ 5 & 6 \end{bmatrix} = \begin{bmatrix} ab_{11} & ab_{12} \\ ab_{21} & ab_{22} \end{bmatrix}$$

Find AB

$$\begin{bmatrix} 2(-1) + 1(5) & 2(4) + 1(6) \\ -1(-1) + 0(5) & -1(4) + 0(6) \end{bmatrix}$$

$$AB = \begin{bmatrix} 3 & 14 \\ 1 & -4 \end{bmatrix}$$

$A = \begin{bmatrix} 2 & 1 \\ -1 & 0 \end{bmatrix}$ $B = \begin{bmatrix} -1 & 4 \\ 5 & 6 \end{bmatrix}$. Find BA

Dimensions of B?	Dimensions of A?	Dimensions of BA?
2×2	2×2	2×2

$$B \cdot A = BA$$

$$\begin{bmatrix} -1 & 4 \\ 5 & 6 \end{bmatrix} \begin{bmatrix} 2 & 1 \\ -1 & 0 \end{bmatrix} = \begin{bmatrix} ba_{11} & ba_{12} \\ ba_{21} & ba_{22} \end{bmatrix}$$

Find BA

$$\begin{bmatrix} -1(2) + 4(-1) & -1(1) + 4(0) \\ 5(2) + 6(-1) & 5(1) + 6(0) \end{bmatrix} = \begin{bmatrix} -6 & -1 \\ 4 & 5 \end{bmatrix}$$

What do you observe about the results of AB vs BA?
they are completely different

1.2—17

$$A = \begin{bmatrix} 0 & 2 \\ 6 & -8 \\ 2 & 4 \end{bmatrix} \quad B = \begin{bmatrix} 1 & 3 \\ -1 & -5 \end{bmatrix}$$

3×2   2×2 = 3×2

$$AB = \begin{bmatrix} ab_{11} & ab_{12} \\ ab_{21} & ab_{22} \\ ab_{31} & ab_{32} \end{bmatrix}$$

Find AB

$$\begin{bmatrix} 0(1)+2(-1) & 0(3)+2(-5) \\ 6(1)+-8(-1) & 6(-1)+-8(-5) \\ 2(1)+4(-1) & 2(3)+4(-5) \end{bmatrix} = \begin{bmatrix} -2 & -10 \\ 14 & 58 \\ -2 & -14 \end{bmatrix}$$

Find BA

$$BA = \begin{bmatrix} ab_{11} & ab_{12} \\ ab_{21} & ab_{22} \\ ab_{31} & ab_{32} \end{bmatrix}$$

$B = 2 \times 2 \quad A = 3 \times 2$

undefined

$$A = \begin{bmatrix} 1 & -1 & 3 \end{bmatrix}, \quad B = \begin{bmatrix} -2 \\ 4 \\ 5 \end{bmatrix}$$

1×3   3×1

$AB = [ab_{11}]$

1×1

Find AB

$$AB = \begin{bmatrix} 1(-2) + -1(4) + 3 \times 5 \end{bmatrix} = \begin{bmatrix} 9 \end{bmatrix}$$

$$C = \begin{bmatrix} 2 & -3 \\ 1 & 5 \end{bmatrix} \quad D = \begin{bmatrix} 1 & -4 \\ 3 & -2 \end{bmatrix}$$
$$2 \times 2 \qquad\qquad 2 \times 2$$

Find CD

$$CD = \begin{bmatrix} cd_{11} & cd_{12} \\ cd_{21} & cd_{22} \end{bmatrix} = \begin{bmatrix} 2(1) - 3(3) & 2(-4) - 3(-2) \\ 1(1) + 5(3) & 1(-4) - 5(-2) \end{bmatrix}$$

$$CD = \begin{bmatrix} -7 & -2 \\ 16 & -14 \end{bmatrix}$$

$$A = \begin{bmatrix} 1 & 2 & 2 \\ 6 & -8 & 6 \\ 2 & 4 & 4 \end{bmatrix} \text{ and } X = \begin{bmatrix} x \\ y \\ z \end{bmatrix} = 3 \times 1$$
$$3 \times 3 \qquad\qquad 3 \times 1$$

Find AX

$$AX = \begin{bmatrix} ax_{11} \\ ax_{21} \\ ax_{31} \end{bmatrix} = \begin{bmatrix} 1x + 2y + 2z \\ 6x + -8y + 6z \\ 2x + 4y + 4z \end{bmatrix}$$

Homework 1.2 Matrix Multiplication

Perform the following matrix multiplications where possible. Write "Undefined" where appropriate.

---

1. $\begin{bmatrix} 4 & -1 \\ -1 & 3 \end{bmatrix} \cdot \begin{bmatrix} -4 & -2 \\ -1 & -6 \end{bmatrix}$
   2×2   2×2   = 2×2

$AB = \begin{bmatrix} ab_{11} & ab_{12} \\ ab_{21} & ab_{22} \end{bmatrix} = \begin{bmatrix} 17 & -2 \\ 1 & -16 \end{bmatrix}$

$\begin{bmatrix} 4(-4)+-1(-1) & 4(-2)+-1(-6) \\ -1(-4)+3(-1) & -1(-2)+3(-6) \end{bmatrix}$

---

2. $\begin{bmatrix} -5 & -5 \\ -5 & 5 \end{bmatrix} \cdot \begin{bmatrix} -3 & -2 \\ -5 & 4 \end{bmatrix}$
   2×2   2×2   = 2×2

$\begin{bmatrix} -5(-3)+-5(-5) & -5(-2)+-5(4) \\ -5(-3)+5(-5) & -5(-2)+5(4) \end{bmatrix}$
 (15)(25) (10)(-20)
 (15)(-25) (10)(20)

$= \begin{bmatrix} 40 & -10 \\ -10 & 30 \end{bmatrix}$

---

3. $\begin{bmatrix} 6 & 3 \\ -2 & -4 \end{bmatrix} \cdot \begin{bmatrix} 6 & -3 \\ -5 & 4 \end{bmatrix}$
   2×2   2×2   = 2×2

$\begin{bmatrix} 6(6)+3(-5) & 6(-3)+3(4) \\ -2(6)+-4(-5) & -2(-3)+-4(4) \end{bmatrix}$
 (36)(-15) (-18)(12)
 (-12)(20) (6)(-16)

$= \begin{bmatrix} 21 & -16 \\ 8 & -10 \end{bmatrix}$

---

4. $\begin{bmatrix} 3 & -5 & 0 & 1 \end{bmatrix} \cdot \begin{bmatrix} -6 & -2 \\ -1 & -6 \end{bmatrix}$
   1×4   2×2

undefined

---

5. $\begin{bmatrix} 4 & 1 \\ 4 & 3 \end{bmatrix} \cdot \begin{bmatrix} 3 & -1 \end{bmatrix}$
   2×2   1×2

undefined

---

6. $\begin{bmatrix} -1 & -4 \\ -4 & 2 \\ -3 & 3 \end{bmatrix} \cdot \begin{bmatrix} -5 & 6 \\ 2 & 2 \end{bmatrix}$
   3×2   2×2   = 3×2

$\begin{bmatrix} -1(-5)+-4(2) & -1(6)+-4(2) \\ -4(-5)+2(2) & -4(6)+2(2) \\ -3(-5)+3(2) & -3(6)+3(2) \end{bmatrix}$
 (5)(-8)   (-6)(-8)
 (20)(4)   (-24)(4)
 (15)(6)   (-18)(6)

$= \begin{bmatrix} -3 & -14 \\ 24 & -20 \\ 21 & -12 \end{bmatrix}$

1.2—20

Homework 1.2 Matrix Multiplication

Perform the following matrix multiplications where possible. Write "Undefined" where appropriate.

---

use calculator!

7. $\begin{bmatrix} 1 & 4 & -5 & 4 \\ 5 & 3 & -6 & 6 \end{bmatrix} \cdot \begin{bmatrix} -2 & 2 \\ 2 & -1 \\ 2 & 6 \\ 4 & -3 \end{bmatrix}$

2×4   4×2

$\begin{bmatrix} 1(-2)+4(2)-5(2)+4(4) & 1(2)+4(-1)-5(6)+4(-3) \\ -28-1016 & 2-4-30-12 \\ 5(-2)+3(2)-6(2)+6(4) & 5(2)+3(-1)-6(6)+6(-3) \\ -106-1224 & 10-3-36-18 \end{bmatrix}$

$= \begin{bmatrix} 12 & -44 \\ 8 & -47 \end{bmatrix}$

---

8. $\begin{bmatrix} -4 & 2 \\ 1 & 4 \\ 1 & 1 \\ -6 & -3 \end{bmatrix} \cdot \begin{bmatrix} -6 & -3 \\ -6 & -6 \end{bmatrix} = 4 \times 2$

4×2   ②×2

$\begin{bmatrix} -4(-6)+2(-6) & -4(-3)+2(-6) \\ 1(-6)+4(-6) & 1(-3)+4(-6) \\ 1(-6)+1(-6) & 1(-3)+1(-6) \\ -6(-6)-3(-6) & -6(-3)-3(-6) \end{bmatrix}$

$= \begin{bmatrix} 12 & 0 \\ -30 & -27 \\ -12 & -9 \\ 54 & 36 \end{bmatrix}$

---

9. $\begin{bmatrix} 0 \\ 1 \\ 2 \\ 3 \end{bmatrix} \cdot \begin{bmatrix} 1 & -1 & 0 & 2 \end{bmatrix}$

4×1   1×4   = 4×4

$\begin{bmatrix} 0 & 0 & 0 & 0 \\ 1(1) & 1(-1) & 0 & 1(2) \\ 2(1) & 2(-1) & 0 & 2(2) \\ 3(1) & 3(-1) & 0 & 3(2) \end{bmatrix} = \begin{bmatrix} 0 & 0 & 0 & 0 \\ 1 & -1 & 0 & 2 \\ 2 & -2 & 0 & 4 \\ 3 & -3 & 0 & 6 \end{bmatrix}$

---

10. $\begin{bmatrix} 1 & 2 & 2 \\ 6 & -8 & 6 \\ 2 & 4 & 4 \end{bmatrix} \cdot \begin{bmatrix} -1 & 0 & 2 \\ 1 & -2 & 1 \\ 3 & 1 & -2 \end{bmatrix}$

3×3   3×3   = 3×3

$\begin{bmatrix} 1(-1)+2(1)+2(3) & 1(0)+2(-2)+2(1) & 1(2)+2(1)+2(-2) \\ 6(-1)+-8(1)+6(3) & 6(0)+-8(-2)+6(1) & 6(2)+-8(1)+6(-2) \\ 2(-1)+4(1)+4(3) & 2(0)+4(-2)+4(1) & 2(2)+4(1)+4(-2) \end{bmatrix}$

$= \begin{bmatrix} 7 & -2 & 0 \\ 4 & 22 & -8 \\ 14 & -4 & 0 \end{bmatrix}$

---

11. $\begin{bmatrix} -1 & 1 & 5 \end{bmatrix} \cdot \begin{bmatrix} 0 \\ 1 \\ 2 \end{bmatrix} = 1 \times 1$ $\begin{bmatrix} -1(0)+1(1)+2(5) \end{bmatrix} = \begin{bmatrix} 11 \end{bmatrix}$

1×3   3×1

Homework 1.2 Matrix Multiplication

CHALLENGE QUESTIONS

1. Consider the matrices below:

Matrix	Dimension
A	$2 \times n$
B	$3 \times m$
C	$4 \times p$
D	3×3

If the product $ABCD$ is defined, find the values of n, m, p and the dimension of $ABCD$.

Can you find any other product of the four matrices where A is not the first matrix in the multiplication? Explain.

2. In matrix theory, a transpose matrix is formed by interchanging the roles of the rows and columns. So, for matrix, A, the first row of A, will become the first column of the transpose of A. We denote the transpose as A^T. A skew-symmetric matrix is one where $A^T = -A$. For matrix $A = \begin{bmatrix} 0 & 2 & -1 \\ -2 & 0 & -4 \\ 1 & 4 & 0 \end{bmatrix}$, find the transpose A^T and show that $A^T = -A$.

1.3. Multiplicative Properties, Determinants and Inverses

How would you answer the following?

$$A = \begin{bmatrix} 2 \\ 3 \end{bmatrix} \quad B = \begin{bmatrix} 1 & 5 \\ 3 & -2 \end{bmatrix} \quad C = \begin{bmatrix} 2 & -3 \\ 7 & 2 \end{bmatrix}$$

Find $A(B - C)$

Find $(B - C)A$

Multiplicative Properties of Matrices

Associative Property of Matrix Multiplication	$A(BC) = (AB)C$
LEFT Distributive Property	$A(B + C) = AB + AC$
RIGHT Distributive Property	$(B + C)A = BA + CA$
Associative Property of Scalar Multiplication	$k(AB) = (kA)B = A(kB)$

Let $A = \begin{bmatrix} 1 & 0 \\ 0 & 1 \end{bmatrix}$ $B = \begin{bmatrix} 2 & -1 \\ -1 & 2 \end{bmatrix}$ $C = \begin{bmatrix} 1 & 2 \\ 3 & 1 \end{bmatrix}$

Prove that $A(B + C) = AB + AC$

$A(B + C)$	$AB + AC$

In addition to its dimensions, the number of rows and columns, matrices have other characteristic values or measures. One of the most important is known as the determinant.

The determinant is a value that only partially describes a **square** matrix...like a person's height. Just as two people can be the same height and look totally different, two matrices can have the same determinant and totally different values. Or they could look very similar and have dramatically different determinants. The determinant is a scalar numerical value that tells us something about the matrix, really much like what volume tells us about a solid in geometry. Though it is beyond the scope of this class, physically, the determinant is a measure of the volume contained by the vectors from the rows of the matrix.

There are transformations that can be applied to matrices, to which the determinant quantifies the extent of the transformation.

For a matrix, A, you might see the notation
$$\det(A) \text{ or } |A|$$
to denote the determinant.

Definition: For a 2x2 matrix, $A = \begin{bmatrix} a_{11} & a_{12} \\ a_{21} & a_{22} \end{bmatrix}$

The ***determinant*** is $\det(A) = a_{11}a_{22} - a_{21}a_{12}$

Or for $A = \begin{bmatrix} a & b \\ c & d \end{bmatrix}$

$$\det(A) = |A| = \det\begin{bmatrix} a & b \\ c & d \end{bmatrix} = ad - bc$$

Find the determinant of $B = \begin{bmatrix} 1 & 5 \\ 3 & -2 \end{bmatrix}$

$\det(B) =$

In the real number system, 1 is the multiplicative identity.
$$1 \cdot x = x \cdot 1$$
Any number multiplied by 1 remains unchanged.

With matrices, there is a multiplicative identity for every square matrix.

Identity Matrices

2x2 identity $\begin{bmatrix} 1 & 0 \\ 0 & 1 \end{bmatrix}$ 3x3 identity $\begin{bmatrix} 1 & 0 & 0 \\ 0 & 1 & 0 \\ 0 & 0 & 1 \end{bmatrix}$ 4x4 identity $\begin{bmatrix} 1 & 0 & 0 & 0 \\ 0 & 1 & 0 & 0 \\ 0 & 0 & 1 & 0 \\ 0 & 0 & 0 & 1 \end{bmatrix}$

Consider the matrix $\begin{bmatrix} a & b \\ c & d \end{bmatrix}$

Find $\begin{bmatrix} 1 & 0 \\ 0 & 1 \end{bmatrix}\begin{bmatrix} a & b \\ c & d \end{bmatrix}$	Find $\begin{bmatrix} a & b \\ c & d \end{bmatrix}\begin{bmatrix} 1 & 0 \\ 0 & 1 \end{bmatrix}$

So, we see that multiplying a matrix by the identity matrix leaves the matrix unchanged, the very definition of a multiplicative identity.

With real numbers, the multiplicative identity tells us that
$$x \cdot 1 = x$$
Also, we see that $\frac{x}{x} = 1$. When you divide x by itself you get the multiplicative identity as your answer. $(x \neq 0)$

We are not able to divide by matrices.

We know that $\begin{bmatrix} 1 & 0 \\ 0 & 1 \end{bmatrix}\begin{bmatrix} a & b \\ c & d \end{bmatrix} = \begin{bmatrix} a & b \\ c & d \end{bmatrix}$

But we cannot expect to divide $\begin{bmatrix} a & b \\ c & d \end{bmatrix}$ by itself to get $\begin{bmatrix} 1 & 0 \\ 0 & 1 \end{bmatrix}$ as we do with real numbers.

This is where we look to the properties of the inverse of a matrix. Most, but not all, square matrices have an inverse.

For a square matrix, A, we will denote its inverse with the notation A^{-1}.

Definition of a matrix inverse:

For a square matrix A, the ***inverse***, A^{-1}, is the unique matrix that has the properties

$$A \cdot A^{-1} = A^{-1} \cdot A = \begin{bmatrix} 1 & \cdots & 0 \\ \vdots & \ddots & \vdots \\ 0 & \cdots & 1 \end{bmatrix} \text{ (The identity matrix)}$$

Consider $A = \begin{bmatrix} 9 & 5 \\ 7 & 4 \end{bmatrix} \; B = \begin{bmatrix} 4 & -5 \\ -7 & 9 \end{bmatrix}$

Prove A and B are inverses

Show that $AB = \begin{bmatrix} 1 & 0 \\ 0 & 1 \end{bmatrix}$	Show that $BA = \begin{bmatrix} 1 & 0 \\ 0 & 1 \end{bmatrix}$

Finding the inverse of a 2x2 matrix

By definition, for a 2x2 matrix, the inverse is as follows.

For $A = \begin{bmatrix} a & b \\ c & d \end{bmatrix}$ We know the determinant is $\det(A) = ad - bc$

The inverse is defined as follows:

$$A^{-1} = \frac{1}{\det(A)} \begin{bmatrix} d & -b \\ -c & a \end{bmatrix}$$

$$\text{or} \quad \frac{1}{\det(A)} * \begin{bmatrix} adjugate \\ matrix \end{bmatrix}$$

When would the inverse of a matrix not exist? _____

Consider $A = \begin{bmatrix} 4 & 19 \\ 2 & 10 \end{bmatrix}$. Find A^{-1}

Step 1: Find the determinant	
Step 2: Using the formula for the inverse, write the expression for A^{-1} $$A^{-1} = \frac{1}{\det(A)} * \begin{bmatrix} adjugate \\ matrix \end{bmatrix}$$	
Step 3: Use scalar multiplication to reach the final answer	$A^{-1} =$

How could you check your answer to make sure you found the correct inverse matrix?

You would have to show that:

$A A^{-1} = \begin{bmatrix} 1 & 0 \\ 0 & 1 \end{bmatrix}$	$A^{-1} A = \begin{bmatrix} 1 & 0 \\ 0 & 1 \end{bmatrix}$

$A = \begin{bmatrix} 4 & 3 \\ 8 & 6 \end{bmatrix}$ Find A^{-1}

$A = \begin{bmatrix} 3 & -1 \\ 2 & 7 \end{bmatrix}$ find A^{-1}

$A = \begin{bmatrix} -1 & 0 \\ 0 & 1 \end{bmatrix}$ find A^{-1}

$A = \begin{bmatrix} 1 & 0 \\ 0 & 1 \end{bmatrix}$ find A^{-1}

Homework 1.3 Multiplicative Properties of Matrices, Determinants, Inverses

1. Find the determinant of each matrix.

$A = \begin{bmatrix} 1 & 3 \\ -3 & 2 \end{bmatrix}$	$B = \begin{bmatrix} 2 & 4 \\ 0 & -3 \end{bmatrix}$	$C = \begin{bmatrix} -3 & 19 \\ -2 & 13 \end{bmatrix}$
$\det A = (1)(2) - (-3)(3)$ $2 + 9$ ⑪	$\det(B) = (2)(-3) - 0$ ⊖6	$|C| = (-3)(13) - (-2)(19)$ $-39 \quad -38$ ⊖1
$D = \begin{bmatrix} 8 & 1 \\ 6 & 4 \end{bmatrix}$	$E = \begin{bmatrix} -3 & -3 \\ -2 & 0 \end{bmatrix}$	$F = \begin{bmatrix} 4 & \frac{1}{2} \\ 3 & \frac{1}{2} \end{bmatrix}$
$|D| = (8 \cdot 4) - (6 \cdot 1)$ $30 - 6$ ㉚	$|E| = (-3 \cdot 0) - (-2 \cdot -3)$ $0 + -6$ ⊖6	$|F| = \frac{1}{2}(3) - \frac{1}{2}(4)$ $\frac{3}{2} - 2 = \dfrac{-1}{2}$

2. Find the inverse of each matrix.

$A = \begin{bmatrix} 1 & 3 \\ -3 & 1 \end{bmatrix}$	$B = \begin{bmatrix} 2 & 4 \\ 0 & -3 \end{bmatrix}$
$|A| = (3 \cdot -3) - (1 \cdot 1) = -9 - 1$ $= \boxed{-10}$ $A^{-1} = -10 \begin{bmatrix} 1 & -3 \\ 3 & 1 \end{bmatrix}$ $A^{-1} = \begin{bmatrix} -10 & 30 \\ -30 & -10 \end{bmatrix}$	$|B| = (2 \cdot -3) - (0 \cdot 4) = \boxed{-6}$ $A^{-1} = -6 \begin{bmatrix} -3 & -4 \\ 0 & 2 \end{bmatrix}$ $A^{-1} = \begin{bmatrix} 18 & 24 \\ 0 & -12 \end{bmatrix}$
$C = \begin{bmatrix} 3 & 2 \\ 37 & 25 \end{bmatrix}$	$D = \begin{bmatrix} 2 & 6 \\ 1 & 3 \end{bmatrix}$
$|C| = (3 \cdot 25) - (37 \cdot 2) = \boxed{-2}$ $72 - 74$ $A^{-1} = -2 \begin{bmatrix} 25 & -2 \\ -37 & 3 \end{bmatrix} = \begin{bmatrix} -50 & 4 \\ 74 & -6 \end{bmatrix}$	$|D| = (2 \cdot 3) - (1 \cdot 6) = 0$ undefined

1.3—30

Homework 1.3 Multiplicative Properties of Matrices, Determinants, Inverses

3. Find the inverse of each matrix.

$A = \begin{bmatrix} -3 & 2 \\ -3 & -8 \end{bmatrix}$

$|A| = (-3 \cdot -8) - (-3 \cdot 2)$
$24 + 6 = \boxed{30}$

$A^{-1} = 30 \begin{bmatrix} -8 & -2 \\ 3 & -3 \end{bmatrix}$

$A^{-1} = \begin{bmatrix} -240 & -60 \\ 90 & -90 \end{bmatrix}$

$B = \begin{bmatrix} 6 & 7 \\ -1 & -8 \end{bmatrix}$

$|B| = (6 \cdot -8) - (-1 \cdot 7)$
$-48 + 7 = \boxed{-41}$

$B^{-1} = -41 \begin{bmatrix} -8 & -7 \\ 1 & 6 \end{bmatrix}$

$B^{-1} = \begin{bmatrix} 328 & 287 \\ -41 & -246 \end{bmatrix}$

4. By performing the matrix multiplications in the equation below, solve for x and y.

$\begin{bmatrix} 7 & -3 \\ -16 & 7 \end{bmatrix} \begin{bmatrix} 7 & 3 \\ 16 & 7 \end{bmatrix} \begin{bmatrix} x \\ y \end{bmatrix} = \begin{bmatrix} 7 & -3 \\ -16 & 7 \end{bmatrix} \begin{bmatrix} 10 \\ 8 \end{bmatrix}$

2×2   2×2 2×1   2×2   2×1

$\begin{bmatrix} x \\ y \end{bmatrix} \begin{bmatrix} 7 \cdot 7 + -3 \cdot 16 & 7 \cdot 3 + -3 \cdot 7 \\ -16 \cdot 7 + 7 \cdot 16 & -16 \cdot 3 + 7 \cdot 7 \end{bmatrix} = \begin{bmatrix} 70 + -24 \\ -160 + 56 \end{bmatrix}$

$\begin{bmatrix} x \\ y \end{bmatrix} \begin{bmatrix} 1 & 0 \\ 0 & 1 \end{bmatrix} = \begin{bmatrix} 46 \\ -104 \end{bmatrix}$

$x = 46$
$y = -104$

Homework 1.3 Multiplicative Properties of Matrices, Determinants, Inverses

5. Given that $A = \begin{bmatrix} 1 & 2 & 3 \\ 2 & 1 & 0 \\ -1 & 2 & 4 \end{bmatrix}$ and $A^{-1} = \frac{1}{3}\begin{bmatrix} 4 & -2 & -3 \\ -8 & 7 & 6 \\ 5 & -4 & -3 \end{bmatrix}$

Prove that $AA^{-1} = \begin{bmatrix} 1 & 0 & 0 \\ 0 & 1 & 0 \\ 0 & 0 & 1 \end{bmatrix}$

$$AA^{-1} = \frac{1}{3}\begin{bmatrix} 1 & 2 & 3 \\ 2 & 1 & 0 \\ -1 & 2 & 4 \end{bmatrix} \begin{bmatrix} 4 & -2 & -3 \\ -8 & 7 & 6 \\ 5 & -4 & -3 \end{bmatrix}$$

$$\frac{1}{3}\begin{bmatrix} 1(4)+2(-2)+3(-3) & 1(-8)+2(7)+3(6) & 1(-3)+2(6)+3(-3) \\ 2(4)+1(-2)+0 & 2(-8)+1(7)+0(6) & 2(-3)+1(6)+0(-3) \\ -1(4)+2(-2)+4(-3) & -1(-8)+2(7)+4(6) & -1(-3)+2(6)+4(-3) \end{bmatrix}$$

(annotations above first row entries: 4, -4, -9)

$$\frac{1}{3}\begin{bmatrix} 3 & 0 & 0 \\ 0 & 3 & 0 \\ 0 & 0 & 3 \end{bmatrix} = \begin{bmatrix} 1 & 0 & 0 \\ 0 & 1 & 0 \\ 0 & 0 & 1 \end{bmatrix}$$

Homework 1.3 Multiplicative Properties of Matrices, Determinants, Inverses

CHALLENGE QUESTIONS

1. If matrices A and B have the same determinants, solve for x.

$$A = \begin{bmatrix} 3 & x \\ 2 & 2x \end{bmatrix} \quad B = \begin{bmatrix} 4x & 1 \\ 5 & 2 \end{bmatrix}$$

2. If $detA = 2detB$ and $detA + detB = 10$, solve for x and y.

$$A = \begin{bmatrix} x & y \\ 1 & 2 \end{bmatrix} \quad B = \begin{bmatrix} y & 1 \\ x & 1 \end{bmatrix}$$

1.4. Matrix Equations

In today's lesson we will begin solving matrix equations, equations where the unknown variable is a matrix.

First let's compare solving for scalar variables versus solving for a matrix.

How do we solve this equation? a, x, b are all real numbers.	How do we solve this equation? A, X, B are all matrices of real numbers.
$$\frac{ax}{a} = \frac{b}{a}$$ $$x = \frac{b}{a}$$	$$AX = B$$ $$A^{-1}AX = A^{-1}B$$ $$IX = A^{-1}B$$ $$\boxed{X = A^{-1}B}$$

Solve the matrix equation: $\begin{bmatrix} 5 & 7 \\ 2 & 3 \end{bmatrix} X = \begin{bmatrix} 4 & 0 \\ 12 & 6 \end{bmatrix}$

$$AX = B$$

$$X = \begin{bmatrix} 5 & 7 \\ 2 & 3 \end{bmatrix}^{-1} \begin{bmatrix} 4 & 0 \\ 12 & 6 \end{bmatrix}$$

$\det(A) = 5 \cdot 3 - 2 \cdot 7$
$= 1$

$$X = \frac{1}{1} \begin{bmatrix} 3 & -7 \\ -2 & 5 \end{bmatrix} \begin{bmatrix} 4 & 0 \\ 12 & 6 \end{bmatrix}$$

$$X = \begin{bmatrix} 3(4) + -7(12) & 3(0) + -7(6) \\ -2(4) + 5(12) & -2(0) + 5(6) \end{bmatrix}$$

$$X = \begin{bmatrix} -72 & -42 \\ 52 & 30 \end{bmatrix}$$

Now solve $\begin{bmatrix} 5 & 7 \\ 2 & 3 \end{bmatrix} X = \begin{bmatrix} 2 & -1 \\ 1 & 4 \end{bmatrix}$ det A = 1

$X = \begin{bmatrix} 5 & 7 \\ 2 & 3 \end{bmatrix}^{-1} \begin{bmatrix} 2 & -1 \\ 1 & 4 \end{bmatrix}$

$X = \frac{1}{1} \begin{bmatrix} 3 & -7 \\ -2 & 5 \end{bmatrix} \begin{bmatrix} 2 & -1 \\ 1 & 4 \end{bmatrix}$

$X = \begin{bmatrix} 3(2) + -7(1) & 3(-1) + -7(4) \\ -2(2) + 5(1) & -2(-1) + 5(4) \end{bmatrix}$

$X = \begin{bmatrix} -1 & -31 \\ 1 & 22 \end{bmatrix}$

By now you should be able to view a matrix equation

$$AX = B$$

And know that multiplying both sides on the left by A^{-1} yields

$$A^{-1}AX = A^{-1}B$$

$$X = A^{-1}B$$

So, you should be comfortable going straight from $AX = B$ to $X = A^{-1}B$ to solve for X.

Let $A = \begin{bmatrix} 2 & 7 \\ 2 & 8 \end{bmatrix}$ $B = \begin{bmatrix} 4 & 12 \\ 10 & 20 \end{bmatrix}$. Solve the matrix equation $AX = B$.

$X = A^{-1}B$

det A = 2·8 − 2·7
    = 2

$X = \frac{1}{2} \begin{bmatrix} 8 & -7 \\ -2 & 2 \end{bmatrix} \begin{bmatrix} 4 & 12 \\ 10 & 20 \end{bmatrix}$

$\frac{1}{2} \begin{bmatrix} 8(4) + -7(10) & 8(12) + -7(20) \\ -2(4) + 2(10) & -2(12) + 2(20) \end{bmatrix}$
    30 − 70   96 − 140
    −8 + 20   −24 + 40

$\frac{1}{2} \begin{bmatrix} -38 & -44 \\ 12 & 16 \end{bmatrix}$

$X = \begin{bmatrix} -19 & -22 \\ 6 & 8 \end{bmatrix}$

1.4—35

1.4—36

Let $A = \begin{bmatrix} 1 & 2 \\ -1 & -1 \end{bmatrix}$ $B = \begin{bmatrix} 4 & -7 & 9 \\ -2 & 4 & -5 \end{bmatrix}$. Solve $AX = B$

2×2 $\quad\quad$ 2×3 $\quad\quad$ $X = 2\times 3$

$X = A^{-1}B$

$\det A = (1\cdot -1)-(-1\cdot 2)$
$\quad\quad -1+2$
$\quad\quad\quad \textcircled{1}$

$X = \dfrac{1}{1}\begin{bmatrix} -1 & -2 \\ 1 & 1 \end{bmatrix}\begin{bmatrix} 4 & -7 & 9 \\ -2 & 4 & -5 \end{bmatrix}$

$\dfrac{1}{1}\begin{bmatrix} -1(4)-2(-2) & -1(-7)-2(4) & -1(9)-2(-5) \\ 1(4)+1(-2) & 1(-7)+1(4) & 1(9)+1(-5) \end{bmatrix}$

$X = \begin{bmatrix} 0 & -1 & 1 \\ 2 & -3 & 4 \end{bmatrix}$

Let $A = \begin{bmatrix} 16 & 4 \\ 8 & 2 \end{bmatrix}$ $B = \begin{bmatrix} -1 & 7 \\ 1 & 0 \end{bmatrix}$. Solve $AX = B$

$X = A^{-1}B$
$\det A = (10\cdot 2)-(4\cdot 8)$
$\quad\quad 32 - 32$
$\quad\quad\quad 0$

undefined

Homework 1.4 Matrix Equations

Solve each matrix equation

1. $\begin{bmatrix} 5 & -6 \\ 0 & -1 \end{bmatrix} Y = \begin{bmatrix} -20 \\ -10 \end{bmatrix}$ 2×2 2×1

   2×1

   $Y = \begin{bmatrix} 5 & -6 \\ 0 & -1 \end{bmatrix}^{-1} \begin{bmatrix} -20 \\ -10 \end{bmatrix}$

   $Y = -\frac{1}{5} \begin{bmatrix} -1 & 6 \\ 0 & 5 \end{bmatrix} \begin{bmatrix} -20 \\ -10 \end{bmatrix}$

   $Y = -\frac{1}{5} \begin{bmatrix} -1(-20) + 6(-10) \\ 0(-20) + 5(-10) \end{bmatrix}$

   $Y = -\frac{1}{5} \begin{bmatrix} -40 \\ -50 \end{bmatrix} = \begin{bmatrix} 8 \\ 10 \end{bmatrix}$

2. $\begin{bmatrix} -21 \\ 18 \end{bmatrix} = \begin{bmatrix} -3 & -2 \\ 2 & 2 \end{bmatrix} X$ $\det A = (-3 \cdot 2) - (-2 \cdot 2)$
   $= -6 + 4$
   $= -2$

   $X = \begin{bmatrix} -3 & -2 \\ 2 & 2 \end{bmatrix}^{-1} \begin{bmatrix} -21 \\ 18 \end{bmatrix}$

   $X = \frac{1}{-2} \begin{bmatrix} 2 & 2 \\ -2 & -3 \end{bmatrix} \begin{bmatrix} -21 \\ 18 \end{bmatrix}$

   $X = \frac{1}{-2} \begin{bmatrix} 2(-21) + 2(18) \\ -2(-21) - 3(18) \end{bmatrix}$ $-42 + 36$

   $X = \frac{1}{-2} \begin{bmatrix} -6 \\ -12 \end{bmatrix} = \begin{bmatrix} 3 \\ 6 \end{bmatrix}$

3. $A = \begin{bmatrix} 0 & -2 \\ 4 & -6 \end{bmatrix}, B = \begin{bmatrix} -12 & 2 \\ -40 & 38 \end{bmatrix}$

   Solve AX=B

   $X = A^{-1} B$
   $\det A = (0 \cdot -6) - (4 \cdot -2)$
   $0 + 8$
   $= \boxed{8}$

   $X = \frac{1}{8} \begin{bmatrix} -6 & 2 \\ -4 & 0 \end{bmatrix} \begin{bmatrix} -12 & 2 \\ -40 & 38 \end{bmatrix}$

   $X = \frac{1}{8} \begin{bmatrix} -6(-12) + 2(-40) & -6(2) + 2(38) \\ -4(-12) + 0(-40) & -4(2) + 0(38) \end{bmatrix}$

   $\frac{1}{8} \begin{bmatrix} -8 & 64 \\ 48 & -8 \end{bmatrix} = \begin{bmatrix} -1 & 8 \\ 6 & -1 \end{bmatrix}$

4. $\begin{bmatrix} 5 & -2 \\ -8 & 3 \end{bmatrix} B = \begin{bmatrix} 5 \\ -8 \end{bmatrix}$ $\det A = (5 \cdot 3) - (-2 \cdot -8)$
   $15 - 16$
   $\boxed{-1}$

   $B = \begin{bmatrix} 5 & -2 \\ -8 & 3 \end{bmatrix}^{-1} \begin{bmatrix} 5 \\ -8 \end{bmatrix}$

   $B = \frac{1}{-1} \begin{bmatrix} 3 & 2 \\ 8 & 5 \end{bmatrix} \begin{bmatrix} 5 \\ -8 \end{bmatrix}$

   $\frac{1}{-1} \begin{bmatrix} 3(5) + 2(-8) \\ 8(5) + 5(-8) \end{bmatrix}$

   $\frac{1}{-1} \begin{bmatrix} -1 \\ 0 \end{bmatrix} = \begin{bmatrix} 1 \\ 0 \end{bmatrix}$

Homework 1.4 Matrix Equations

Solve each matrix equation

5. $\begin{bmatrix} 1 & -1 \\ 1 & 1 \end{bmatrix} X = \begin{bmatrix} 3 & -8 & 8 \\ -9 & 2 & 0 \end{bmatrix}$

$\det A = (1 \cdot 1) - (-1 \cdot 1)$
$= 1 + 1$
$= \boxed{2}$

$2 \times 2 \quad 2 \times 3 = 2 \times 3$

$X = \dfrac{1}{2} \begin{bmatrix} 1 & 1 \\ -1 & 1 \end{bmatrix} \begin{bmatrix} 3 & -8 & 8 \\ -9 & 2 & 0 \end{bmatrix}$

$\dfrac{1}{2} \begin{bmatrix} 1(3)+1(-9) & 1(-8)+1(2) & 1(8) \\ -1(3)+1(-9) & -1(-8)+1(2) & -1(8) \end{bmatrix}$

$\dfrac{1}{2} \begin{bmatrix} -6 & -6 & 8 \\ -12 & 10 & -8 \end{bmatrix} = \begin{bmatrix} -3 & -3 & 4 \\ -6 & 5 & 4 \end{bmatrix}$

6. $A = \begin{bmatrix} 0 & 5 \\ -1 & -3 \end{bmatrix} \quad B = \begin{bmatrix} 35 & 30 \\ -14 & -25 \end{bmatrix}$

Solve $AX = B$

$\det A = (0 \cdot -3) - (5 \cdot -1)$
$= 0 + 5$
$= \boxed{5}$

$X = \begin{bmatrix} 0 & 5 \\ -1 & -3 \end{bmatrix}^{-1} \begin{bmatrix} 35 & 30 \\ -14 & -25 \end{bmatrix}$

$X = \dfrac{1}{5} \begin{bmatrix} -3 & -5 \\ 1 & 0 \end{bmatrix} \begin{bmatrix} 35 & 30 \\ -14 & -25 \end{bmatrix}$

$\dfrac{1}{5} \begin{bmatrix} -3(35)+-5(-14) & -3(30)-5(-25) \\ 1(35)+0(-14) & 1(30)-0(25) \end{bmatrix}$

$\dfrac{1}{5} \begin{bmatrix} -35 & 35 \\ 35 & 30 \end{bmatrix} = \begin{bmatrix} -7 & 7 \\ 7 & 6 \end{bmatrix}$

7. Let $A = \begin{bmatrix} 1 & 2 & 1 \\ 3 & -1 & -1 \\ 1 & 0 & 1 \end{bmatrix}, B = \begin{bmatrix} 8 \\ -2 \\ 4 \end{bmatrix}$ & $A^{-1} = \dfrac{1}{8}\begin{bmatrix} 1 & 2 & 1 \\ 4 & 0 & -4 \\ -1 & -2 & 7 \end{bmatrix}$

Solve $AX = B$

$X = A^{-1} B$

$3 \times 3 \quad 3 \times 1 = 3 \times 1$

$X = \dfrac{1}{8} \begin{bmatrix} 1 & 2 & 1 \\ 4 & 0 & -4 \\ -1 & -2 & 7 \end{bmatrix} \begin{bmatrix} 8 \\ -2 \\ 4 \end{bmatrix}$

$\dfrac{1}{8} \begin{bmatrix} 1(8)+2(-2)+1(4) \\ 4(8)+0(-2)+-4(4) \\ -1(8)+-2(-2)+7(4) \end{bmatrix}$

$\dfrac{1}{8} \begin{bmatrix} 8 \\ 16 \\ 24 \end{bmatrix} = \begin{bmatrix} 1 \\ 2 \\ 3 \end{bmatrix}$

Homework 1.4 Matrix Equations

CHALLENGE QUESTIONS

1. For matrices larger than 2 × 2 it can be more difficult to find the inverse when working with matrix equations. There are a variety of methods that may be employed manually or computationally to invert large matrices.

Consider the equation
$$\begin{bmatrix} 2 & 1 & 3 \\ 4 & -1 & 3 \\ -2 & 5 & 5 \end{bmatrix} \begin{bmatrix} x_1 \\ x_2 \\ x_3 \end{bmatrix} = \begin{bmatrix} 1 \\ -1 \\ 3 \end{bmatrix}$$

This is an equation of the form
$$AX = B$$

We can rewrite A as the product of two matrices, L and U, where L has non-zero entries below the diagonal and U has non-zero entries above the diagonal.

$$A = LU$$
$$\begin{bmatrix} 1 & 2 & 0 \\ 3 & 6 & -1 \\ 1 & 2 & 1 \end{bmatrix} = \begin{bmatrix} 1 & 0 & 0 \\ 2 & 1 & 0 \\ -1 & -2 & 1 \end{bmatrix} \begin{bmatrix} 2 & 1 & 3 \\ 0 & -3 & -3 \\ 0 & 0 & 2 \end{bmatrix}$$

So, $A = \begin{bmatrix} 1 & 2 & 0 \\ 3 & 6 & -1 \\ 1 & 2 & 1 \end{bmatrix}, L = \begin{bmatrix} 1 & 0 & 0 \\ 2 & 1 & 0 \\ -1 & -2 & 1 \end{bmatrix}$ and $U = \begin{bmatrix} 2 & 1 & 3 \\ 0 & -3 & -3 \\ 0 & 0 & 2 \end{bmatrix}$

So, take our original equation
$$AX = B$$

And substitute LU for A.
$$LUX = B$$

Let's create a matrix Y such that $Y = UX$

Now solve $LY = B$ for Y. You will get the values for $Y = \begin{bmatrix} y_1 \\ y_2 \\ y_3 \end{bmatrix}$	Now remember we defined Y as $Y = UX$. Now take the values you have for Y and solve $$Y = UX \quad or \quad UX = Y$$ for X.

This is called the "LU Decomposition" method for solving matrix equations. From a computational standpoint it is more efficient to use the LU Decomposition than to find the inverse of the original matrix A.

1.5. Solving Systems of Linear Equations with Matrices

Look at the equation $\begin{bmatrix} 5 & -7 \\ 2 & 3 \end{bmatrix} \begin{bmatrix} x \\ y \end{bmatrix} = \begin{bmatrix} 3 \\ 7 \end{bmatrix} \Rightarrow \begin{bmatrix} x \\ y \end{bmatrix} = \begin{bmatrix} 5 & -7 \\ 2 & 3 \end{bmatrix}^{-1} \begin{bmatrix} 3 \\ 7 \end{bmatrix}$

Left Hand Side	Right Hand Side
$\begin{bmatrix} 5 & -7 \\ 2 & 3 \end{bmatrix} \begin{bmatrix} x \\ y \end{bmatrix}$	$= \begin{bmatrix} 3 \\ 7 \end{bmatrix}$
Is this matrix operation valid? What will be the dimensions of the answer matrix? Perform the multiplication $\quad$ 2×2 $\quad$ 2×1 → 2×1 $\begin{bmatrix} 5 & -7 \\ 2 & 3 \end{bmatrix} \begin{bmatrix} x \\ y \end{bmatrix}$	

Putting it all back together:

$\begin{bmatrix} 5x - 7y \\ 2x + 3y \end{bmatrix} = \begin{bmatrix} 3 \\ 7 \end{bmatrix}$ $\quad$ $5x - 7y = 3$
 $\quad\quad\quad\quad\quad\quad\quad\quad\quad\quad$ $2x + 3y = 7$

This should look familiar; you should have solved systems of linear equations in your prior studies of Algebra.

$$5x - 7y = 3$$
$$2x + 3y = 7$$

You have learned to solve these by substitution or elimination. We will now use matrices and inverses to solve systems of linear equations.

Linear Systems of Equations	Matrix Equation
$5x - 7y = 3$ $2x + 3y = 7$	$\begin{bmatrix} 5 & -7 \\ 2 & 3 \end{bmatrix} \begin{bmatrix} x \\ y \end{bmatrix} = \begin{bmatrix} 3 \\ 7 \end{bmatrix}$

$\det A = 5 \cdot 3 - 2 \cdot -7 = \boxed{29}$

How do we solve $\begin{bmatrix} 5 & -7 \\ 2 & 3 \end{bmatrix} \begin{bmatrix} x \\ y \end{bmatrix} = \begin{bmatrix} 3 \\ 7 \end{bmatrix}$

$\begin{bmatrix} x \\ y \end{bmatrix} = \begin{bmatrix} 5 & -7 \\ 2 & 3 \end{bmatrix} \begin{bmatrix} 3 \\ 7 \end{bmatrix} = \frac{1}{29} \begin{bmatrix} 3 & 7 \\ -2 & 5 \end{bmatrix} \begin{bmatrix} 3 \\ 7 \end{bmatrix} = \frac{1}{29} \begin{bmatrix} 9+49 \\ -6+35 \end{bmatrix} = \frac{1}{29} \begin{bmatrix} 58 \\ 29 \end{bmatrix} = \begin{bmatrix} 2 \\ 1 \end{bmatrix}$

$\boxed{x = 2 \quad y = 1}$

So, if we are given a linear system

$$ax + by = c$$
$$dx + ey = f$$

We can rewrite as a matrix equation

$$\begin{bmatrix} a & b \\ d & e \end{bmatrix} \begin{bmatrix} x \\ y \end{bmatrix} = \begin{bmatrix} c \\ f \end{bmatrix}$$

coefficient ← → variable ← → solution

Definitions

Matrix $\begin{bmatrix} a & b \\ d & e \end{bmatrix}$ is the **coefficient matrix**.	Matrix $\begin{bmatrix} x \\ y \end{bmatrix}$ is the **variable matrix**.
Matrix $\begin{bmatrix} c \\ f \end{bmatrix}$ is the **solution matrix**.	

So, from our previous work with matrices and inverses, we know that the solution to

$$\begin{bmatrix} a & b \\ d & e \end{bmatrix} \begin{bmatrix} x \\ y \end{bmatrix} = \begin{bmatrix} c \\ f \end{bmatrix}$$

is found by multiplying the inverse matrix, by the solution matrix.

$$\begin{bmatrix} x \\ y \end{bmatrix} = \begin{bmatrix} a & b \\ d & e \end{bmatrix}^{-1} \begin{bmatrix} c \\ f \end{bmatrix}$$

1.5—42

Consider the linear system $$-2x + 3y = -11$$ $$5x + y = 19$$	Rewrite as a matrix equation $\begin{bmatrix} -2 & 3 \\ 5 & 1 \end{bmatrix} \begin{bmatrix} x \\ y \end{bmatrix} = \begin{bmatrix} -11 \\ 19 \end{bmatrix}$

det A =
(-2·1) − (5·3)
= −17

Solve the matrix equation

$\begin{bmatrix} x \\ y \end{bmatrix} = \begin{bmatrix} -2 & 3 \\ 5 & 1 \end{bmatrix}^{-1} \begin{bmatrix} -11 \\ 19 \end{bmatrix} = \frac{1}{-17} \begin{bmatrix} 1 & -3 \\ -5 & -2 \end{bmatrix} \begin{bmatrix} -11 \\ 19 \end{bmatrix}$

$= \frac{1}{-17} \begin{bmatrix} -11 + -57 \\ 55 - 38 \end{bmatrix} = \frac{1}{-17} \begin{bmatrix} -68 \\ 17 \end{bmatrix} = \begin{bmatrix} 4 \\ -1 \end{bmatrix}$

x = 4
y = −1

x = 4	y = −1

Consider the linear system $$4x + 5y = 23$$ $$5x - y = 7$$	Rewrite as a matrix equation $\begin{bmatrix} 4 & 5 \\ 5 & -1 \end{bmatrix} \begin{bmatrix} x \\ y \end{bmatrix} = \begin{bmatrix} 23 \\ 7 \end{bmatrix}$

det A =
(4·−1) − (5·5)
= −29

Solve the matrix equation

$\begin{bmatrix} x \\ y \end{bmatrix} = \begin{bmatrix} 4 & 5 \\ 5 & -1 \end{bmatrix}^{-1} \begin{bmatrix} 23 \\ 7 \end{bmatrix} = \frac{1}{-29} \begin{bmatrix} -1 & -5 \\ -5 & 4 \end{bmatrix} \begin{bmatrix} 23 \\ 7 \end{bmatrix}$

$\frac{1}{-29} \begin{bmatrix} -23 - 35 \\ -115 + 28 \end{bmatrix} = \frac{1}{-29} \begin{bmatrix} -58 \\ -87 \end{bmatrix} = \begin{bmatrix} 2 \\ 3 \end{bmatrix}$

x = 2	y = 3

Consider the linear system $3x + 7y = 31$ $9x + 21y = 19$	Rewrite as a matrix equation $\begin{bmatrix} 3 & 7 \\ 9 & 21 \end{bmatrix} \begin{bmatrix} x \\ y \end{bmatrix} = \begin{bmatrix} 31 \\ 19 \end{bmatrix}$
Solve the matrix equation $\begin{bmatrix} x \\ y \end{bmatrix} = \begin{bmatrix} 3 & 7 \\ 9 & 21 \end{bmatrix}^{-1} \begin{bmatrix} 31 \\ 19 \end{bmatrix}$	$\det A = (3 \cdot 21) - (7 \cdot 9)$ $\det A = 0$ undefined
$x =$	$y =$

Consider the linear system $4x + y = 10$ $3x + 5y = -1$	Rewrite as a matrix equation $\begin{bmatrix} 4 & 1 \\ 3 & 5 \end{bmatrix}$
Solve the matrix equation	
$x =$	$y =$

Rewrite as a matrix equation

$$x + 2y + z = 4$$
$$2x - 3y + 7z = 4$$
$$x - y - 4z = 7$$

$$\begin{bmatrix} 1 & 2 & 1 \\ 2 & -3 & 7 \\ 1 & -1 & 4 \end{bmatrix} \begin{bmatrix} x \\ y \\ z \end{bmatrix} = \begin{bmatrix} 4 \\ 4 \\ 7 \end{bmatrix} \qquad \begin{bmatrix} x \\ y \\ z \end{bmatrix} = \begin{bmatrix} 1 & 2 & 1 \\ 2 & -3 & 7 \\ 1 & -1 & 4 \end{bmatrix}^{-1} \begin{bmatrix} 4 \\ 4 \\ 7 \end{bmatrix}$$

*need calculator

In order to solve this, we would need to know how to take the inverse of a __3 × 3__ matrix. For this class, we will only find inverses manually of 2x2 matrices. Our next lesson will detail how to take the inverse of a 3x3 matrix with the TI-Nspire.

Homework 1.5 Solving Systems of Linear Equations with Matrices

Write each linear system of equations as a matrix equation, then solve each matrix equation using inverse matrices.

1. $\begin{aligned} 3x + 5y &= -1 \\ -3x - 3y &= 3 \end{aligned}$ $\begin{bmatrix} 3 & 5 \\ -3 & -3 \end{bmatrix} \begin{bmatrix} x \\ y \end{bmatrix} = \begin{bmatrix} -1 \\ 3 \end{bmatrix}$

$\det A = (3 \cdot -3) - (5 \cdot -3) = -9 + 15 = 6$

$\begin{bmatrix} x \\ y \end{bmatrix} = \begin{bmatrix} 3 & 5 \\ -3 & -3 \end{bmatrix}^{-1} \begin{bmatrix} -1 \\ 3 \end{bmatrix} = \frac{1}{6} \begin{bmatrix} -3 & -5 \\ 3 & 3 \end{bmatrix} \begin{bmatrix} -1 \\ 3 \end{bmatrix}$

$\frac{1}{6} \begin{bmatrix} 3 + -15 \\ -3 + 9 \end{bmatrix} = \frac{1}{6} \begin{bmatrix} 12 \\ 6 \end{bmatrix} = \begin{bmatrix} 2 \\ 1 \end{bmatrix}$

x = 2, y = 1

2. $\begin{aligned} -4x - 2y &= 8 \\ -3x + 2y &= 6 \end{aligned}$ $\begin{bmatrix} -4 & -2 \\ -3 & 2 \end{bmatrix} \begin{bmatrix} x \\ y \end{bmatrix} = \begin{bmatrix} 8 \\ 6 \end{bmatrix}$

$\det A = (-4 \cdot 2) - (-3 \cdot -2) = -8 - 6 = -14$

$\begin{bmatrix} x \\ y \end{bmatrix} = \begin{bmatrix} -4 & -2 \\ -3 & 2 \end{bmatrix}^{-1} \begin{bmatrix} 8 \\ 6 \end{bmatrix} = \frac{1}{-14} \begin{bmatrix} 2 & 2 \\ 3 & -4 \end{bmatrix} \begin{bmatrix} 8 \\ 6 \end{bmatrix}$

$\frac{1}{-14} \begin{bmatrix} 16 + 12 \\ 24 - 24 \end{bmatrix} = \begin{bmatrix} 28 \\ 0 \end{bmatrix} = \begin{bmatrix} -2 \\ 0 \end{bmatrix}$

x = -2, y = 0

3. $\begin{aligned} 4x + 5y &= 9 \\ 12x + 15y &= 20 \end{aligned}$ $\begin{bmatrix} 4 & 5 \\ 12 & 15 \end{bmatrix} = \begin{bmatrix} 9 \\ 20 \end{bmatrix}$

$\det A = (4 \cdot 15) - (12 \cdot 5) = 60 - 60 = 0$

undefined

4. $\begin{aligned} x + 3y &= 16 \\ x - y &= 0 \end{aligned}$ $\begin{bmatrix} 1 & 3 \\ 1 & -1 \end{bmatrix} \begin{bmatrix} x \\ y \end{bmatrix} = \begin{bmatrix} 16 \\ 0 \end{bmatrix}$

$\det A = (1 \cdot -1) - (1 \cdot 3) = -1 - 3 = -4$

$\begin{bmatrix} x \\ y \end{bmatrix} = \frac{1}{-4} \begin{bmatrix} -1 & -3 \\ -1 & 1 \end{bmatrix} \begin{bmatrix} 16 \\ 0 \end{bmatrix} = \frac{1}{-4} \begin{bmatrix} -16 + 0 \\ -16 + 0 \end{bmatrix}$

$\frac{1}{-4} \begin{bmatrix} -16 \\ -16 \end{bmatrix} = \begin{bmatrix} 4 \\ 4 \end{bmatrix}$

x = 4, y = 4

5. $\begin{aligned} 5x - 8y &= 4 \\ -x + 2y &= 0 \end{aligned}$ $\begin{bmatrix} 5 & -8 \\ -1 & 2 \end{bmatrix} \begin{bmatrix} x \\ y \end{bmatrix} = \begin{bmatrix} 4 \\ 0 \end{bmatrix}$ $\det A = (5 \cdot 2) - (-8 \cdot -1) = 10 - 8 = 2$

$\begin{bmatrix} x \\ y \end{bmatrix} = \frac{1}{2} \begin{bmatrix} 2 & 8 \\ 1 & 5 \end{bmatrix} \begin{bmatrix} 4 \\ 0 \end{bmatrix} = \frac{1}{2} \begin{bmatrix} 8 + 0 \\ 4 + 0 \end{bmatrix} = \begin{bmatrix} 4 \\ 2 \end{bmatrix}$

x = 4, y = 2

Homework 1.5 Solving Systems of Linear Equations with Matrices

Write each linear system of equations as a matrix equation, then solve each matrix equation using inverse matrices.

$\det A =$
$(-5 \cdot 3) - (-4 \cdot 3)$
$-15 + 12 = \boxed{-3}$

6. $-5x - 4y = 17$
   $3x + 3y = -9$

$\begin{bmatrix} -5 & -4 \\ 3 & 3 \end{bmatrix} = \begin{bmatrix} 17 \\ -9 \end{bmatrix}$

$\begin{bmatrix} x \\ y \end{bmatrix} = -\frac{1}{3} \begin{bmatrix} 3 & 4 \\ -3 & -5 \end{bmatrix} \begin{bmatrix} 17 \\ -9 \end{bmatrix}$

$\begin{bmatrix} x \\ y \end{bmatrix} = \frac{1}{-3} \begin{bmatrix} 51 + -36 \\ -51 + 45 \end{bmatrix} = \begin{bmatrix} 15 \\ -6 \end{bmatrix}$

$\begin{bmatrix} -5 \\ 2 \end{bmatrix}$

$\boxed{x = -5 \\ y = 2}$

7. $5x + 6y = 11$
   $3x + 4y = 7$

$\begin{bmatrix} 5 & 6 \\ 3 & 4 \end{bmatrix} \begin{bmatrix} x \\ y \end{bmatrix} = \begin{bmatrix} 11 \\ 7 \end{bmatrix}$

$\begin{bmatrix} x \\ y \end{bmatrix} = \begin{bmatrix} 5 & 6 \\ 3 & 4 \end{bmatrix} \begin{bmatrix} 11 \\ 7 \end{bmatrix}$

$\det A = (5 \cdot 4) - (6 \cdot 3)$
$20 - 18 = \boxed{2}$

$\begin{bmatrix} x \\ y \end{bmatrix} = \frac{1}{2} \begin{bmatrix} 4 & -6 \\ -3 & 5 \end{bmatrix} \begin{bmatrix} 11 \\ 7 \end{bmatrix} = \frac{1}{2} \begin{bmatrix} 44 - 42 \\ -33 + 35 \end{bmatrix}$

$\frac{1}{2} \begin{bmatrix} 2 \\ 2 \end{bmatrix} = \begin{bmatrix} 1 \\ 1 \end{bmatrix} = \boxed{x = 1 \\ y = 1}$

8. $ax + 7y = 1$
   $2ax + 15y = -1$

   a is a real number, $a \neq 0$.

$\det A = (1a \cdot 15) - (2a \cdot 7)$
$15 - 14 = \boxed{1}$

$\begin{bmatrix} 1a & 7 \\ 2a & 15 \end{bmatrix} \begin{bmatrix} x \\ y \end{bmatrix} = \begin{bmatrix} 1 \\ -1 \end{bmatrix}$

$\begin{bmatrix} x \\ y \end{bmatrix} = \begin{bmatrix} 15 & -7 \\ -2a & 1a \end{bmatrix} \begin{bmatrix} 1 \\ -1 \end{bmatrix} = \begin{bmatrix} 15 + 7 \\ -2a - 1a \end{bmatrix} = \begin{bmatrix} 22 \\ -3a \end{bmatrix}$

$\boxed{x = 22 \\ y = -3a}$

$\det A =$
$(3 \cdot 3) - (4 \cdot 2)$
$9 - 8 = \boxed{1}$

9. $3x + 2y = a$
   $4x + 3y = b$

   a, b real numbers

$\begin{bmatrix} 3 & 2 \\ 4 & 3 \end{bmatrix} \begin{bmatrix} x \\ y \end{bmatrix} = \begin{bmatrix} a \\ b \end{bmatrix} = \begin{bmatrix} 3 & -2 \\ -4 & 3 \end{bmatrix} \begin{bmatrix} a \\ b \end{bmatrix} = \begin{bmatrix} 3a + -2b \\ -4a + 3b \end{bmatrix}$

$\boxed{x = 3a - 2b \\ y = -4a + 3b}$

Homework 1.5 Solving Systems of Linear Equations with Matrices

CHALLENGE QUESTIONS

1. Sophomore Algebra students are selling tickets to 7th and 8th graders to view an exhibition of their homework and allow them to connect on SnapChat. If you charge $2 for 7th graders and $3 for eighth graders, the revenue is $29. If you charge $2 for eighth graders and $3 for seventh, the revenue is $26. Let x be the number of 7th graders and y the number of 8th graders. Set up a system of equations in x and y, then solve using matrices. How many 7th and 8th graders bought tickets?

1.6. Solving Systems of Linear Equations with the TI-Nspire

So far, all of the linear systems we have solved by hand have been 2x2 systems, systems of 2 equations and 2 variables.

The resulting matrix equation yielded a 2x2 matrix of coefficients that required you to know how to find the inverse.

Systems of linear equations modeling real life scenarios are typically much larger than 2x2 and make use of computational power to solve the system.

We will not be solving 3x3 systems of equations by hand but will be using the TI-Nspire.

Some TI-Nspire basics first.

Let's enter the following matrices on the TI-Nspire. $A = \begin{bmatrix} 6 & 1 \\ -5 & -2 \end{bmatrix}$ and $B = \begin{bmatrix} -1 & 3 \\ 0 & 8 \end{bmatrix}$

Instructions	Display
Home 1: New Document 1: Add Calculator MENU [7: Matrix and Vector] [1: Create] [1: Matrix] Number of Rows: 2 Number of columns: 2 [OK]	
Let's enter the values of matrix A now. You can tab to move between elements. After all numbers are entered, use the right arrow or tab to position the cursor outside the matrix. To store the matrix in the variable, A, CTRL VAR A Enter Repeat these same steps to store the matrix B.	

Now we have stored the following matrices on our calculator.
$$A = \begin{bmatrix} 6 & 1 \\ -5 & -2 \end{bmatrix} \text{ and } B = \begin{bmatrix} -1 & 3 \\ 0 & 8 \end{bmatrix}$$

Use your calculator to find the answer to the following matrix operations.

Operation	Calculator Entry	Answer
$A + B$	[a] + [b] [enter]	$\begin{bmatrix} 5 & 4 \\ 3 & 6 \end{bmatrix}$
$A - B$	[a] − [b] [enter]	$\begin{bmatrix} 7 & -2 \\ -13 & -10 \end{bmatrix}$
$3B$	[3][b][enter]	$\begin{bmatrix} -3 & 9 \\ 24 & 24 \end{bmatrix}$
AB	[a] × [b][enter]	$\begin{bmatrix} 2 & 26 \\ -11 & -31 \end{bmatrix}$
BA	[b] × [a][enter]	$\begin{bmatrix} -21 & -7 \\ 8 & -8 \end{bmatrix}$
A^{-1}	[a][^ − 1] [enter]	$\begin{bmatrix} \frac{2}{7} & \frac{1}{7} \\ \frac{-5}{7} & \frac{-6}{7} \end{bmatrix}$
$\|A\|$ or det A	[menu] [7: Matrix & Vector] [3: Determinant][a][enter] Or [d] [e] [t] (a) [enter]	-7

Now let's solve a 2x2 system of linear equations using the TI-Nspire.

$$2x + 7y = 34$$
$$3x + 17y = 77$$

Rewrite as a matrix equation, $AX = B$.

$$\begin{bmatrix} 2 & 7 \\ 3 & 17 \end{bmatrix} \begin{bmatrix} x \\ y \end{bmatrix} = \begin{bmatrix} 34 \\ 77 \end{bmatrix}$$

Now to solve $AX = B$, we know our solution will be $X = A^{-1}B$.

Enter the 2x2 matrix A, and 2x1 matrix B into your calculator, and find the solution.

$$\begin{bmatrix} x \\ y \end{bmatrix} = \begin{bmatrix} 2 & 7 \\ 3 & 17 \end{bmatrix}^{-1} \begin{bmatrix} 34 \\ 77 \end{bmatrix}$$

Solution $\begin{bmatrix} x \\ y \end{bmatrix} = \begin{bmatrix} 3 \\ 4 \end{bmatrix}$	$X = \begin{bmatrix} \\ \end{bmatrix}$
$x = 3$	$y = 4$

Manually, it is much harder to solve a 3x3 than it is a 2x2. We have not learned how to find the determinant and inverse of 3x3 matrices. On the TI-Nspire, it is the same level of difficulty to solve both 2x2s, 3x3s, etc.

Solve the following system:
$$2x + 3y - z = 3$$
$$-x + 5y + 3z = 24$$
$$5y - 4z = -10$$

First write a matrix equation of the form $AX = B$	$\begin{bmatrix} 2 & 3 & -1 \\ -1 & 5 & 3 \\ 0 & 5 & -4 \end{bmatrix} \begin{bmatrix} x \\ y \\ z \end{bmatrix} = \begin{bmatrix} 3 \\ 24 \\ -10 \end{bmatrix}$
Write the matrix equation solving for X $X = A^{-1}B$	$\begin{bmatrix} x \\ y \\ z \end{bmatrix} = \begin{bmatrix} 2 & 3 & -1 \\ -1 & 5 & 3 \\ 0 & 5 & -4 \end{bmatrix}^{-1} \begin{bmatrix} 3 \\ 24 \\ -10 \end{bmatrix}$
Solve for X using your calculator	$\begin{bmatrix} 1 \\ 2 \\ 5 \end{bmatrix}$

$x = 1$	$y = 2$	$z = 5$

Homework 1.6 Solving Systems of Linear Equations with the TI-Nspire

Use your calculator to evaluate the following:

1. $\begin{bmatrix} -4 & -1 & -5 \\ 1 & 0 & 5 \end{bmatrix} - \begin{bmatrix} 5 & -1 & -3 \\ -1 & -3 & -1 \end{bmatrix}$	$\begin{bmatrix} -9 & 0 & -2 \\ 2 & 3 & 6 \end{bmatrix}$
2. $\begin{bmatrix} 5 & 6 & -5 \end{bmatrix} \begin{bmatrix} -3 & -3 \\ -2 & -3 \\ 6 & -1 \end{bmatrix}$	$\begin{bmatrix} -57 & -28 \end{bmatrix}$
3. $\begin{vmatrix} -2 & -3 \\ 4 & 5 \end{vmatrix}$	det = 2
4. $\det \begin{bmatrix} 1 & 3 & 4 \end{bmatrix}$	undefined
5. $\begin{vmatrix} 0 & 5 & 0 \\ 1 & 5 & -5 \\ 4 & 2 & 0 \end{vmatrix}$	−100
6. $\begin{bmatrix} -7 & 11 \\ -4 & 10 \end{bmatrix}^{-1}$	$\begin{bmatrix} \frac{-5}{13} & \frac{11}{26} \\ \frac{-2}{13} & \frac{7}{26} \end{bmatrix}$
7. $\begin{bmatrix} 3 & -4 & 3 \\ 1 & -2 & -1 \\ -1 & -4 & -4 \end{bmatrix}^{-1}$	$\begin{bmatrix} \frac{-2}{13} & \frac{14}{13} & \frac{-5}{13} \\ \frac{-5}{26} & \frac{9}{26} & \frac{-3}{13} \\ \frac{3}{13} & \frac{-8}{13} & \frac{1}{13} \end{bmatrix}$
8. $\begin{bmatrix} 1 & 2 \\ 0 & -1 \\ 4 & 7 \end{bmatrix}^{-1}$	undefined

Homework 1.6 Solving Systems of Linear Equations with the TI-Nspire

Solve the matrix equation.

9. $\begin{bmatrix} 1 & -2 \\ -5 & 7 \end{bmatrix} A = \begin{bmatrix} -6 \\ 30 \end{bmatrix}$	$A = \begin{bmatrix} -6 \\ 0 \end{bmatrix}$

Solve the following systems of linear equations.

10. $\begin{matrix} 5x + 4y = 10 \\ -2x - 3y = -11 \end{matrix}$	Write the matrix equation $\begin{bmatrix} 5 & 4 \\ -2 & -3 \end{bmatrix} \begin{bmatrix} x \\ y \end{bmatrix} = \begin{bmatrix} 10 \\ -11 \end{bmatrix}$ $\begin{bmatrix} 5 & 4 \\ -2 & -3 \end{bmatrix}^{-1} \times \begin{bmatrix} 10 \\ -11 \end{bmatrix}$
Solve the matrix equation $\begin{bmatrix} x \\ y \end{bmatrix} = \begin{bmatrix} -2 \\ 5 \end{bmatrix}$	$x = -2$ $y = 5$
11. $\begin{matrix} -8x + 8y = 16 \\ -7x + 9y = 22 \end{matrix}$	Write the matrix equation $\begin{bmatrix} -8 & 8 \\ -7 & 9 \end{bmatrix} \begin{bmatrix} x \\ y \end{bmatrix} = \begin{bmatrix} 16 \\ 22 \end{bmatrix}$ $\begin{bmatrix} -8 & 8 \\ -7 & 9 \end{bmatrix}^{-1} \begin{bmatrix} 16 \\ 22 \end{bmatrix}$
Solve the matrix equation $\begin{bmatrix} x \\ y \end{bmatrix} = \begin{bmatrix} 2 \\ 4 \end{bmatrix}$	$x = 2$ $y = 4$

Homework 1.6 Solving Systems of Linear Equations with the TI-Nspire

Solve the following systems of linear equations.

12.	$6x - 3y - 3z = 6$ $-4x + 5y - 5z = 0$ $-5x + y + 5z = -6$	Write the matrix equation $\begin{bmatrix} 6 & -3 & -3 \\ -4 & 5 & -5 \\ -5 & 1 & 5 \end{bmatrix}^{-1} \begin{bmatrix} 6 \\ 0 \\ -6 \end{bmatrix}$
Solve the matrix equation $\begin{bmatrix} x \\ y \\ z \end{bmatrix} = \begin{bmatrix} 0 \\ -1 \\ -1 \end{bmatrix}$		$x = 0$ $y = -1$ $z = -1$
13.	$-6r + 3s - t = -1$ $-r + s + t = 2$ $5r + 4s - t = 12$	Write the matrix equation $\begin{bmatrix} -6 & 3 & -1 \\ -1 & 1 & 1 \\ 5 & 4 & -1 \end{bmatrix}^{-1} \begin{bmatrix} -1 \\ 2 \\ 12 \end{bmatrix}$
Solve the matrix equation $\begin{bmatrix} r \\ s \\ t \end{bmatrix} = \begin{bmatrix} 1 \\ 2 \\ 1 \end{bmatrix}$		$r = 1$ $s = 2$ $t = 1$

Homework 1.6 Solving Systems of Linear Equations with the TI-Nspire
CHALLENGE QUESTIONS

The triangle given by the points $(1,2), (4,5)$ and $(3,-4)$ can be written in a matrix as: $\begin{bmatrix} 1 & 4 & 3 \\ 2 & 5 & -4 \end{bmatrix}$. Moving the triangle to the right 4 and down 3 is represented by
$$\begin{bmatrix} 1 & 4 & 3 \\ 2 & 5 & -4 \end{bmatrix} + \begin{bmatrix} 4 & 4 & 4 \\ -3 & -3 & -3 \end{bmatrix}$$

Multiplying by the matrices below will perform rotations of the triangle.

90°	180°	270°	360°
$\begin{bmatrix} 0 & -1 \\ 1 & 0 \end{bmatrix}$	$\begin{bmatrix} -1 & 0 \\ 0 & -1 \end{bmatrix}$	$\begin{bmatrix} 0 & 1 \\ -1 & 0 \end{bmatrix}$	$\begin{bmatrix} 1 & 0 \\ 0 & 1 \end{bmatrix}$

Consider the triangle below.

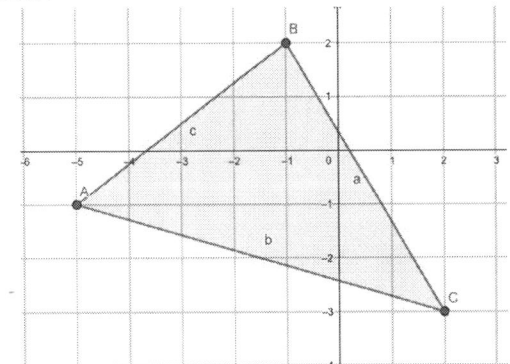

Using transformation matrices and your TI-Nspire, give the coordinates of the requested triangles.

Matrix that represents the original triangle $\triangle ABC$.	
Matrix expression that indicates a rotation of 270°, and a vertical translation of 4 unit upward. Give the coordinates of the translated triangle.	
Matrix expression that indicates moving the triangle down 1 and to the left 4, then rotating by 180°. Give the coordinates of the transformed triangle.	

1.7. Review

Assessment Checklist. Below are the competencies one should master in preparation for an assessment on matrices.

- [ ] Identify and classify matrices by their dimensions
- [ ] Based on matrix dimension, know when an operation is undefined
- [ ] Solve simple equations based on the equality of matrices
- [ ] Addition and Subtraction of matrices
- [ ] Multiplication of matrices by a scalar value
- [ ] Multiplication of matrices
- [ ] Find the determinant of a 2x2 matrix
- [ ] Find the inverse of a 2x2 matrix
- [ ] Solve a 2x2 matrix equation, $AX = B$
- [ ] Solve a 2x2 system of linear equations with matrices
- [ ] TI-Nspire: Enter matrices and store them on the TI-Nspire
- [ ] TI-Nspire: Perform all matrix operations we learned: addition, subtraction, scalar multiplication, matrix multiplication
- [ ] TI-Nspire: Find inverses, determinants of 2x2 and 3x3 matrices
- [ ] TI-Nspire: Solve 2x2 and 3x3 matrix equations
- [ ] TI-Nspire: Solve 2x2 and 3x3 systems of linear equations

Review 1.7

Use your TI-Nspire for 1-7. Use the following matrices to solve problems 1-6.

✷ Store

$$A = \begin{bmatrix} 1 & 5 & -4 \\ 2 & 6 & 3 \\ 1 & 8 & 4 \end{bmatrix}_{3\times 3} \quad B = \begin{bmatrix} 3 & -9 \\ 14 & 2 \end{bmatrix}_{2\times 2} \quad C = \begin{bmatrix} 3 & 4 \end{bmatrix}_{1\times 2} \quad D = \begin{bmatrix} 3 & 6 & 1 \\ 9 & -7 & 4 \end{bmatrix}_{2\times 3}$$

1. AD 3×3 2×3 **undefined**	2. CB $\begin{bmatrix} 65 & -19 \end{bmatrix}$	3. $\frac{1}{2}DA$ $\begin{bmatrix} 8 & \frac{69}{2} & 5 \\ -\frac{1}{2} & \frac{35}{2} & \frac{-41}{2} \end{bmatrix}$
4. D^2 (not square) **undefined**	5. $\det(A)$ -65	6. A^{-1} $\begin{bmatrix} 0 & \frac{4}{5} & \frac{-3}{5} \\ \frac{1}{13} & \frac{-8}{65} & \frac{11}{65} \\ \frac{-2}{13} & \frac{3}{65} & \frac{4}{65} \end{bmatrix}$

7. Given the system of equations: $\begin{cases} 2x - 3y = -18 \\ x + 4y - z = -4 \\ 2x + y + 2z = 32 \end{cases}$

   a. Write a matrix equation that represents the system above.

$$\begin{bmatrix} 2 & 3 & 0 \\ 1 & 4 & -1 \\ 2 & 1 & 2 \end{bmatrix} \begin{bmatrix} x \\ y \\ z \end{bmatrix} = \begin{bmatrix} -18 \\ -4 \\ 32 \end{bmatrix} \qquad \begin{bmatrix} x \\ y \\ z \end{bmatrix} = \begin{bmatrix} 2 & -3 & 0 \\ 1 & 4 & -1 \\ 2 & 1 & 2 \end{bmatrix}^{-1} \begin{bmatrix} -18 \\ -4 \\ 32 \end{bmatrix}$$

$$\begin{bmatrix} x \\ y \\ z \end{bmatrix} = \begin{bmatrix} -3 \\ 4 \\ 17 \end{bmatrix}$$

   b. Solve the matrix equation.

   $x = -3$ $\qquad y = 4 \qquad$ $z = 17$

Review 1.7

For 8-13, you should be able to do each problem WITHOUT the TI-Nspire. For problems 8-12, use the following matrices.

$$A = \begin{bmatrix} 3 & 2 \\ 19 & 13 \end{bmatrix} \quad (2\times 2) \qquad B = \begin{bmatrix} -1 & 2 & -1 \\ 0 & 3 & -2 \end{bmatrix} \quad (2\times 3) \qquad C = \begin{bmatrix} -4 & 2 \\ 1 & -1 \end{bmatrix} \quad (2\times 2)$$

---

8. Find AB $2\times 2 \cdot 2\times 3 = 2\times 3$

$$\begin{bmatrix} 3 & 2 \\ 19 & 13 \end{bmatrix} \begin{bmatrix} -1 & 2 & -1 \\ 0 & 3 & -2 \end{bmatrix}$$

$$\begin{bmatrix} 3(-1)+2(0) & 3(2)+2(3) & 3(-1)+2(-2) \\ 19(-1)+13(0) & 19(2)+13(3) & 19(-1)+13(-2) \end{bmatrix}$$

$$\begin{bmatrix} -3 & 12 & -7 \\ -19 & 77 & -45 \end{bmatrix}$$

---

9. BA $2\times 3 \quad 2\times 2$

undefined

---

10. A^{-1}

$$\begin{bmatrix} 3 & 2 \\ 19 & 13 \end{bmatrix}^{-1}$$

$\det A = (3\cdot 13)-(2\cdot 19) = 1$

$$A^{-1} = \frac{1}{1}\begin{bmatrix} 3 & 2 \\ 19 & 13 \end{bmatrix}$$

---

11. $\det(C)$

$$\begin{bmatrix} -4 & 2 \\ 1 & -1 \end{bmatrix}$$

$\det C = (-4\cdot -1)-(2\cdot 1)$

$\det C = \boxed{2}$

---

12. $\frac{1}{2}(A+C)$ $2\times 2 \quad 2\times 2$

$$\frac{1}{2}\left(\begin{bmatrix} 3 & 2 \\ 19 & 13 \end{bmatrix} + \begin{bmatrix} -4 & 2 \\ 1 & -1 \end{bmatrix}\right) = \frac{1}{2}\begin{bmatrix} -1 & 4 \\ 20 & 12 \end{bmatrix} = \begin{bmatrix} -\frac{1}{2} & 2 \\ 10 & 6 \end{bmatrix}$$

Review 1.7

13. For the system of equations, $\begin{cases} 4x + 3y = -2 \\ 14x + 11y = -8 \end{cases}$

   a. Write the system of equations as a matrix equation.

   $$\begin{bmatrix} 4 & 3 \\ 14 & 11 \end{bmatrix} \begin{bmatrix} x \\ y \end{bmatrix} = \begin{bmatrix} -2 \\ -8 \end{bmatrix}$$

   b. Solve the matrix equation.

   $$\begin{bmatrix} x \\ y \end{bmatrix} = \begin{bmatrix} 4 & 3 \\ 14 & 11 \end{bmatrix}^{-1} \begin{bmatrix} -2 \\ -8 \end{bmatrix}$$

   $\det A = (4 \cdot 11) - (3 \cdot 14)$
   $\det A = \boxed{2}$

   $$= \frac{1}{2} \begin{bmatrix} 11 & -3 \\ -14 & 4 \end{bmatrix} \begin{bmatrix} -2 \\ -8 \end{bmatrix} = \frac{1}{2} \begin{bmatrix} 11(-2) + -3(-8) \\ -14(-2) + 4(-8) \end{bmatrix} = \frac{1}{2} \begin{bmatrix} 2 \\ -4 \end{bmatrix} = \begin{bmatrix} 1 \\ -2 \end{bmatrix}$$

   $x = 1 \qquad y = -2$

2. LINEAR PROGRAMMING

Linear programming (LP, also called linear optimization) is a method to achieve the best outcome (such as maximum profit or lowest cost) in a mathematical model whose requirements are represented by linear relationships.

LP first gained notice during WWII. Soviet mathematicians worked problems to reduce expenditure, while maximizing loss to the enemy. Air Force mathematicians worked on problems to maximize the effectiveness of 70 people in 70 different jobs.

2.1. Systems of Inequalities, Feasible Regions, Objective Functions

A key skill in working linear programming (LP) problems is the ability to graph systems of linear inequalities. This section is a quick review of the linear graphing skills needed for the LP unit.

Graphing a line in Standard Form
The equation of a line in Standard Form is given by $Ax + By = C$.

Graph the line $2x + 4y = 12$

x	y
0	3	(0,3)
6	0	(6,0)

You need two points to graph a line. Easiest thing to do is graph the intercepts.

Set $x = 0$, solve for y.
Set $y = 0$, solve for x.

2.1—62

Graphing a linear inequality
The prior example was graphing a linear equation. The solution was a line. To graph a linear inequality, our solution is a region.

Graph the inequality
$2x + 4y \geq 12$

x	y
0	3
6	0

Notice in this case, we graph the line and shade above the line as the solution to the inequality.

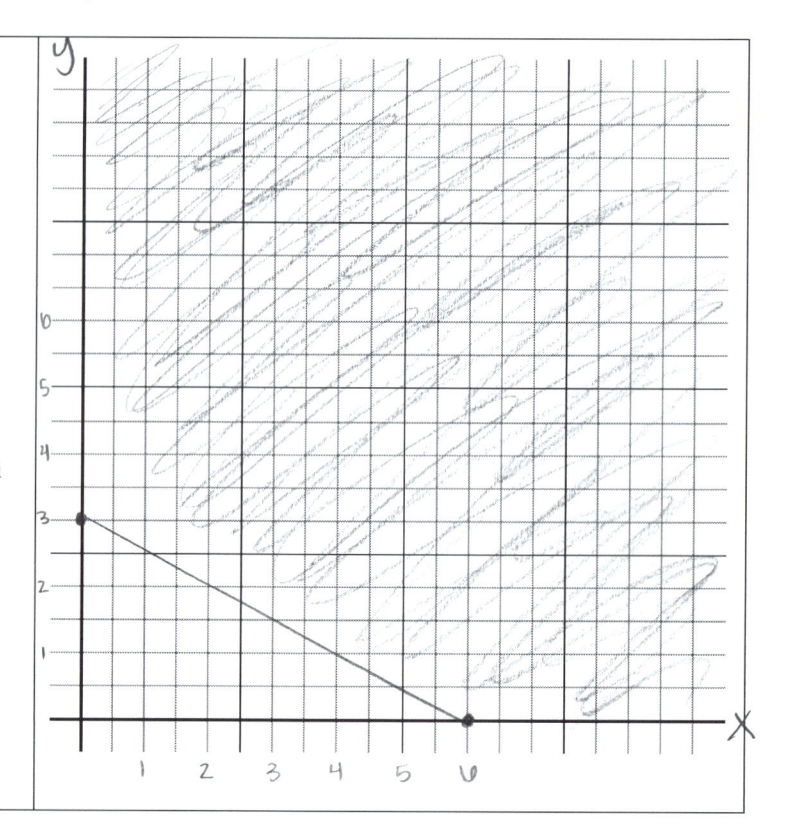

Graph the inequality
$2x + 4y < 12$

x	y
0	3
6	0

Notice in this case, we graph a dotted line and shade below the line as the solution to the inequality.

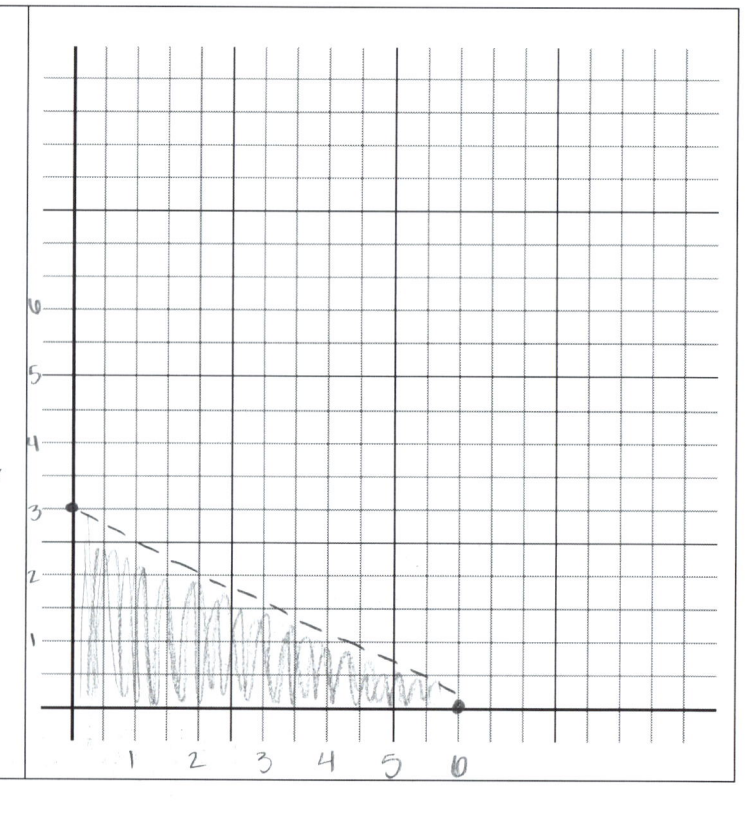

2.1—63

Graphing a system of linear inequalities

When we have more than one inequality to graph, and we wish to find the region that satisfies each inequality simultaneously, we are then graphing a system of linear inequalities.

Graph the following system of linear inequalities. Break it down into 4 problems:

1st quadrant
$$x \geq 0$$
$$y \geq 0$$
$$2x + 3y \leq 18$$
$$x + y \geq 4$$

$x \geq 0$	$y \geq 0$	$2x + 3y \leq 18$	$x + y \geq 4$
Shading to the right of the y-axis.	Shading above the x-axis.	Graph the line $2x + 3y = 18$ | x | y | | 0 | 6 | | 9 | 0 | Shade where?	Graph the line $x + y = 4$ | x | y | | 0 | 4 | | 4 | 0 | Shade where?

$<$ = shade below; dotted

$\leq$ = shade below; solid

$>$ = shade above; dotted

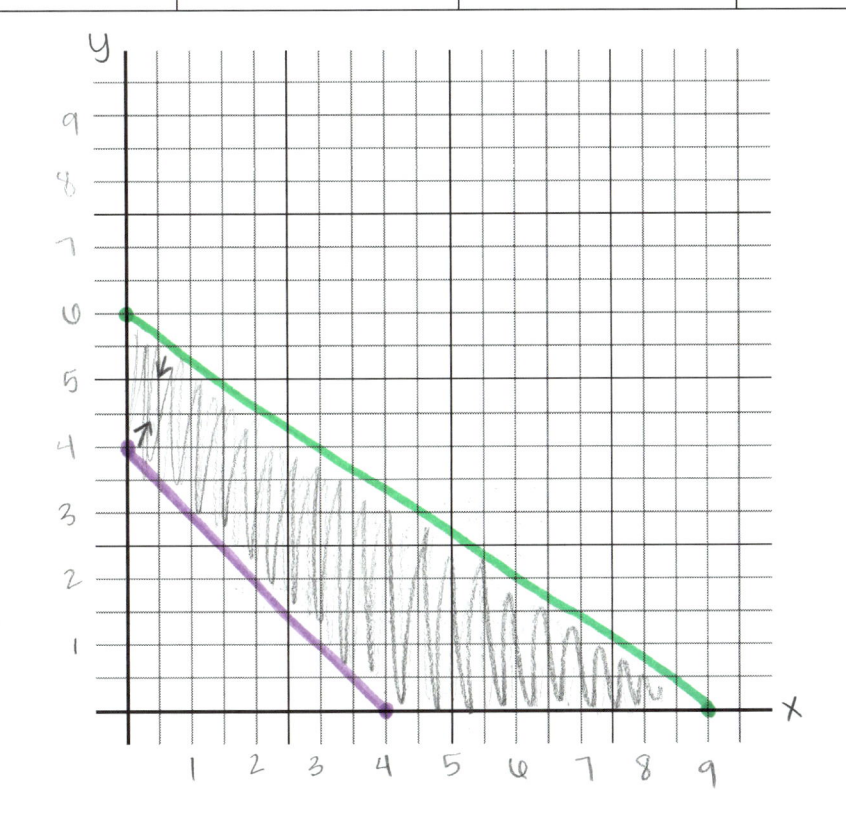

$\geq$ = shade above; solid

So, in the prior example
$$x \geq 0$$
$$y \geq 0$$
$$2x + 3y \leq 18$$
$$x + y \geq 4$$

The solution of the graph is given below.

The shaded region that satisfies each inequality is called the **feasible region**. Each of the inequalities is considered a **constraint**. Constraints: $x \geq 0$ $y \geq 0$ $2x + 3y \leq 18$ $x + y \geq 4$ At the intersection of each boundary line of the feasible region are the **vertices** of the feasible region.	*(graph showing feasible region with vertices)*
Vertices (9,0) (0,6) (0,4) (4,0)	

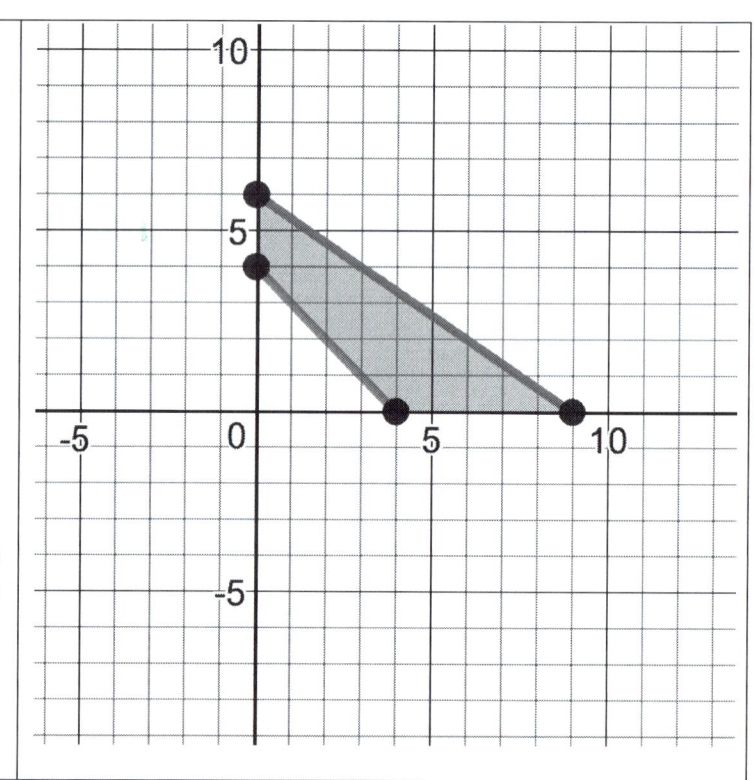

Now to go one step further, the goal of linear programming or linear optimization is not just to graph a feasible region of constraints. The goal is to optimize a particular function. For example, if the constraints dealt with limitations that a baker faces in baking different pastries, the goal would be to maximize profit, or minimize costs.

Let's revisit the same feasible region from our original problem. But let's add the condition that we wish to maximize the function

$$P = 20x + 15y$$

The function P is our **objective function**. It is the condition that we wish to optimize (either maximize or minimize) while still meeting all the constraints of the feasible region.

For an interactive look at this, visit https://tinyurl.com/Algebra2LP
The slider for the value of B shows the objective function intersecting our feasible region.

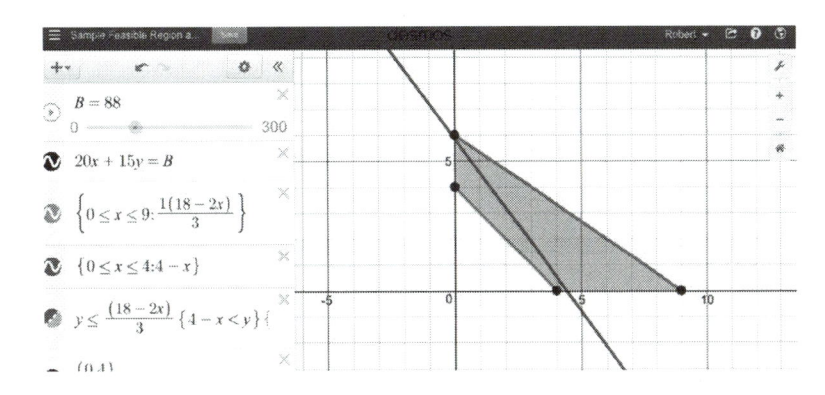

Use the slider for the value of B to see where the objective function P is maximized.

Change the objective function from $P = 20x + 15y$ to $P = 20x + 35y$. How does that change where the slider meets the feasible region?

Evaluate the original objective function at each of the vertices of the feasible region.

Vertices	Objective Function $P = 20x + 15y$	Max or Min?
(9,0)	20(9) + 15(0) =	180
(0,6)	20(0) + 15(6) =	90
(0,4)	20(0) + 15(4) =	60
(4,0)	20(4) + 15(0) =	80

max value = 180 @ (9,0)

min value = 60 @ (0,4)

2.1—65

If you revisit the interactive graph at the link above, you should see the max and min of our objective function graphically.

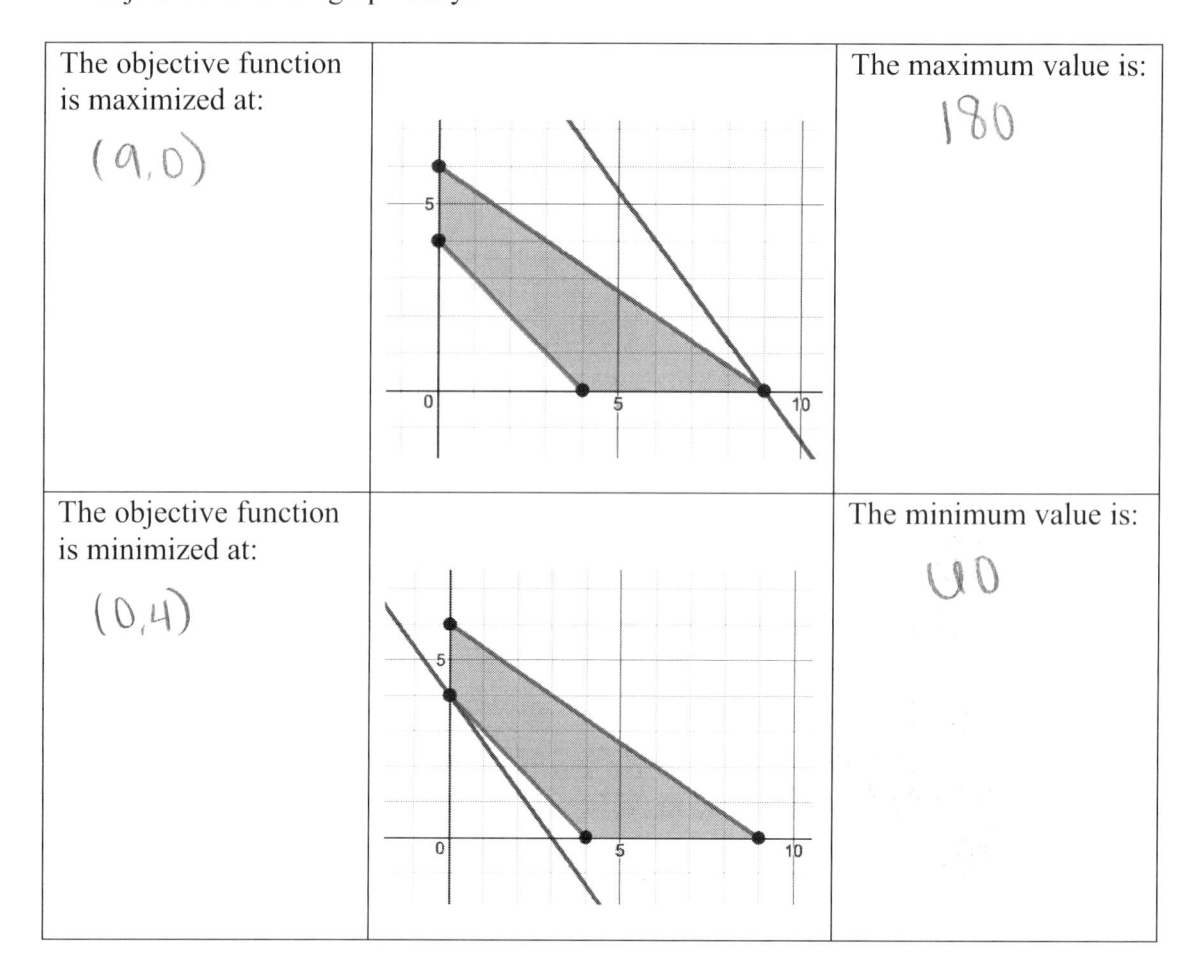

The objective function is maximized at: (9,0)		The maximum value is: 180
The objective function is minimized at: (0,4)		The minimum value is: 40

2.1—67

Maximize the objective function $S = x + 2y$ subject to the contraints: $\begin{cases} 2x + y \leq 7 \\ y - x \geq 1 \\ x \geq 0, y \geq 0 \end{cases}$

Steps: Graph the feasible region, find the vertices, test the objective function at each of the vertices.

$2x + y \leq 7$

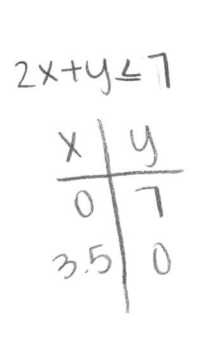

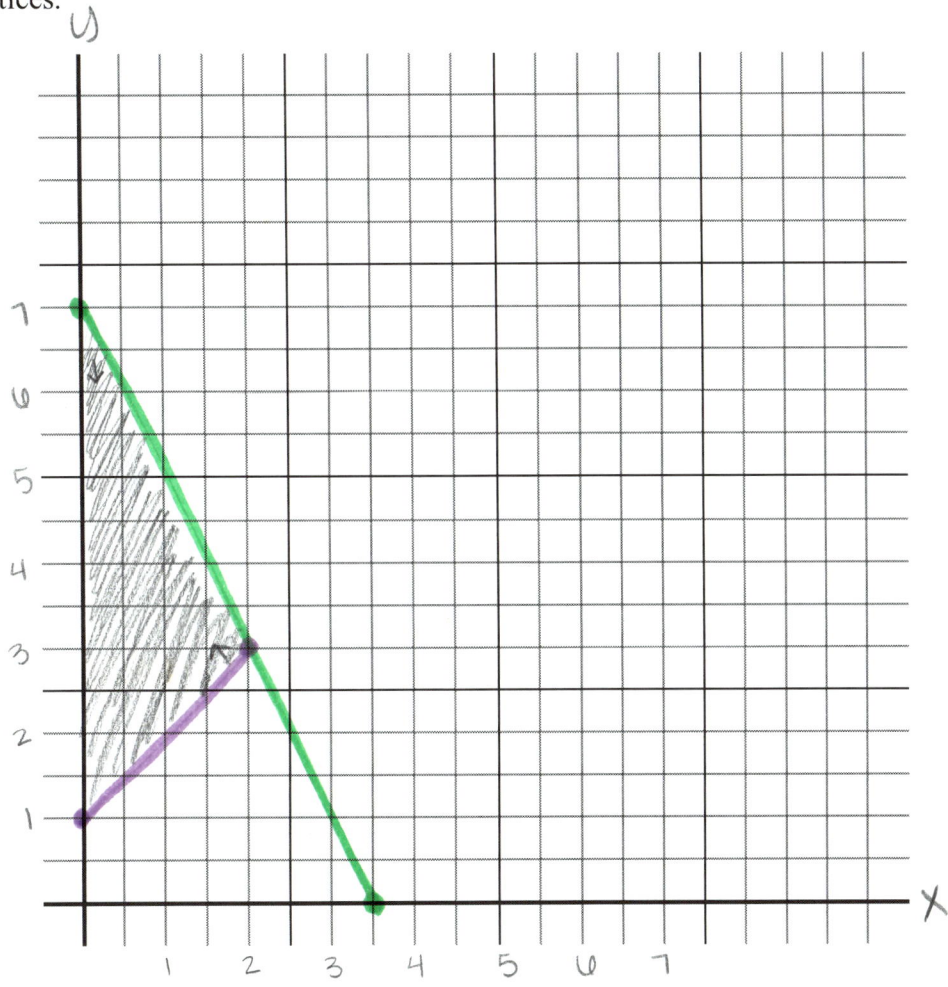

x	y
0	7
3.5	0

$y - x \geq 1$

x	y
0	1
-1	0
↓	↓
2	3

vertices =
(0,7) (0,1) (2,3)

For your reference.

In the prior problem, we found ourselves needing to find the point of intersection of the two lines:

$$2x + y = 7$$
$$y - x = 1$$

You can use any of the methods below to find the intersection point (solution to the system of equations.)

Substitution 1	Matrices & TI-Nspire 2	TI-Nspire, Solve System of Equations 3
Solve one equation for one of the variables: $$y = x + 1$$ Substitute that into the other equation. $$2x + (x + 1) = 7$$ Solve for the variable. $$2x + x + 1 = 7$$ $$3x + 1 = 7$$ $$3x = 6$$ $$x = 2$$ Substitute $x = 2$ into either original equation to find that $y = 3$.	Write as a matrix equation. $$\begin{bmatrix} 2 & 1 \\ -1 & 1 \end{bmatrix} \begin{bmatrix} x \\ y \end{bmatrix} = \begin{bmatrix} 7 \\ 1 \end{bmatrix}$$ Then solve for $\begin{bmatrix} x \\ y \end{bmatrix}$ on your calculator. $$\begin{bmatrix} x \\ y \end{bmatrix} = \begin{bmatrix} 2 & 1 \\ -1 & 1 \end{bmatrix}^{-1} \begin{bmatrix} 7 \\ 1 \end{bmatrix}$$	MENU 3: Algebra 7: Solve System of Equations 2: Solve System of Equations $$\text{solve}\left(\begin{cases} 2 \cdot x + y = 7 \\ y - x = 1 \end{cases}, \{x, y\}\right)$$ x=2 and y=3

2.1—69

Homework 2.1 Systems of Inequalities, Feasible Regions, Objective Functions

1. Consider the feasible region graphed below.

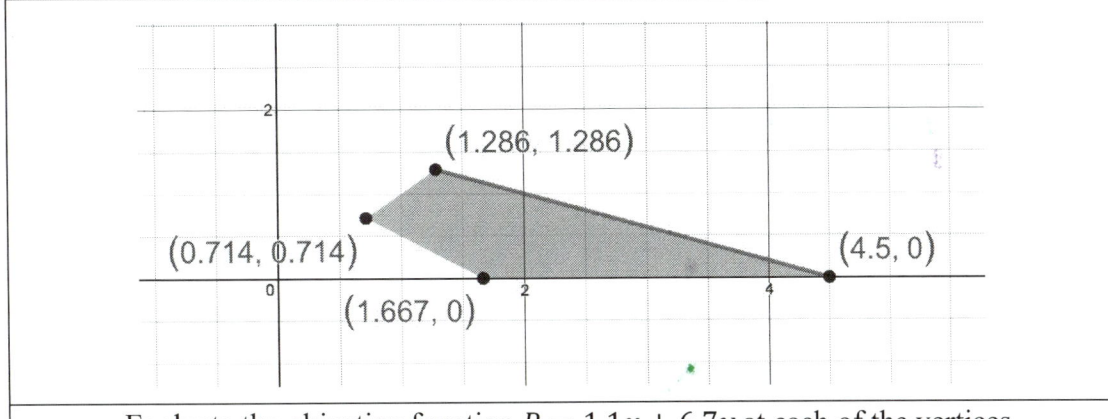

Vertices shown: (0.714, 0.714), (1.286, 1.286), (1.667, 0), (4.5, 0)

Evaluate the objective function $P = 1.1x + 6.7y$ at each of the vertices			
1.1(0.714) + 6.7(0.714) = 5.5692 ✓	1.1(1.667) + 6.7(0) = 1.8337 ✓	1.1(1.286) + 6.7(1.286) = 10.0308 ✓	1.1(4.5) + 6.7(0) = 4.95 ✓
Identify where the maximum occurs and its value.	(1.286, 1.286) 10.0308 ✓	Identify where the minimum occurs and its value.	(1.667, 0) 1.8337 ✓

$x = 1.286$ $y = 1.286$ $x = 1.667$ $y = 0$

2. Maximize the objective function $U = 12p + 10q$

Constraints: $\begin{cases} p + q \leq 10 \\ 2p - q \leq 2 \\ p \geq 0, q \geq 0 \end{cases}$

$p + q \leq 10$
$p + q = 10$

p	q
0	10 ✓
10	0

$2p - q \leq 2$
$2p - q = 2$

p	q
0	-2
1	0

(2, 2)

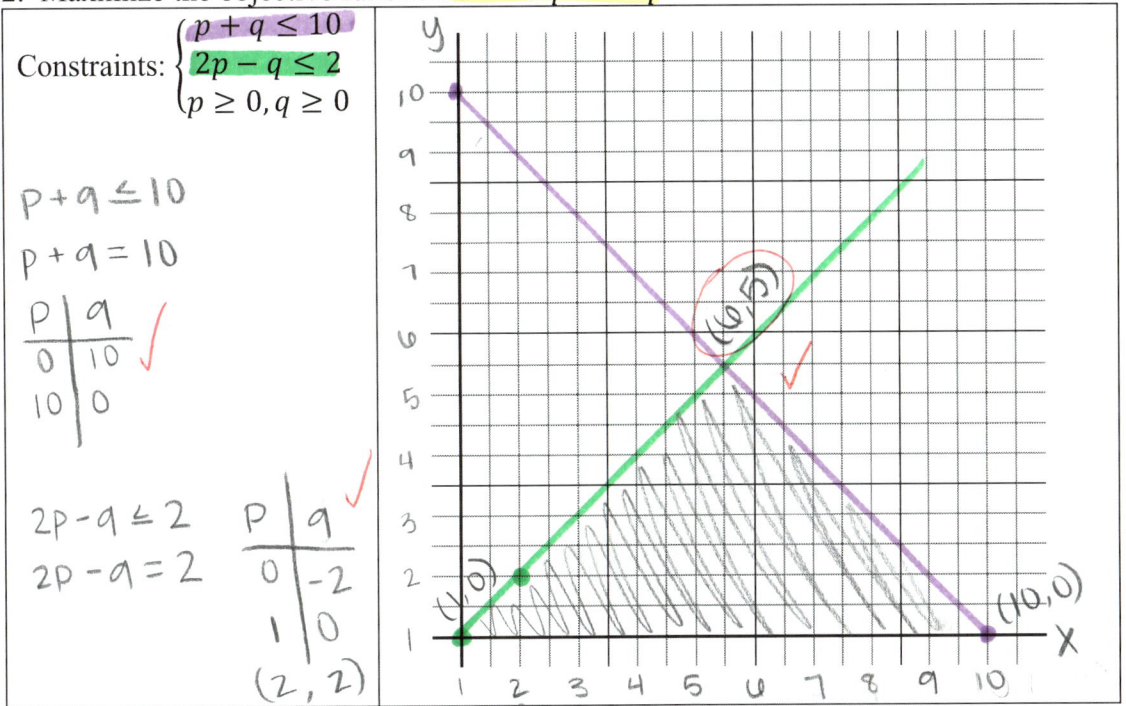

Vertices: (1, 0) (10, 0) (6, 5)

$12(1) + 10(0) = 12$ ← min
$12(10) + 10(0) = 120$
$12(6) + 10(5) = 122$ ← max

Homework 2.1 Systems of Inequalities, Feasible Regions, Objective Functions

3. Minimize the objective function $Q = 4A + 5B$.
Draw the feasible region, find the vertices, and evaluate the objective function at each vertex.

Constraints: $\begin{cases} A + B \geq 8 \\ 2A + B \geq 10 \\ A \geq 0, B \geq 0 \end{cases}$

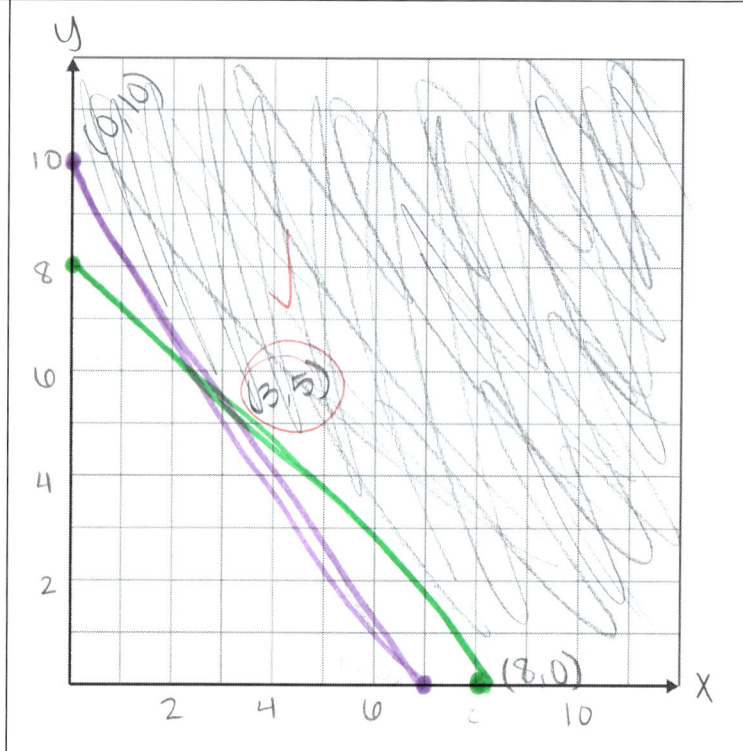

$A + B = 8$

A	B
0	8
8	0	✓

$2A + B = 10$

A	B
0	10
5	0	✓

Vertices:
(2, 6) (0, 10) (8, 0)

Evaluate Objective Function
$4(2) + 5(6) = 38$
$4(0) + 5(10) = 50$
$4(8) + 5(0) = 32$

Solution
Max = 50 (0, 10)
Min = 32 (8, 0)

Homework 2.1 Systems of Inequalities, Feasible Regions, Objective Functions

CHALLENGE PROBLEMS

1. Consider the feasible region graphed below. If the objective function passes through the point (5,0) and has a slope of -2, at what point is the objective function maximized and what is the maximum value?

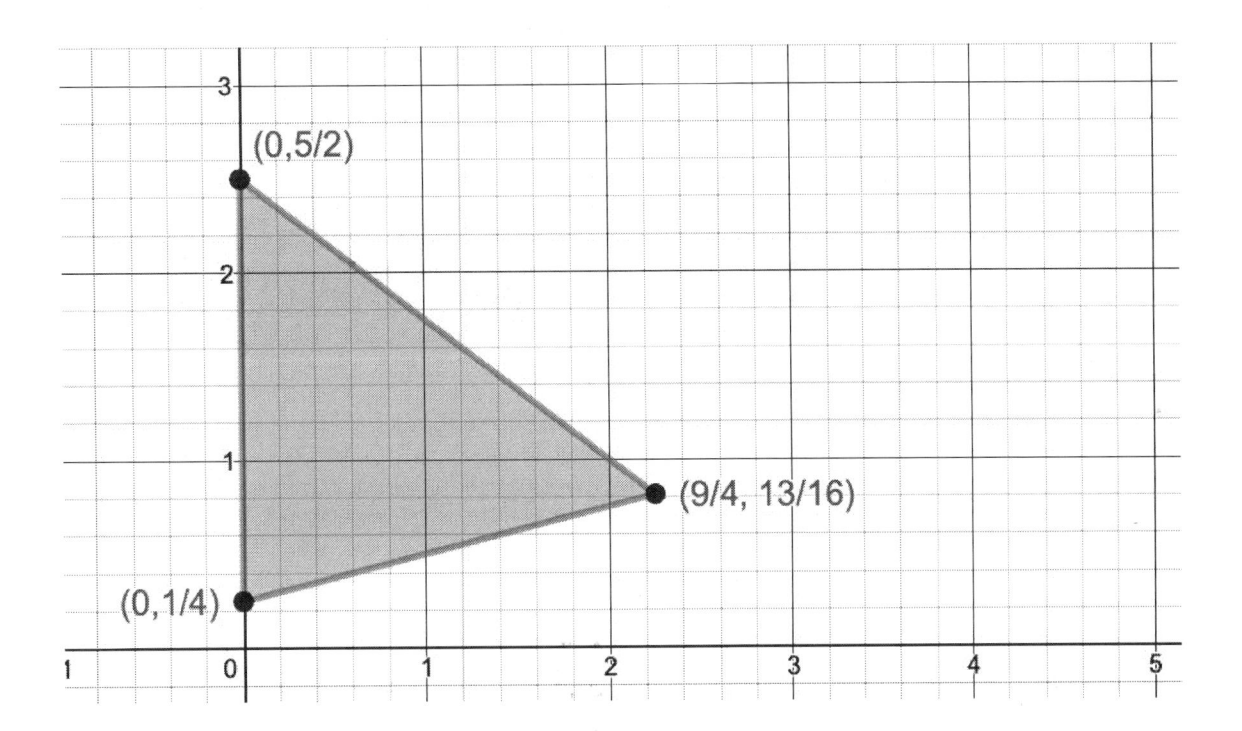

2. In the same feasible region above, a different objective function has an x-intercept of (5,0) and it is maximized at the point (0,5/2). What is a possible objective function?

2.2. Linear Programming Application: Maximization

Perfect Parking Company has 600 square meters of parking available to customers. The hourly rates for parking are $2.50 for cars and $7.50 for buses. The average car will require 6 square meters, and a bus will take 30 square meters of space. If the lot is only allowed to handle 60 vehicles how many of each vehicle type should be accepted to maximize the income? What is the maximum income?

In order to answer this, we need to determine:

Variables	Constraints	Objective Function
Goal of the optimization? Max or min?	Feasible region	Vertices
Value of the objective function at each vertex		

To define the variables, there will be a question in the problem that nicely sets up what the variables should be.

If the lot is only allowed to handle 60 vehicles how many of each vehicle type should be accepted to maximize the income? What is the maximum income?	Define the variables B = number of buses C = number of cars

There will also be a question that helps define the goal of the optimization (maximum or minimum), what is being optimized, and other supporting information to express the objective function in terms of our variables.

The hourly rates for parking are $2.50 for cars and $7.50 for buses. .... What is the maximum income?	What is being optimized? income Max or Min? Max
The hourly rates for parking are $2.50 for cars and $7.50 for buses ...... What is the maximum income?	Define objective function $I = 7.5B + 2.5C$

Next to determine the constraints – typically contained a few sentences. It is best to organize these in a grid before writing the inequalities.

Perfect Parking Company has 600 square meters of parking available to customers. The hourly rates for parking are $2.50 for cars and $7.50 for buses. The average car will require 6 square meters, and a bus will take 30 square meters of space. If the lot is only allowed to handle 60 vehicles how many of each vehicle type should be accepted to maximize the income? What is the maximum income?

Constraints	Buses B	Cars C	Limit
Amount of parking space	30B	6B	≤ 600
Number of vehicles	B	C	≤ 60

Objective Function: $I = 7.5B + 2.5C$

Constraints:

$B \geq 0$ $30B + 6C \leq 600$
$C \geq 0$ $B + C \leq 60$

B	C
0	100
20	0

B	C
0	60
60	0

Vertices:

$(0, 60)$
$(20, 0)$
calc. $\rightarrow (10, 50)$

Evaluate the objective function $I = 7.5B + 2.5C$

$7.5(0) + 2.5(60) = 150$ $7.5(20) + 2.5(0) = 150$
$7.5(10) + 2.5(50) = 200$

Solution

max income is $200, 10 buses & 50 cars

Some key points:
- There are multiple steps to each problem.
- Concentration and organization are key.
- Colors can be helpful when drawing the feasible region.
- Be consistent with your variables and the order in which they appear in the objective function and constraints. (find the question)
- If you define your variables as B and C, it doesn't matter which is the horizontal and which is the vertical, you decide.
- If B is horizontal and C is vertical, then write all constraints with B first and C next.
- Write your objective function with B first and C second.
- The original problem contains words; therefore, your answer should contain words, not just numbers. Look to the original questions and make sure your answer fulfills the question.
    - For example, in the problem we just completed "If the lot is only allowed to handle 60 vehicles how many of each vehicle type should be accepted to maximize the income? What is the maximum income?"
    - The answer should be "Allow __10__ buses and __50__ cars in the parking lot for a maximum income of __$200__"

War Eagle Bakery makes decorated sheet cakes for parties in two sizes, a full sheet and a half sheet. A batch of full-sheet cakes takes 4 hours of oven time and 2 hours of decorating time, while a batch of half-sheet cakes takes 2 hours of oven time and 4 hours of decorating time. The oven and decorating rooms are each available 24 hours per day. The bakery makes a profit of $30 on each batch of full-sheet cakes and $35 on each batch of half-sheet. How many batches of each type should they bake? What is their profit?

Variables full + half	Objective Function (max profit)
F = number of full H = number of half	P = 30F + 35H

Determine Constraints

	F	H	Limit
oven time	4 hours	2 hours	≤ 24 hours
decorating time	2 hours	4 hours	≤ 24 hours

Constraints

F ≥ 0   4F + 2H ≤ 24
H ≥ 0   2F + 4H ≤ 24

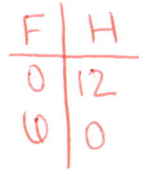

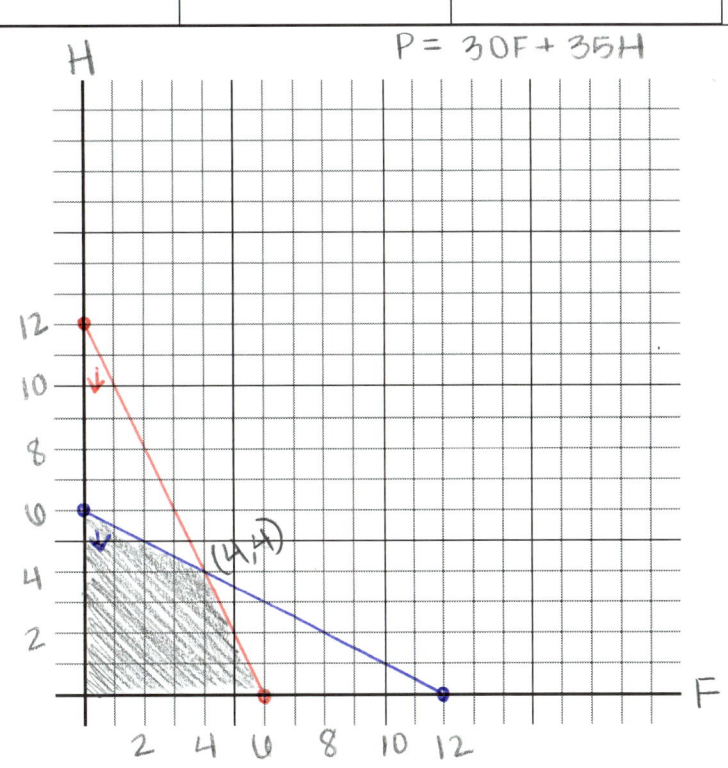

F	H
0	12
6	0

F	H
0	6
12	0

P = 30F + 35H

Vertices
(0, 6) (6, 0) (4, 4)

30(0) + 35(6) = 210
30(6) + 35(0) = 180
30(4) + 35(4) = 260 — max

Solution

max profit of $260 with 4 batches of each.

2.2—76

The Elite Tweet Pottery Shoppe makes two kinds of birdbaths: fancy glazed and simple unglazed. The unglazed birdbath requires 1/2 hour on the pottery wheel and 1 hour in the kiln. The glazed birdbath takes 1 hour on the wheel and 5 hours in the kiln. The company's one pottery wheel is available for at most 8 hours per day. The six kilns can be used a total of at most 25 hours per day. The company has a standing order for 5 unglazed birdbaths per day, so it must produce at least that many. The pottery shop's profit on each unglazed birdbath is $10, while its profit on each glazed birdbath is $40. How many of each kind of birdbath should be produced in order to maximize profit? What is the maximum profit?

Constraints	Glazed	Unglazed	Limit
Pottery Wheel	$1G$	$\frac{1}{2}U$	≤ 8
Kiln	$5G$	$1U$	≤ 25
#per day		U	≥ 5

Objective Function: $P = 40G + 10U$

Constraints:

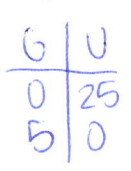

$G \geq 0$
*$U \geq 5$

$1G + \frac{1}{2}U \leq 8$
$5G + 1U \leq 25$

G	U
0	16
8	0

G	U
0	25
5	0

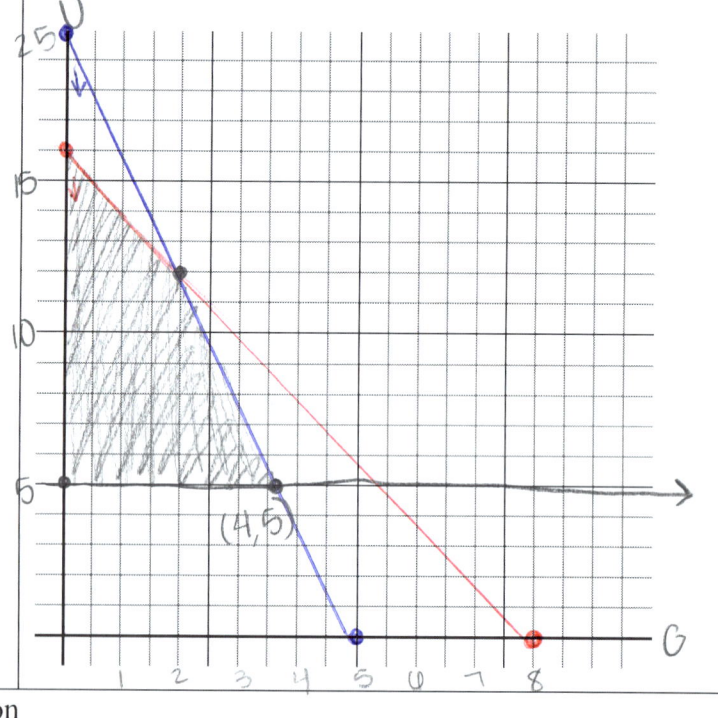

Vertices:
(0, 5)
(0, 10)
(4, 5)
(3, 10)

Evaluate the objective function
$40(0) + 10(5) = 50$ $40(4) + 10(5) = 210$
$40(0) + 10(10) = 100$ $40(3) + 10(10) = 220$

Solution

3 glazed & 10 unglazed should be made to get a maximum profit of $220.

Homework 2.2 Linear Programming Application: Maximization

Piñatas are made to sell at a craft fair. It takes 2 hours to make a mini piñata and 3 hours to make a regular- sized piñata. The owner of the craft booth will make a profit of $12 for each mini piñata sold and $24 for each regular-sized piñata sold. If the craft booth owner has no more than 30 hours available to make piñatas and wants to have at least 12 piñatas to sell, how many of each size piñata should be made to maximize profit?

maximize profit

Variables	Objective Function
M = # of mini R = # of regular	$P = 12M + 25R$

Determine Constraints

	M	R	Limit
hours	$2M$	$3R$	≤ 30
# of piñatas	M	R	≥ 12

Constraints

$M \geq 0$ $2M + 3R \leq 30$
$R \geq 0$ $M + R \geq 12$

M	R
0	10
15	0

M	R
0	12
12	0

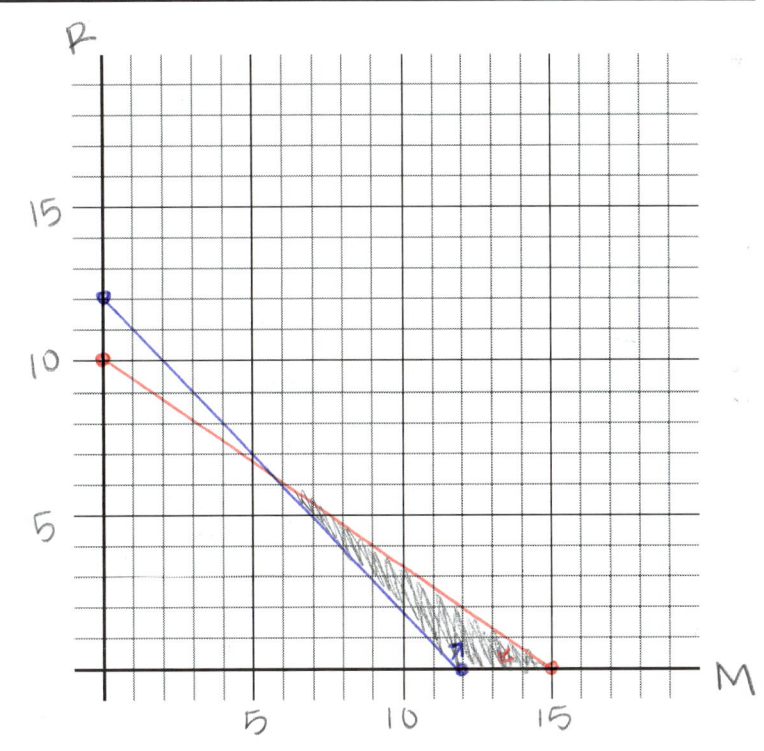

Vertices

$(12, 0) \ (15, 0) \ (6, 6)$

$12(12) + 25(0) = 144$
$12(15) + 25(0) = 180$
$12(6) + 25(6) = 222$ — max

Solution

To make a max profit of $222, you need to make 6 minis and 6 regulars.

Homework 2.2 Linear Programming Application: Maximization

2. Your candy factory is making chocolate-covered peanuts and chocolate-covered pretzels. For each case of peanuts, you make $40 profit; for each case of pretzels, you make $55 profit. The table below shows the number of machine hours and man hours needed to make one case of each type of candy. It also shows the maximum number of hours available. How many cases of each should you produce to maximize profits?

Production Hours	Peanuts	Pretzels	Maximum Hours
Machine Hours	2	6	150
Man Hours	5	4	155

Variables	Objective Function
P = # peanut cases Z = # pretzel cases	P = 40P + 55Z

Determine Constraints

constraints	P	Z	Limits
machine hrs	2P	6Z	≤ 150
staff hrs	5P	4Z	≤ 155

Constraints

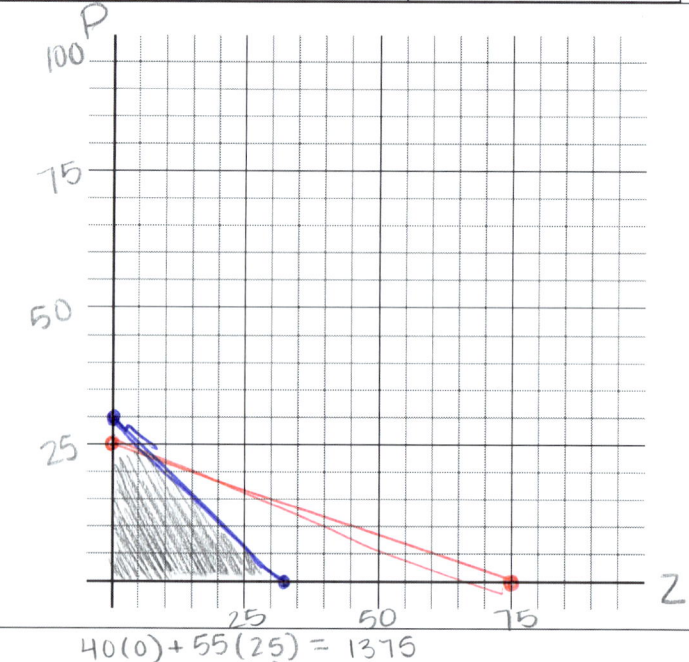

$P \geq 0$ $2P + 6Z \leq 150$
$Z \geq 0$ $5P + 4Z \leq 155$

P	Z
0	25
75	0

P	Z
0	38.75
31	0

Vertices

(0, 25) (31, 0) (15, 20)

40(0) + 55(25) = 1375
40(31) + 55(0) = 1240
40(15) + 55(20) = 1100 — max

Solution

To max profit of $1700, you need to make 15 cases of peanuts and 20 cases of pretzels.

Homework 2.2 Linear Programming Application: Maximization

For all of the challenge problems, use the feasible region below.

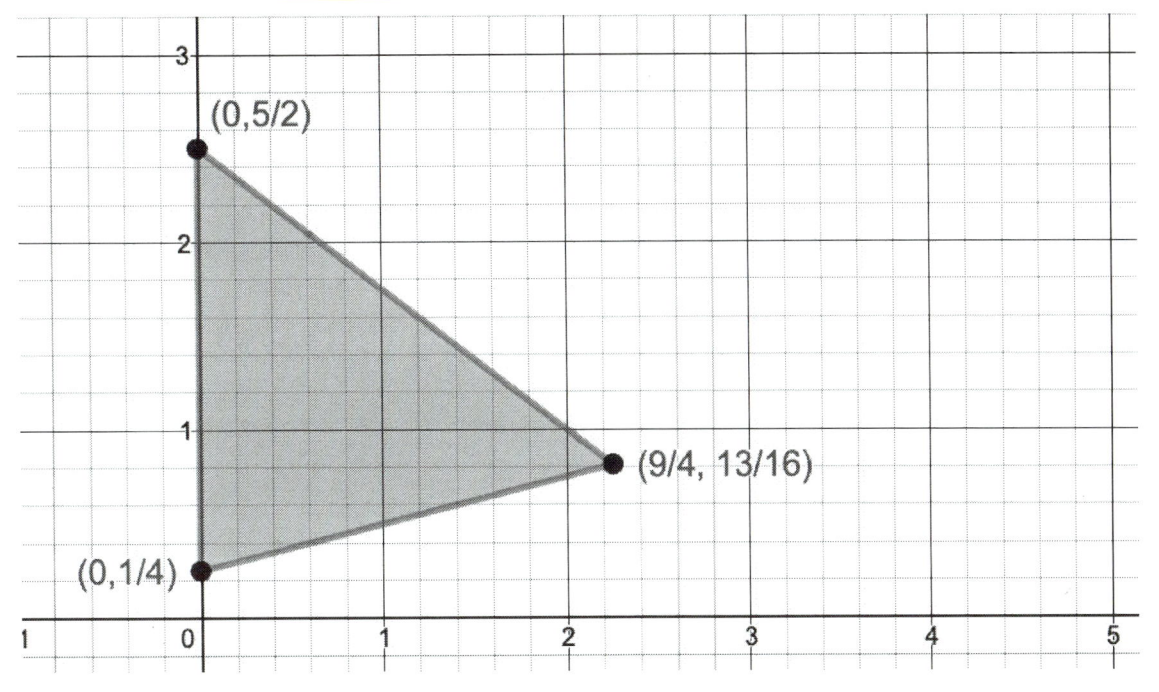

[Feasible region triangle with vertices (0, 5/2), (0, 1/4), and (9/4, 13/16)]

1. Write an objective function that will give the same maximum value for each pair of vertices.

   a. (0, 5/2) and (9/4, 13/16)

   b. (0, 1/4) and (0, 5/2)

   c. (0, 1/4) and (9/4, 13/16)

2. Is it possible to write an objective function that gives the same maximum value at all three vertices? If so, what is the objective function? If not, why not?

2.3. Linear Programming Application: Minimization

What are the general steps to solve a linear programming problem?

define variables (question)	objective function	max or min?
constraints	graphing	feasible region
verticies	solve problem	

What are the challenges you encounter with linear programming problems?

setting the problem up	answer the correct?	where to shade ↓ verticies wrong

In the prior lesson and homework, we considered linear programming problems that maximized an objective function. Today's lesson will continue the work in linear programming but focus on problems that ask for an objective function to be minimized.

The minimum daily requirements from a powdered dietary supplement are 1200 units of vitamin A, 800 units of vitamin B and 1000 units of vitamin C. An ounce of dietary powder X costs $3 and an ounce of dietary powder Y costs $2. Use the table below to find **how many ounces of each powder should be consumed each day to minimize the cost and still meet the minimum daily requirements?**

	Serving	Vitamin A	Vitamin B	Vitamin C
Dietary Powder X	1 ounce	500 units	100 units	100 units
Dietary Powder Y	1 ounce	200 units	200 units	400 units

First, define the variables.

To do this, we look to the questions in the problem to assist. "how many pounds of each powder should be consumed each day to minimize the cost and still meet the minimum daily requirements?"	$x =$ oz of powder x $y =$ oz of powder y

Next, once the variables are defined you can write the objective function.

What are we asked to optimize? COST Is it a (minimization) or maximization?	Look to the problem for details about the entity you are asked to optimize. This will enable you to write the objective function. $C = 3x + 2y$

Define the constraints.

Constraint	X	Y	Limit
amt. Vitamin A	500x	200y	≥ 1200
amt. Vitamin B	100x	200y	≥ 800
amt. Vitamin C	100x	400y	≥ 1000

| Variables x = oz powder x y = oz powder y | Objective Function $C = 3x + 2y$ |

Constraints

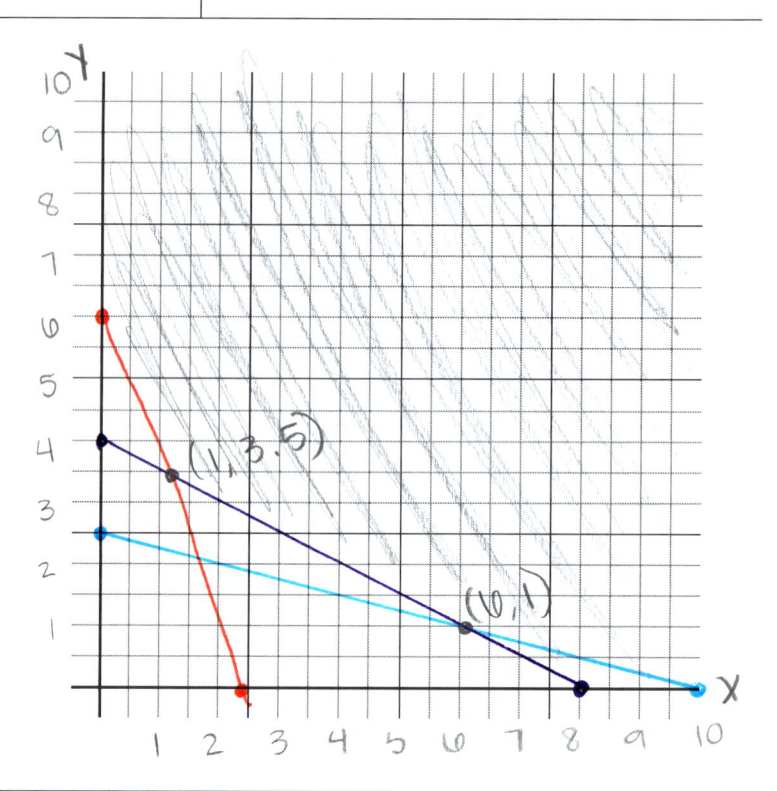

$x \geq 0$
$y \geq 0$

$500x + 200y \geq 1200$
$100x + 200y \geq 800$
$100x + 400y \geq 1000$

x	y
0	6
2.4	0

x	y
0	4
8	0

x	y
0	2.5
10	0

Vertices

$(0,6)$ $3(0) + 2(6) = 12$
$(10,0)$ $3(10) + 2(0) = 30$
$(1, \frac{7}{2})$ $3(1) + 2(3.5) = \boxed{10}$
$(6,1)$ $3(6) + 2(1) = 20$

Solution

1 once of powder x and 3.5 onces of powder y for minimum cost of $10.

Woodward has contracted with a rock group, U3 to perform at the school. The school gym can accommodate at most 1000 people. The school has guaranteed the ticket receipts would be at least $4800. The tickets are $4 for students and $6 for non-students. U3 will be paid a fee of $2.50 for each student ticket and $4.50 for each non-student ticket. **What is the minimum pay for U3, and how many of each ticket type most be sold for this minimum payoff?** minimum

Variables S = # student tickets N = # non-student tickets	Objective Function $P = 2.5S + 4.5N$

Determine Constraints

	S	N	Limits
capacity	S	N	≤ 1000
receipts	4S	6N	≥ 4800

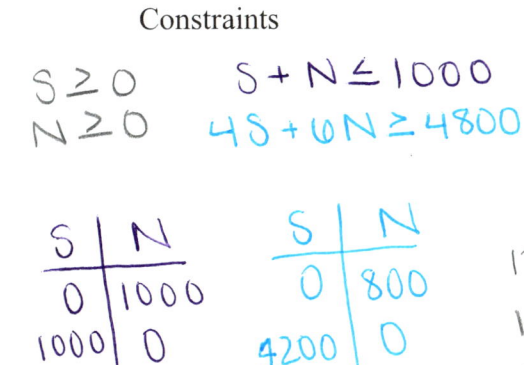

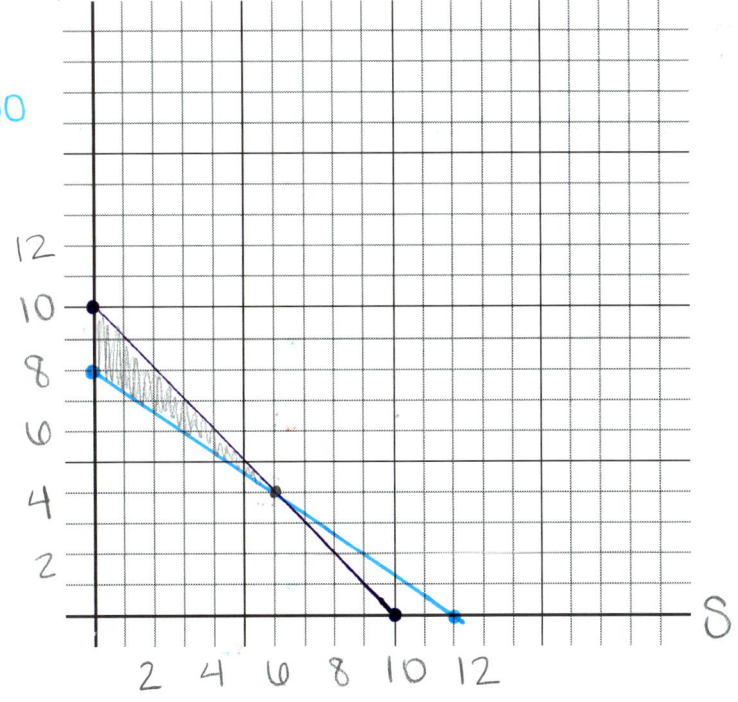

Constraints

$S \geq 0$ $S + N \leq 1000$
$N \geq 0$ $4S + 6N \geq 4800$

S	N
0	1000
1000	0

S	N
0	800
1200	0

Vertices

$(0, 1000)$ $2.5(0) + 4.5(1000) = 4500$
$(0, 800)$ $2.5(0) + 4.5(800) = 3600$
$(600, 400)$ $2.5(600) + 4.5(400) = \boxed{3300}$

Solution

The minimum pay is $3300 and 600 student tickets, 400 non-students.

Homework 2.3 Linear Programming Application: Minimization

1. The officers of the senior class are planning to rent busses and vans for a class trip. Each bus can transport 40 students, requires 3 chaperones, and costs $1200 to rent. Each van can transport 8 students, requires 1 chaperone, and costs $100 to rent. The officers want to be able to accommodate at least 400 students with no more than 36 chaperones. How many vehicles of each type should they rent in order to minimize the transportation costs? What are the minimum transportation costs?

Variables B = buses V = vans	Objective Function C = 1200B + 100V

Determine Constraints	B	V	Limits
chaperones	3B	1V	≤ 36
students	40B	8V	≥ 400

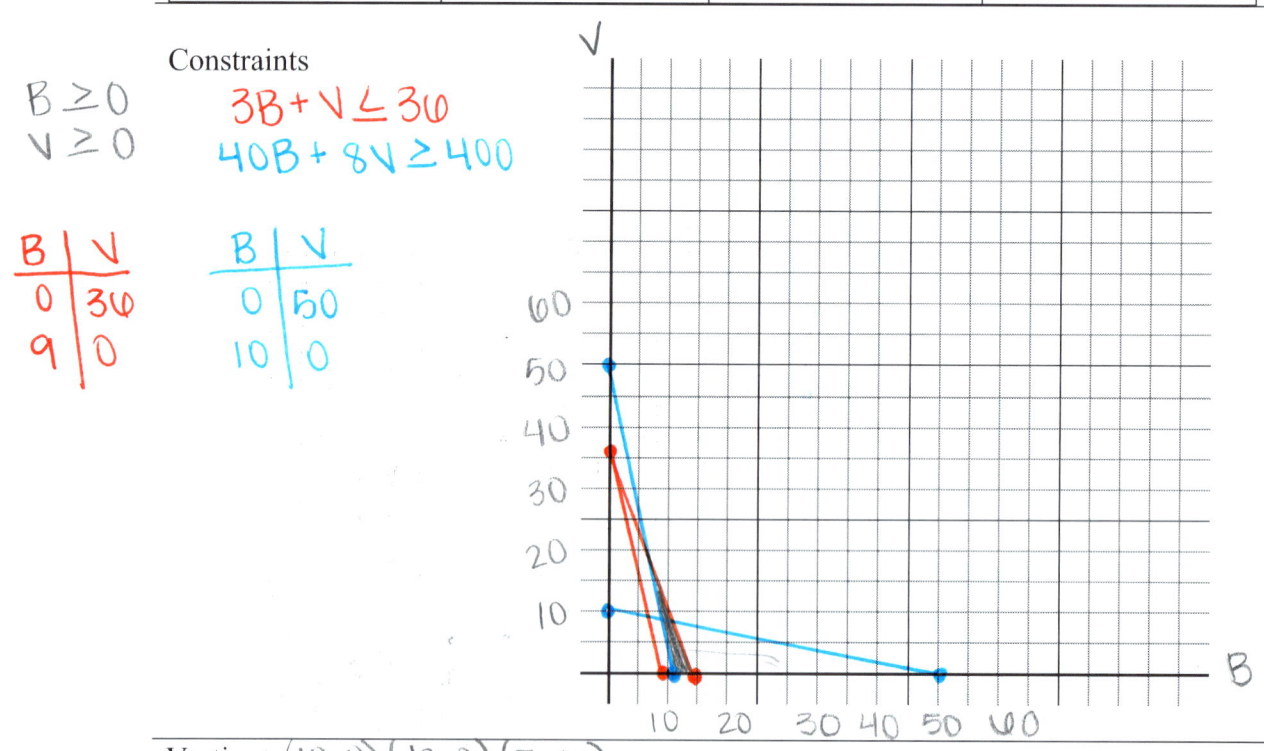

Constraints

B ≥ 0   3B + V ≤ 36
V ≥ 0   40B + 8V ≥ 400

B	V
0	36
9	0

B	V
0	50
10	0

Vertices (10, 0) (12, 0) (7, 15)

1200(10) + 100(0) = 12000
1200(12) + 100(0) = 14400
1200(7) + 100(15) = (9900)

Solution

Use 7 buses + 15 vans to minimize a cost of $9900.

Homework 2.3 Linear Programming Application: Minimization

2. A construction firm employs two levels of tile installers: craftsmen and apprentices. Craftsmen make $200 per day and apprentices make $120. Craftsmen install 500 square feet of specialty tile, 100 square feet of plain tile, and 100 linear feet of trim in one day. An apprentice installs 100 square feet of specialty tile, 200 square feet of plain tile, and 100 linear feet of trim in one day. The firm has a job that requires at least 2000 square feet of specialty tile, at least 1600 square feet of plain tile, and at least 1200 linear feet of trim completed per day. How many craftsmen and how many apprentices should be assigned to this job so that it can be completed with minimum labor cost? What is the minimum labor cost?

Variables C = craftsmen A = apprentices	Objective Function C = 200C + 120A

Determine Constraints

	C	A	Limit
Specialty Tile	500C	100A	≥ 2000
Plain Tile	100C	200A	≥ 1600
Trim	100C	100A	≥ 1200

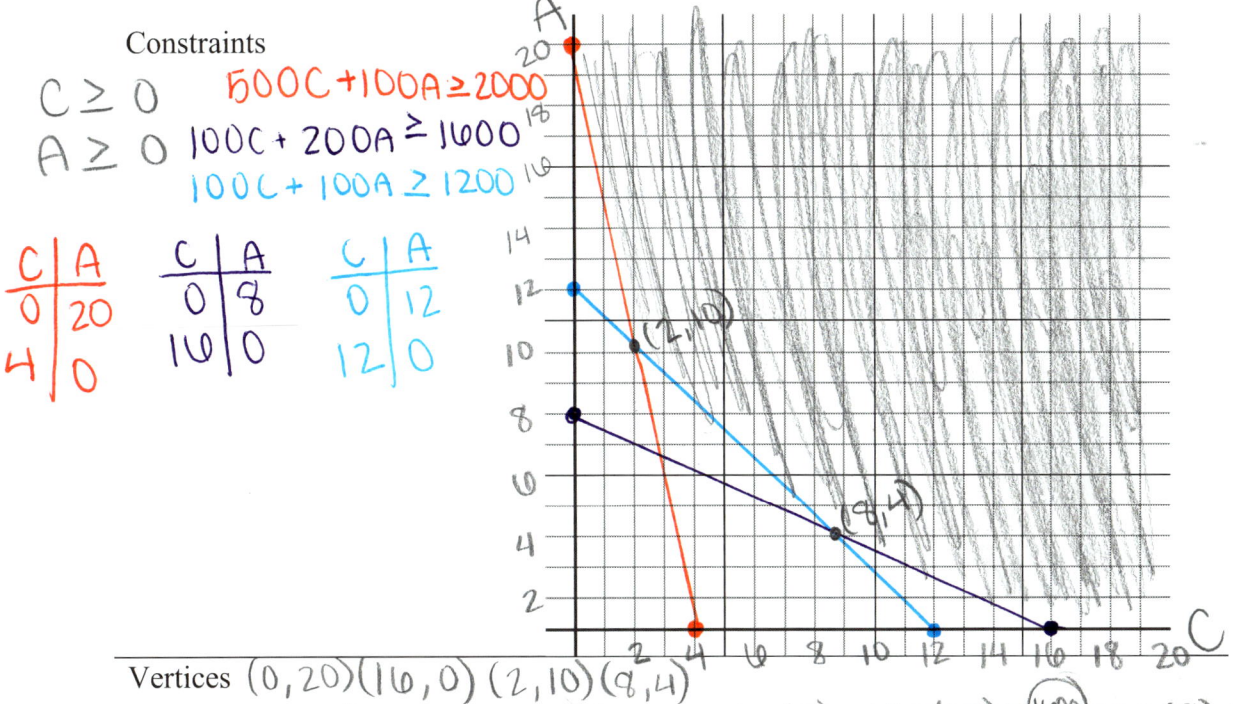

Constraints

$C \geq 0$ $500C + 100A \geq 2000$
$A \geq 0$ $100C + 200A \geq 1600$
              $100C + 100A \geq 1200$

C	A
0	20
4	0

C	A
0	8
16	0

C	A
0	12
12	0

Vertices (0, 20) (16, 0) (2, 10) (8, 4)
200(0) + 120(20) = 2400   200(16) = 3200   200(2) + 120(10) = 1600   200(8) + 120(4) = 2080

Solution
In order to achieve the min cost of $1600, hire 2 craftsmen & 10 apprentises.

Homework 2.3 Linear Programming Application: Minimization

CHALLENGE PROBLEM

1. The Big Bad Boat Company produces canoes, kayaks and jet skis. They must produce at least 5000 canoes and 12,000 kayaks each year; they can produce at most 30,000 jet skis in a year. The company has two factories: one in Georgia, and one in Alabama; each factory is open for a maximum of 240 days per year. The Georgia factory makes 20 cabin cruisers, 40 pontoons, and 60 jet skis per day. The Alabama factory makes 10 cruisers, 30 pontoons, and 50 jet skis per day. The cost to run the Georgia factory per day is $960,000; the cost to run the Alabama factory per day is $750,000. How many days of the year should each factory run in order to meet the boat production, yet do so at a minimum cost? What is the minimum cost?

Variables	Objective Function

Determine Constraints

Constraints

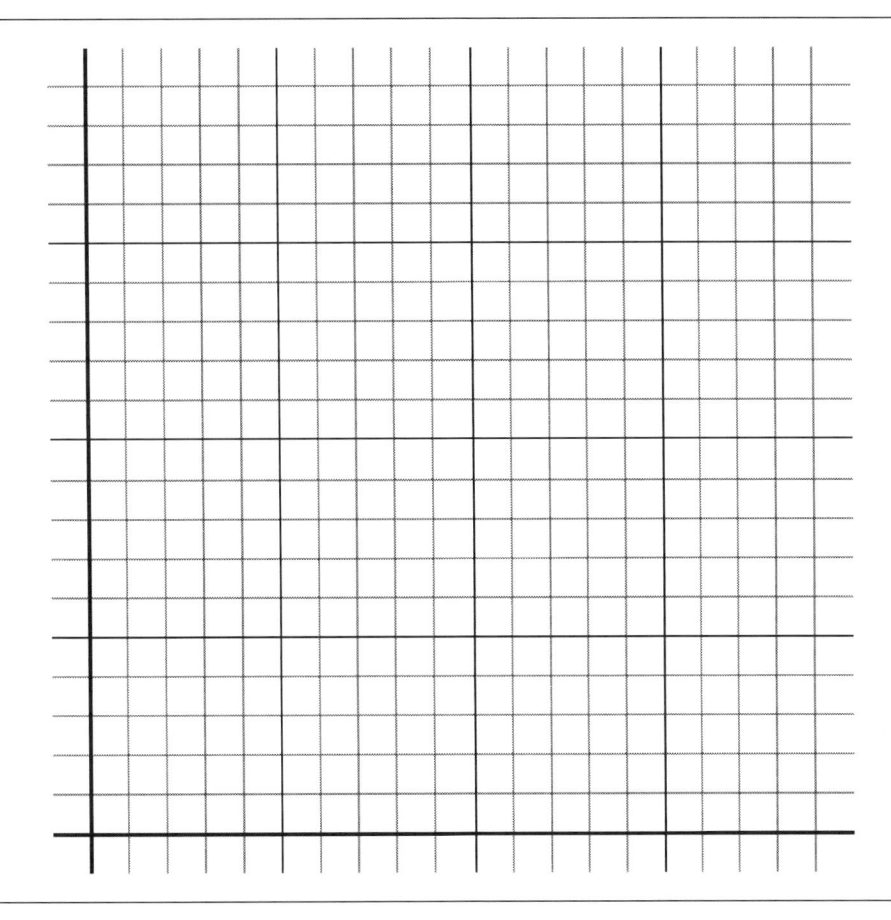

Vertices

Solution

2.3—87

Homework 2.3 Linear Programming Application: Minimization

CHALLENGE PROBLEM
2. Miranda is mixing two sports drinks together to create one that has the electrolytes she needs for her workouts. She needs a total of at least 600 mg sodium, 630 mg potassium, and 400 mg calcium. "Blue Wave" has 20 mg sodium, 60 mg potassium, and 20 mg calcium and it costs 10 cents per ounce. "Yellow Thunder" has 40 mg sodium, 30 mg potassium, and 20 mg calcium and it costs 15 cents per ounce. How many ounces of each sports drink should Miranda mix together to minimize her total cost?

Variables	Objective Function

Determine Constraints

Constraints

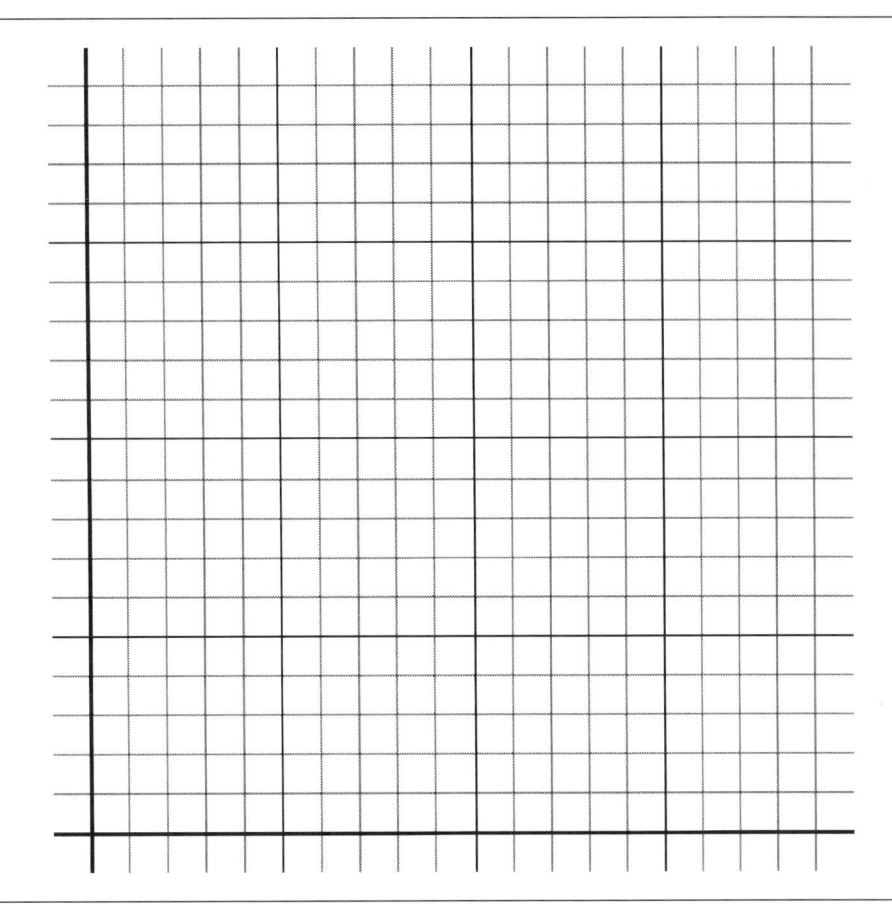

Vertices

Solution

2.4. Review

Assessment Checklist. Below are the competencies one should master in preparation for an assessment on matrices.

- [ ] Know how to graph linear inequalities, systems of linear inequalities
- [ ] Know how a system of linear inequalities will create a feasible region
- [ ] Find the vertices of a feasible region
- [ ] When necessary, find the point at which two lines intersect, either graphically, algebraically, or with your calculator
- [ ] How to evaluate an objective function to determine its max/min in a feasible region
- [ ] LP Applications: defining variables
- [ ] LP Applications: listing constraints
- [ ] LP Applications: writing the objective function
- [ ] LP Applications: graphing the feasible region
- [ ] LP Applications: identifying vertices
- [ ] LP Applications: testing the objective function at the vertices
- [ ] LP Applications: answering the question completely, with both words and numbers

Review 2.4

1. An ice cream supplier has two machines. Each machine produces vanilla and chocolate ice cream. Machine A produces 4 gallons of vanilla per hour and 5 gallons of chocolate per hour. Machine B produces 3 gallons of vanilla per hour and 10 gallons of chocolate per hour. To meet contractual obligations, the company must produce <mark>at least 60 gallons</mark> of vanilla and <mark>100 gallons of</mark> chocolate per hour. It costs $28 per hour to run Machine A and $25 to run Machine B. <mark>How many hours should each machine be operated to fulfill the contract at minimal expense?</mark>   min.

4V 5C
3V 10C

Variables A = machine A B = machine B	Objective Function C = 28A + 25B

Determine Constraints

	A	B	Limits
vanilla	4A	3B	≥ 60
chocolate	5A	10B	≥ 100

Constraints

a ≥ 0
b ≥ 0

4A + 3B ≥ 60
5A + 10B ≥ 100

A	B
0	20
15	0

A	B
0	10
20	0

Vertices (0, 20)   (20, 0)   (12, 4)

28(0) + 25(20) = 500   28(20) = 560   [28(12) + 25(4) = 436]

Solution

Machine A should be operated for 12 hours; Machine B for 4 hours. Min cost = $436.

Review 2.4

2. Gauss has started a lawn service. He has two different lawn mowers. The regular lawn mower can mow 4 acres/hour at an hourly cost of $4. The super lawn mower can mow 6 acres/hour at an hourly cost of $3. You have accepted a job that requires you to mow at least 12 acres of a pasture and provide the clippings for livestock. The regular mower can generate 4 bags of clippings per hour. The super mower only generates 2 bags of clippings per hour. You have guaranteed that you will produce at least 8 bags of clippings per hour. <u>How many hours should you run each mower to minimize hourly costs?</u>

Variables	Objective Function
Let R= # of hours of the regular mower. Let S= # hours of the Super mower	$C = 4R + 3S$

Determine Constraints

	R	S	Limits
acres/hour	4R	6S	≥ 12
clippings	4R	2S	≥ 8

Constraints

$S \geq 0$ $4R + 6S \geq 12$
$R \geq 0$ $4R + 2S \geq 8$

R	S
0	2
3	0

R	S
0	4
2	0

Vertices

$(0, 4)\ (3, 0)\ \left(\frac{3}{2}, 1\right)$

$4(0) + 3(4) = 12$
$4(3) + 3(0) = 12$
$\boxed{4\left(\frac{3}{2}\right) + 3(1) = 9}$

Solution

For min cost of $9, regular = $\frac{3}{2}$ hr, super = 1 hr

Review 2.4

3. Solve the following matrix equation for x, y, and z.

$$2\begin{bmatrix} 2 & x \\ 5 & 9 \end{bmatrix} + \begin{bmatrix} y+5 & 7 \\ -4 & -9 \end{bmatrix} = \begin{bmatrix} 7 & 15 \\ 6 & z \end{bmatrix} \quad \begin{bmatrix} 4 & 2x \\ 10 & 18 \end{bmatrix} + \begin{bmatrix} y+5 & 7 \\ -4 & -9 \end{bmatrix} = \begin{bmatrix} 7 & 15 \\ 6 & z \end{bmatrix}$$

$4 + y + 5 = 7$ $2x + 7 = 15$ $18 - 9 = z$

$\boxed{y = -2}$ $2x = 8$ $\boxed{z = 9}$

                $\boxed{x = 4}$

4. For the feasible region shown below, where is the objective function maximized? What is the maximum value? Show appropriate work to receive credit!

(plug in)

$P = 3x + 4y$

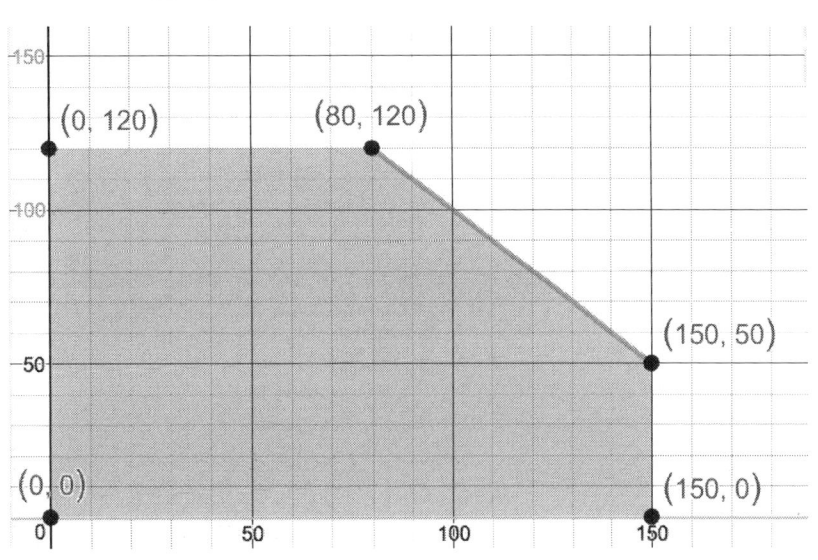

Vertices: (0, 120), (80, 120), (150, 50), (150, 0), (0, 0)

$3(0) + 4(120) = 480$

$\boxed{3(80) + 4(120) = 720}$

$3(150) + 4(50) = 650$

$3(150) + 4(0) = 450$

$3(0) + 4(0) = 0$

Max = 720
(80, 120)

Review 2.4

CHALLENGE PROBLEM (Linear Programming)

1. Ella is baking cookies for a bake sale to raise money for her favorite charity. She plans to make two types of cookies: ooey-gooey chococale chip and gluten-free oatmeal-raisin. A box of chocolate chip cookies costs $2 to make and takes 15 minutes to make; a box of oatmeal-raisin costs $4 to make and takes 15 minutes to make. Ella has $40 she can spend on supplies and 240 minutes she can set aside for preparing and baking. From past experience, Ella knows that she needs to make at least twice as many boxes of chocolate chip as she does of oatmeal-raisin. She will make $5 per box of chocolate chip and $7 per box of oatmeal-raisin. How many boxes of each type of cookie should Ella make in order to maximize the amount of money she makes for charity?

Variables	Objective Function

Determine Constraints

Constraints

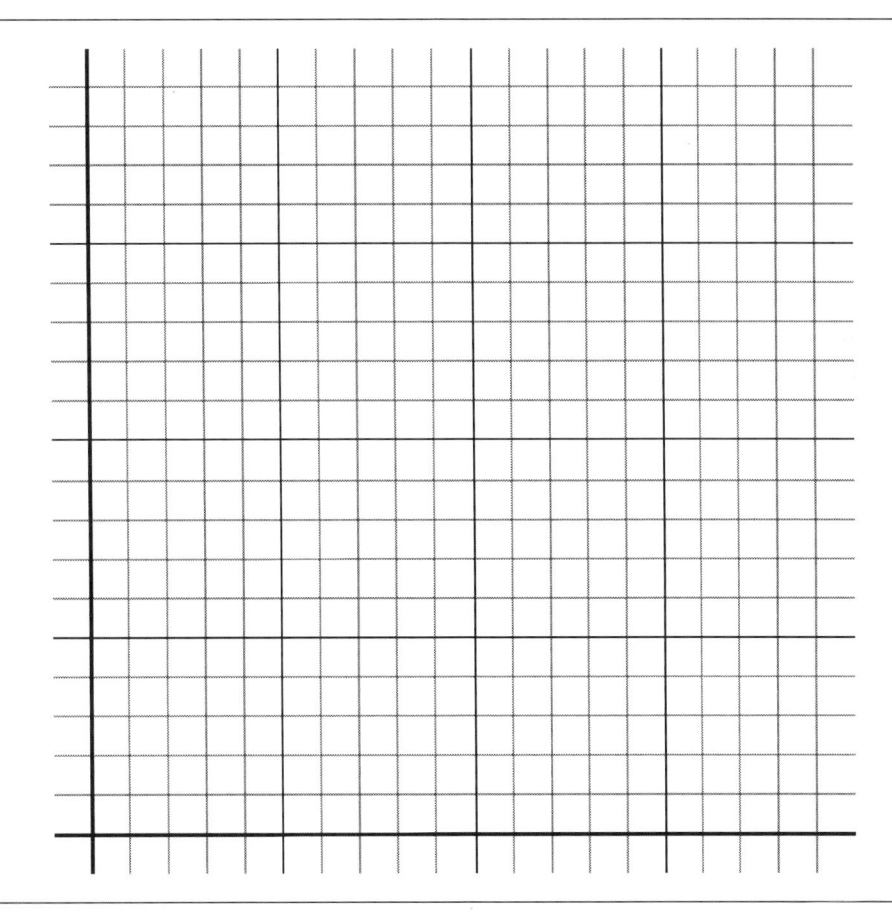

Vertices

Solution

2.5. Review Additional Linear Programming Problems

1. Every year for Homecoming, Woodward Academy has a cookout that requires <u>at least</u> 1000 hamburgers and 550 hot dogs. Woodward purchases from two vendors – "BBQ Brothers" and "Grill Masters". "BBQ Brothers" sells packs of 200 hamburgers and 50 hot dogs for $250. "Grill Masters" sells packs of 50 hamburgers and 50 hot dogs for $100. <mark>If Woodward wants to minimize the total cost, how many packs should be purchased from each vendor?</mark>

Variables B = packs sold by BBQ Bros G = packs Grill Masters	Objective Function $C = 250B + 100G$

Determine Constraints

	B	G	Limits
hamburgers	200B	50G	≥ 1000
hot dogs	50B	50G	≥ 550

Constraints

$B \geq 0$ $200B + 50G \geq 1000$
$G \geq 0$ $50B + 50G \geq 550$

B	G
0	20
5	0

B	G
0	11
11	0

(3, 8)

Vertices
(0, 20) (11, 0) (3, 8)

$250(0) + 100(20) = 2000$
$250(11) + 100(0) = 2750$
$250(3) + 100(8) = 1550$

Solution To minimize cost, 3 packs from BBQ Bros, 8 packs from Grill Master, to cost of $1550.

Review 2.5 Additional Linear Programming Problems

2. A patient in a hospital is required to have **at least 84** units of drug A and **120 units** of drug B each day (assume that an overdose would not be harmful). Each gram of substance M contains 10 units of drug A and 8 units of drug B, and each gram of substance N contains 2 units of drug A and 4 units of drug B. **What is the least total amount of substances M and N that can be taken to meet the minimum daily requirements of drugs A and B?**

Variables M = amt of subs M N = amt of subs N	Objective Function min. Amt = M + N

Determine Constraints

	M	N	Limits
drug A	10M	2N	≥ 84
drug B	8M	4N	≥ 120

Constraints

$M \geq 0$
$N \geq 0$

$10M + 2N \geq 84$
$8M + 4N \geq 120$

M	N
0	42
8.4	0

M	N
0	30
15	0

Vertices (0, 42) (15, 0) (4, 22)

0 + 42 = 42
15 + 0 = 15
4 + 22 = 26

Solution

To meet the minimum daily requirement, 15 units of M + 0 units of N.

Review 2.5 Additional Linear Programming Problems

3. A manufacturer of sleeping bags makes a <mark>standard model</mark> and a <mark>sub-zero model</mark>. Each standard sleeping bag requires <mark>1 labor-hour</mark> from the cutting department and <mark>3 labor-hours</mark> from the assembly department. Each sub-zero sleeping bag requires <mark>2 labor hours</mark> from the cutting department and <mark>4 labor-hours</mark> from the assembly department. The maximum labor-hours available per week in the cutting and assembly departments are <mark>32</mark> and <mark>84</mark>, respectively. In addition, because of demand, the distributor <mark>will not take more than 12 sub-zero sleeping bags</mark> per week. If the company makes a profit of $50 on each standard sleeping bag and $80 on each sub-zero sleeping bag, <mark>how many sleeping bags of each type should be manufactured each week to maximize the total weekly profit?</mark>

Variables S = # of standard Z = # of sub-zero	Objective Function max. P = 50S + 80Z

Determine Constraints

	S	Z	LIMITS
standard	1S	2Z	≤ 32
subzero	3S	4Z	≤ 84
		Z	≤ 12

Constraints

S ≥ 0   S + 2Z ≤ 32
Z ≥ 0   3S + 4Z ≤ 84

S	Z
0	16
32	0

S	Z
0	21
28	0

Z ≤ 12

Vertices
(28,0) (12,0)
(20,6) (8,12)

50(28) + 80(0) = 1400
50(12) + 80(0) = 600
50(20) + 80(6) = 1480
50(8) + 80(12) = 1360

Solution
Max = $1480 ; 20 standard, 0 sub-zero

Review 2.5 Additional Linear Programming Problems

4. The Cruiser Bicycle Company makes two styles of bicycles: the Stylish Traveler, which sells for $500, and the Luxury Tourister, which sells for $800. Each bicycle has the same frame and tires, but the assembly and painting time required for the Stylish Traveler is only 1 hour, while it is 3 hours for the Luxury Tourister. There are 300 frames and 360 hours of labor available for production. <mark>How many bicycles of each model should be produced to maximize revenue?</mark>

Variables	S = # of Stylish Traveler L = # of Luxury Tourister	Objective Function	max. $R = 500S + 800L$

Determine Constraints

	S	L	Limits
Frames	S	L	≤ 300
assembly & painting time	$1S$	$3L$	≤ 360

Constraints

$S \geq 0$ $S + L \leq 300$
$L \geq 0$ $S + 3L \leq 360$

S	L		S	L
0	300		0	120
300	0		360	0

Vertices $(0, 120)$ $(300, 0)$ $(270, 30)$

$500(0) + 800(120) = 96000$ $500(300) = 150000$ $500(270) + 800(30) = \boxed{159000}$

Solution Max profit = $159000 w/ 270 Stylish Travelers + 30 Luxury Touristers

Review 2.5 Additional Linear Programming Problems

5. The Bee in Your Bonnet Hat Company makes two popular styles of hats: berets and toboggans. Each beret requires 5 feet of yarn and 2 hours of sewing machine time. Each toboggan requires 12 feet of yarn and 1 hour of machine time. The company makes a $4 profit on each beret and a $3 profit on each toboggan. Due to supply limitations, the company has just 184 feet of yarn and 48 hours of sewing machine time available today. How many of each hat should they make to maximize profits? What will be their maximum profit?

Variables B = # of berets	Objective Function Max
T = # of toboggans	P = 4B + 3T

Determine Constraints

	B	T	Limits
yarn	5B	12T	≤ 184
sewing Time	2B	1T	≤ 48

Constraints

B ≥ 0   5B + 12T ≤ 184
T ≥ 0   2B + T ≤ 48

B	T
0	15.3
36.8	0

B	T
0	48
24	0

Vertices (20.6, 6.7) (0, 15.3) (24, 0)

$\boxed{4(20.6) + 3(6.7) = 102.5}$ 4(0) + 3(15.3) = 45.9   4(24) + 3(0) = 96

Solution

Max profit = $102.50; 20.6 berets & 6.7 toboggans

Review 2.5 Additional Linear Programming Problems

6. You have 140 tomatoes and 10 onions left over from your garden. You want to use these to make jars of tomato sauce and jars of salsa to sell at a farm stand. A jar of tomato sauce requires 10 tomatoes and ½ of an onion. A jar of salsa requires 5 tomatoes and 1 onion. You will make a profit of $5 on every jar of tomato sauce and $4 on each jar of salsa sold. How many jars of tomato sauce and salsa should you make to ~~maximize~~ minimize your profits?

Variables T = # of tomato sauce S = # of salsa	Objective Function min $P = 5T + 4S$

Determine Constraints

	T	S	Limits
tomatoes	10T	5S	≤ 140
onions	½ T	1S	≤ 10

Constraints

$T \geq 0$ $10T + 5S \leq 140$
$S \geq 0$ ✱ $\frac{1}{2}T + S \leq 10$

T	S
0	28
14	0

T	S
0	10
20	0

Vertices (14, 0) (0, 10) (12, 4)

$5(14) + 4(0) = 70$ $\boxed{5(0) + 4(10) = 40}$ $5(12) + 4(4) = 76$

Solution

Min profit = $40; 0 tomato sauces; 10 salsas
Max profit = $76; 12 tomato sauces; 4 salsas

3. DATA ANALYSIS

Knowing what a collection of data really means and learning from data are skills becoming more in demand in the higher education and corporate careers.

Nate Silver, a well-known statistician at the *New York Times*, and editor of the website fivethirtyeight.com admonishes us to always question the data. He withstood a lot of scrutiny in the 2016 presidential election for his polling that defied the odds and gave our current President higher odds of winning than any other poll.

This is currently a hot field - statisticians work in most every company now, heavy in consumer industries. Companies are discovering how much valuable information they have, especially in a world where so much of our interaction happens online.

Data Science, as a college major or area of study, is a recent addition to many schools, but a popular and lucrative field of study that was non-existent 20 years ago.

How to glean intelligence from "Big Data" datasets and applying human intelligence to data is a growing field. There are news reports of how data was used in the 2016 USA presidential election to help candidates know precisely what message to target to what audiences.

There is a classic story about Target and their targeting of customers regarding vitamins – the full story is found here. https://tinyurl.com/Algebra2Data

For this unit, we will use the TI-Nspire extensively. Make sure you have it with you **and fully charged** every day. (This is important even when we finish the current unit!)

3.1. One variable statistics, Dot Plots, & Box Plots

Let's start out by creating a simple dataset. How many years, including this one, have you attended Woodward Academy?

3	5	10	5	5
4	3	4	2	4
5	4	1		

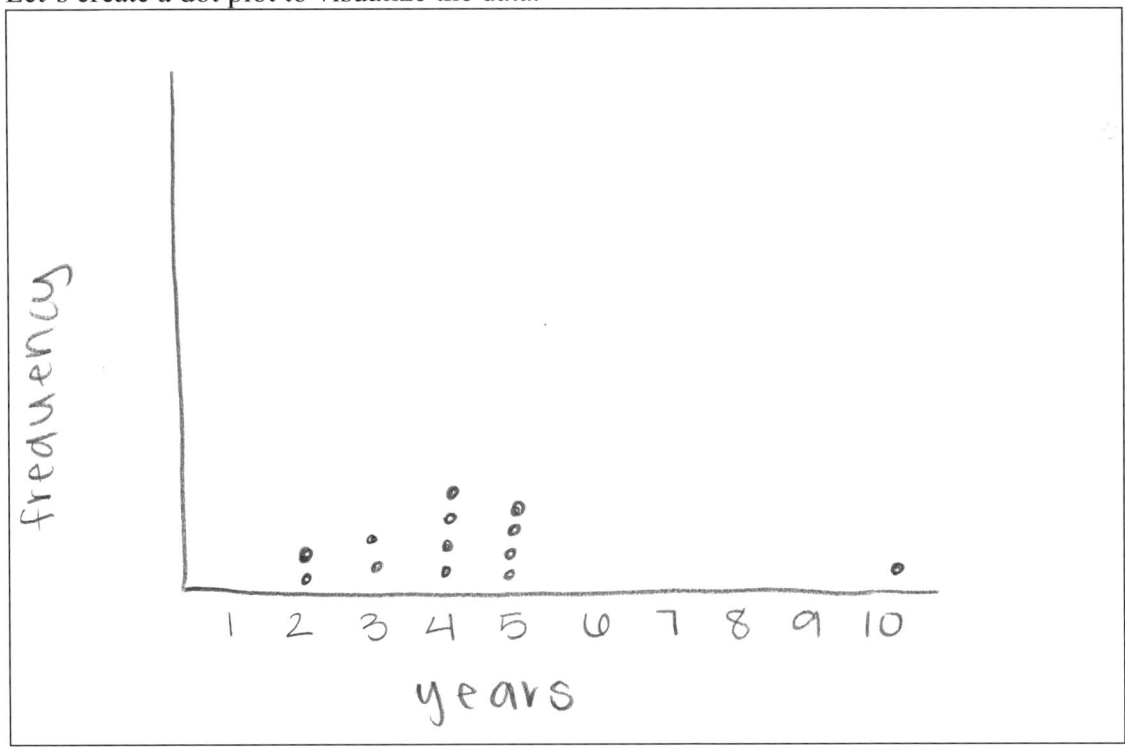

Let's create a dot plot to visualize the data.

You are likely familiar with a few measures we have for datasets. What does each measure mean, what intelligence or use might you or Woodward find for the data?

Measure	Meaning	Possible Usage
Minimum	1	Targeting new families about onboarding
Maximum	10	Retention why have you stayed? what are positives?
Mean or Average	4	Why might we watch what the average value does over time? If it goes up - positive If it goes down - neg.
Median	7	middle 1/2 less 1/2 greater
Mode	(bimodal) 4 & 5	most frequently occurring number

These are but a few simple measures about a dataset, but you should get the idea that the data can provide a lot of information about the data, and you should see the value of human insight coupled with the data. The two go hand in hand.

Let's involve our TI-Nspire to find some information about the dataset we started with – the number of years classmates have attended Woodward.

3	5	10	5	5
4	3	4	2	4
5	4	1		

Calculating common One-variable Statistics

First to enter the data: Doc 1: File 1: New Document 4: Lists & Spreadsheets Label the column WA_years, and then complete the column with the data from above.	*(screenshot of Lists & Spreadsheets showing column A "wa_years" with values 5, 3, 2, 1)*
Use the calculator to provide some useful one-variable statistics. MENU 4: Statistics 1: Stat Calculations 1: One-Variable Statistics When prompted for Number of Lists, leave as 1. (We only have one list or column of data). OK	*(screenshot of One-Variable Statistics dialog with "Num of Lists: 1", OK / Cancel)*

	When prompted for the X1 list, choose the value WA_years, or whatever name you chose for your data list. Select OK.	
	You should now notice your spreadsheet has been populated with various measures or statistics about the dataset.	

We will not drill into each of these, but a few are worth noting at this time. Complete the following with data from your dataset.

$\bar{x}$	Mean or average	4.23077
$\sum x$	Sum of all values	55
$MedianX$	Median of the dataset	4
$MinX$	Minimum value	1
$MaxX$	Maximum value	10
n	Number of elements in dataset	13
sx	Standard Deviation	2.12736

The **standard deviation**, sx, is a measure of how dispersed the dataset is. It is a measure of how far the data elements are scattered from the mean or average.

The formula for standard deviation is

$$sx = \sqrt{\frac{\sum_{i=1}^{n}(x_i - \bar{x})^2}{N-1}}$$

We will not be using this formula directly for calculations, but you can see how it represents the distance of each element from the mean, $\bar{x}$.

The closer the standard deviation is to zero, the more centered the dataset is. The higher the value, the more dispersed the dataset is.

You may have heard of normal curves or normal distributions – the illustration below shows a normal bell curve and the relationship to standard deviation.

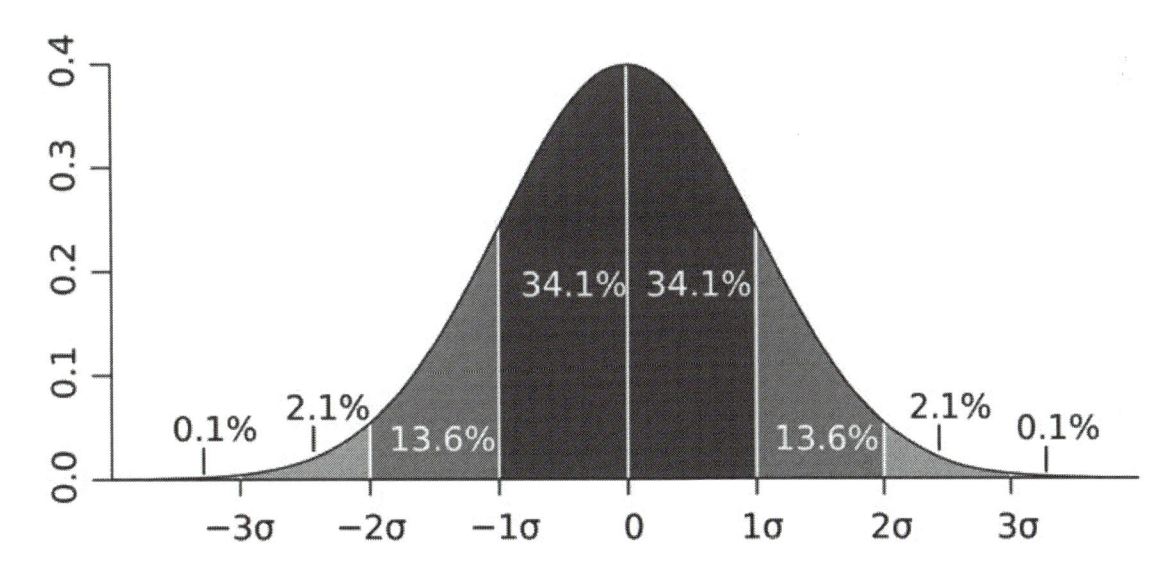

For now, we will just know the meaning of standard deviation at a high level, and how to find it on the calculator. Your future stat class will deal with this all in more depth.

Woodward security is always concerned about the traffic surrounding campus, especially with so many sophomores learning to drive. (sarcasm) Consider below a sampling of card speeds on main street.

Car Speeds on Main Street (mph)

45	47	48	39	46
40	35	51	47	49
31	50	41	40	46

Use your calculator to create a dot plot:

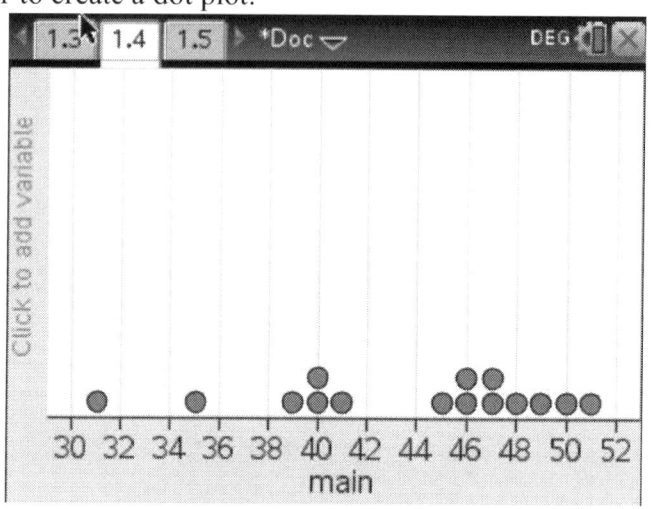

Calculate the one-variable statistics on this dataset.

Average $\bar{x}$	Minimum MinX	Maximum MaxX	Median MX	Standard SX Deviation
43.66	31	51	46	5.77703

What do the statistics tell you about this experiment?

people are driving too fast

Is the mean a good representation of what is happening?

It does not show more than 1 number

What does the median tell you about car speeds?

46 = ½ are going slower
½ are going faster — most above speed limit

Now let's consider car speeds on Cambridge Avenue, between the upper and middle schools.

Car Speeds on Cambridge Avenue (mph)

15	15	15	14	13
39	35	13	37	36
12	32	35	38	40

Let's look at the dot plot.

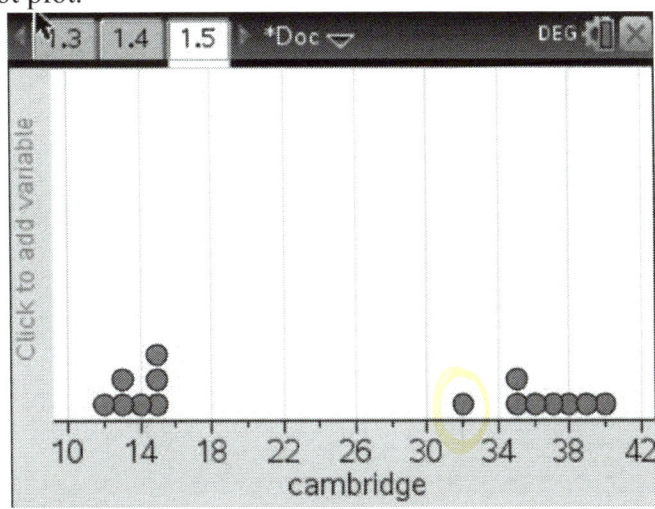

Calculate the one-variable statistics on this dataset.

Average	Minimum	Maximum	Median	Standard Deviation
25.933	12	40	32	11.8591

How do you interpret the mean? Does it tell you something useful about the data?

What do you infer from the standard deviation of Main Street versus Cambridge Avenue?

Using your intelligence about Cambridge Avenue, what do you think can explain the speeds? (This data might be more useful if instead of one variable (speed) there was another variable)

In the Cambridge Avenue data, the box plot is a good way to visualize the data. The mean doesn't tell me much about this data. It is a singular, numerical measure. But the box plot can tell me ranges of what is happening in the data set.

A **Box Plot** is another useful way to represent data. It is sometimes called a *box and whisker* plot.

A box plot is useful to quickly give a visual representation of five key elements of the dataset:

Minimum Value	First Quartile Q1	Median	Third Quartile Q3	Maximum Value
The lowest value in the data	The cut off that defines the lower 25% of the data. 75% of the data is above the value.	The middle of the dataset, the 50th percentile.	The cut off that shows the 75th percentile, 75% of the data below and 25% above the value.	The maximum value of the dataset.

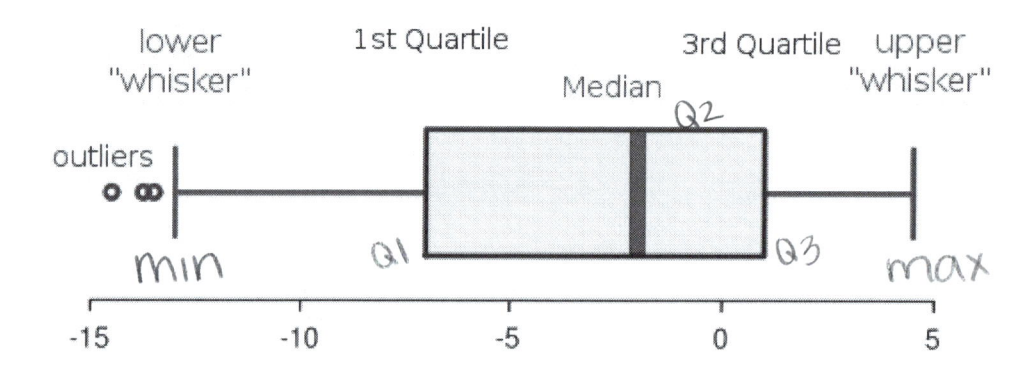

The box plot will also show outliers, data elements that are out of bound, or too extreme to be considered statistically significant in the dataset.

Car Speeds on Cambridge Avenue (mph)

15	15	15	14	13
39	35	13	37	36
12	32	35	38	40

Create a box plot and give the 5-number summary.

min = 12
Q1 = 14
median = 32
Q3 = 37
max = 40

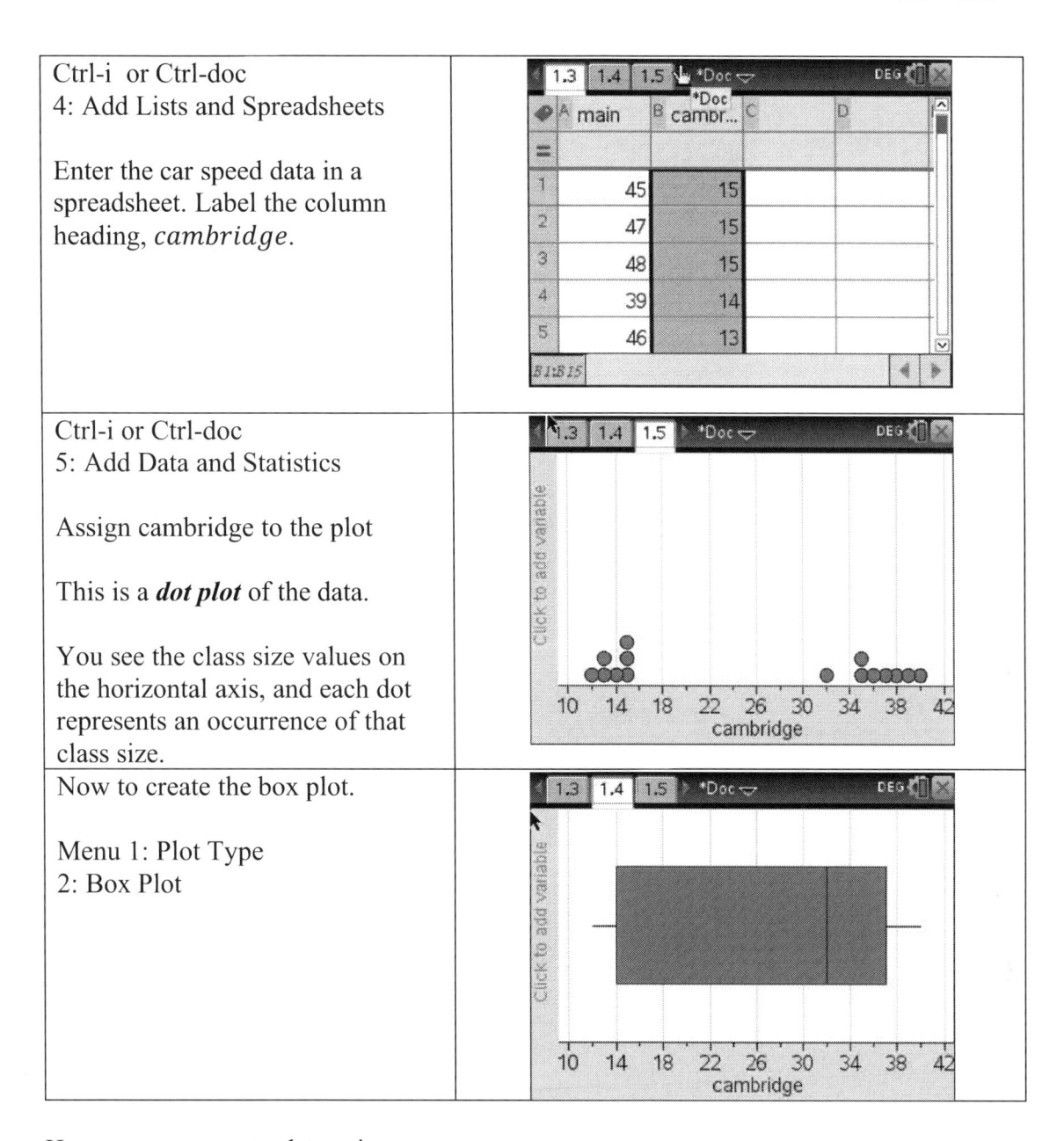

Ctrl-i or Ctrl-doc 4: Add Lists and Spreadsheets Enter the car speed data in a spreadsheet. Label the column heading, *cambridge*.	
Ctrl-i or Ctrl-doc 5: Add Data and Statistics Assign cambridge to the plot This is a ***dot plot*** of the data. You see the class size values on the horizontal axis, and each dot represents an occurrence of that class size.	
Now to create the box plot. Menu 1: Plot Type 2: Box Plot	

Hover your cursor to determine

Minimum	Q1	Median	Q3	Maximum
12	14	32	37	40

What percentage of cars are between 12 and 14 mph?

25%

What percentage of cars are between 32 and 37 mph?

25%

What percentage of cars are between 37 and 40 mph?

25%

What percentage of cars are between 32 and 40 mph?

50%

If we had a car speed of 75 mph, would that be an outlier?

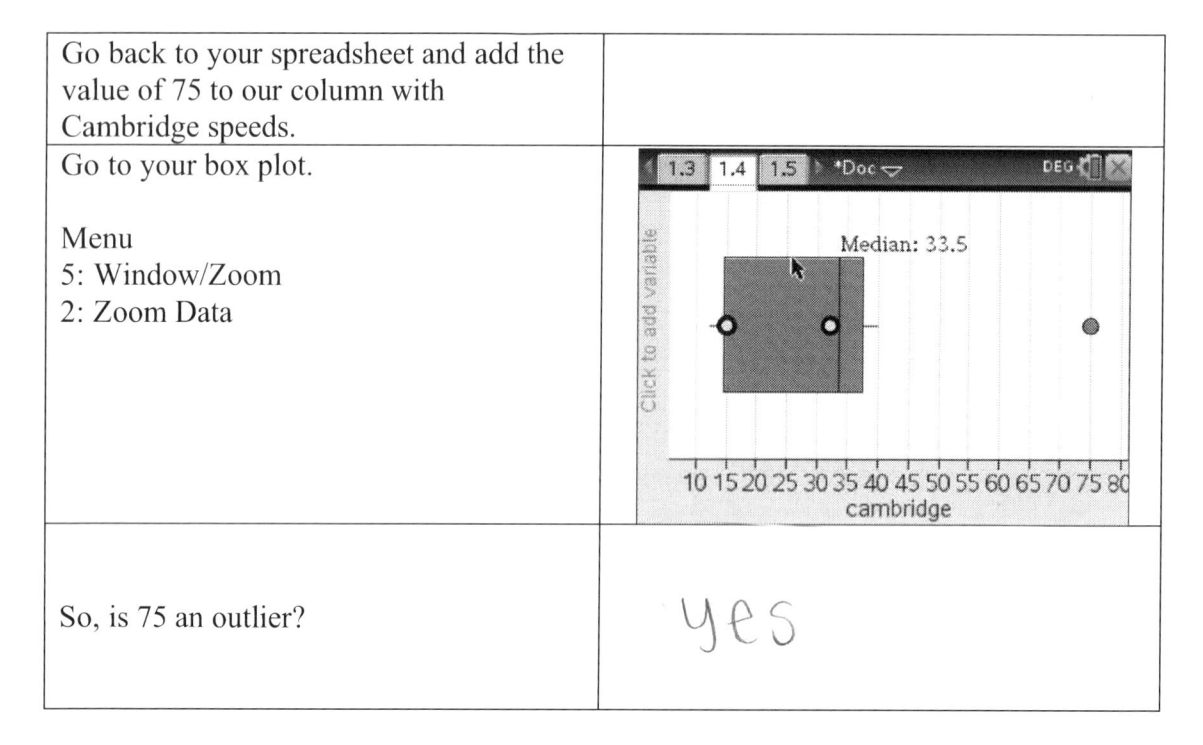

Go back to your spreadsheet and add the value of 75 to our column with Cambridge speeds.	
Go to your box plot. Menu 5: Window/Zoom 2: Zoom Data	
So, is 75 an outlier?	yes

You can also get this 5-number summary (Min, Q1, Median, Q3, Max) WITHOUT making a box plot.

Ctrl-i or Ctrl-Doc 1: Add Calculator Menu 6: Statistics 1: Stat Calculations 1: One-Variable Statistics Num Lists =1 OK X1 List = "cambridge" OK	

Homework 3.1 One variable Statistics, Box Plots

1. Woodward Academy students and alums were surveyed and asked at what age did they have their first real, paying job. Results are shown below

14 16 17 23 17 14 14 12 12 17 17 14 16 16 18 16 17

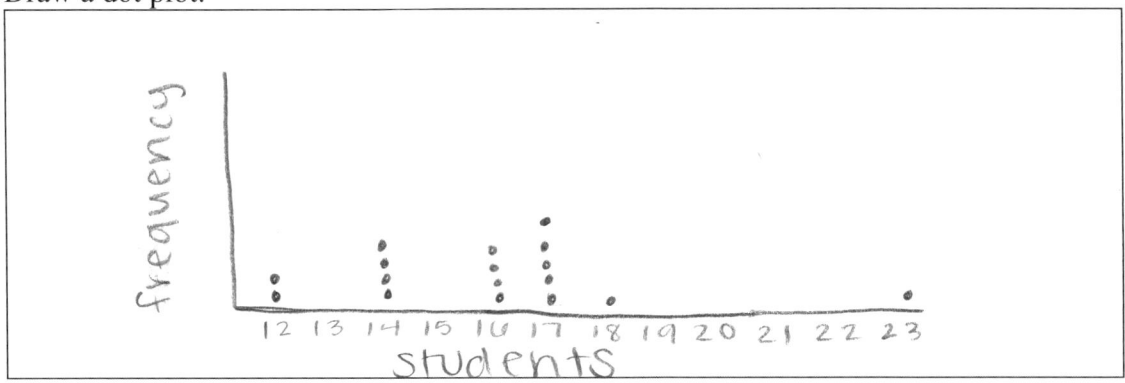

Draw a dot plot.

Create a Box plot on your calculator. Use it to label the box plot below with the elements of a box plot, e.g. Maximum, Q3, etc.

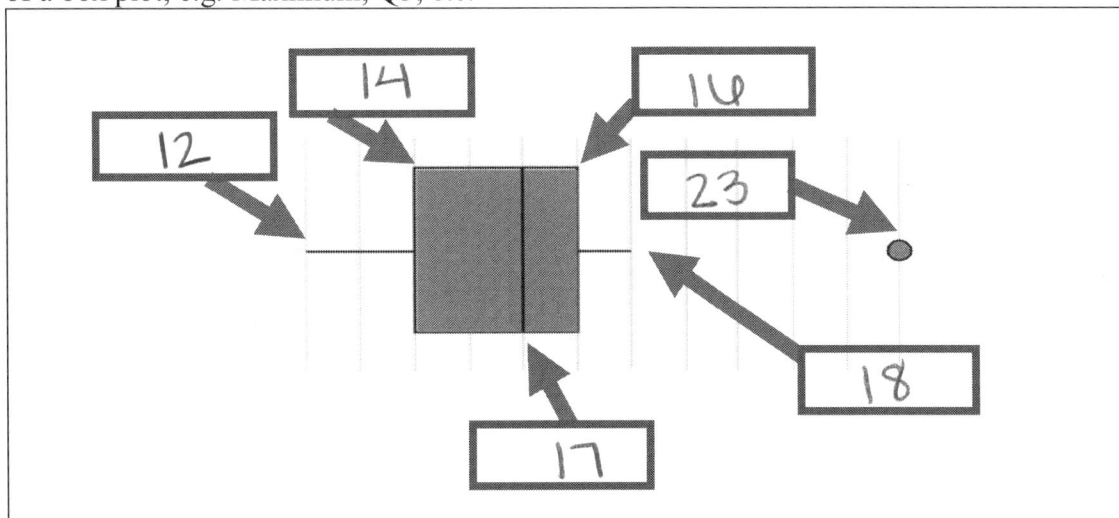

Labels on box plot: 12, 14, 16, 23, 17, 18

Find the value of the following:

Minimum age	Maximum age	Q1	Median	Q3
12	23	14	16	17
Mean Age 15.8824	Standard Deviation 2.57105	Age range of the oldest 25% 17-18	Is there an outlier? If so, what is it? 23	Teens in the range of 14-17 make up what percentage? 50%

Homework 3.1 One variable Statistics, Box Plots

2. The dataset below shows the top 20 finishes for the 2017 women's Peachtree Road Race.

2017 Peachtree Road Race Women's Times in minutes				
32.8	33.0	33.1	33.3	33.6
33.7	33.8	33.8	34.1	34.3
34.3	34.4	34.4	34.5	34.7
34.8	35	35.3	35.3	35.8

What is their average time?	What is the standard deviation?	What is the sum of all the times?
34.2167 min	0.804696	821.2 min
What range of times would put you in the top 25% of this list?	What is the median time?	Half of these women were faster than what time?
32.8 – 33.05 min	34.3	34.3 min

For problems 3-5, refer to the data which shows gender and height for a class of Algebra 2 students.

Gender	Height (cm)	Gender	Height (cm)
Male	168	Female	168
Male	181	Male	183
Male	72	Female	152
Female	159	Female	173
Female	165	Female	162
Male	170	Male	163
Female	60	Female	164
Male	178	Male	168
Male	165	Male	176
Female	149	Male	213

Homework 3.1 One variable Statistics, Box Plots

3. From the data on the prior page, find the 5-number summary for student height. Draw a box plot. Are there any outliers?

Min	Q1	Median	Q3	Max
60	160.5	166.5	174.5	213

outliers → 72, 60, 213

(box plot drawn with x-axis "height" from 60 to 220)

4. Find the 5-number summary and standard deviation for **male** heights.

Min	Q1	Median	Q3	Max	Standard Deviation
72	165	170	181	213	34.389

5. Find the 5-number summary and standard deviation for **female** heights.

Min	Q1	Median	Q3	Max	Standard Deviation
60	150.5	162	166.5	173	34.64

Does the data show either males or females have a greater dispersion of heights? Explain.

the standard deviation is more for male heights, therefore greater dispersion of the male heights

Homework 3.1 One variable Statistics, Box Plots

5. In 2018 the average time of an elite male runner in the Peachtree Road Race was 28.83 minutes. The average time of an elite female runner was 32.93 minutes. Comparing the top ten finishes of each, what can you conclude about the standard deviation of the data sets when compared to each other? Explain. *Girls times vary from mean more than boy ones → Standard dev. is less*

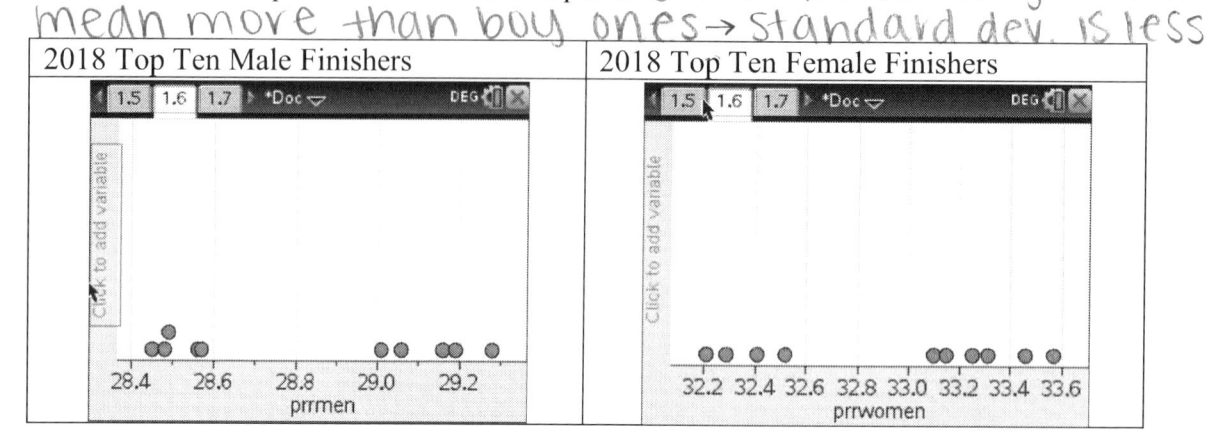

2018 Top Ten Male Finishers	2018 Top Ten Female Finishers

6. The data below shows top finishes of the 2014 Boston Marathon. The time is measured in hours.

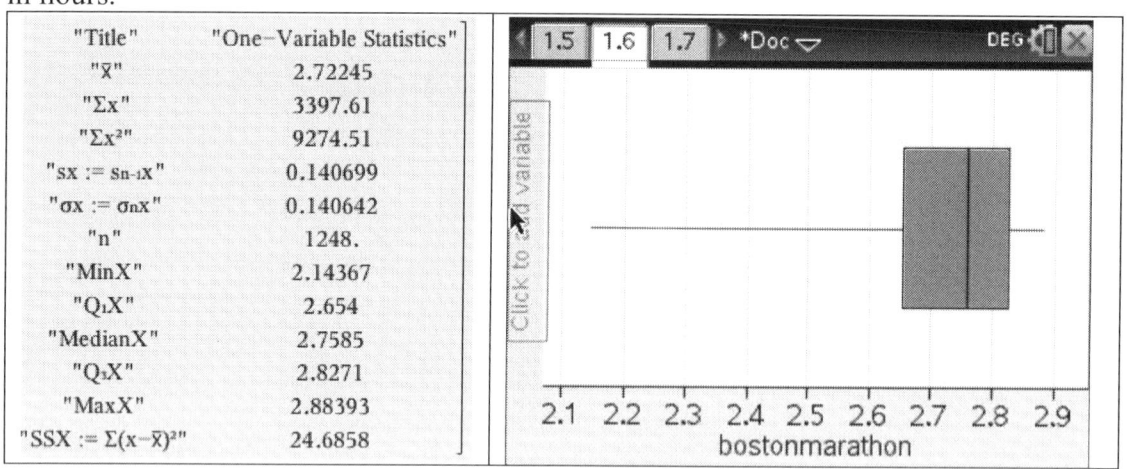

"Title"	"One-Variable Statistics"
"x̄"	2.72245
"Σx"	3397.61
"Σx²"	9274.51
"sx := Sn-1x"	0.140699
"σx := σnx"	0.140642
"n"	1248.
"MinX"	2.14367
"Q₁X"	2.654
"MedianX"	2.7585
"Q₃X"	2.8271
"MaxX"	2.88393
"SSX := Σ(x−x̄)²"	24.6858

How many racers are represented in the data set? **1248**

What is the range of times for the fastest 50% of runners? **2.14367 − 2.7585**

What is the range of times for the slowest 25% of runners? **2.8271 − 2.88393**

What is the average time for the runners?
x̄ = 2.72246

Half of the runners are slower than what time? **2.7585**

How can you explain the widest interval being 2.14 hours to 2.65 hours and all other quartile intervals being smaller ranges of time?

Greater time spread among fastest 25%. Slower 75% times are more tightly grouped

Homework 3.1 One variable Statistics, Box Plots

CHALLENGE PROBLEMS

1. Your math assessment scores are given as

$$75, 80, 85, 90, x, y$$

You know that your median score is 85 and your mean score is 84. Find the two missing test scores.

2. Consider the data values

$$25, 35, 45, 55, 65$$

If you add the value of 20 to the data set does the mean or median have a larger percentage change in value? Will the standard deviation increase or decrease?

3.2. Histograms, Interpreting Two-variable data, and Scatter Plots

So far, the data sets we have seen have all be using raw data sets. You had a data set with n elements, and we used our visualizations of dot plots and box plots, along with the one-variable statistics to make inferences about the data set.

More often, you will be presented with summary data, either a chart or graph that summarizes or groups the data. This is helpful in taking large data sets and being able to work with the information more concisely.

Consider the data set of the scores of Algebra 2 students on the Data Analysis Assessment:

79	66	92	68	74	68	70	74	75	74
71	68	73	69	98	74	80	71	77	71
72	65	71	86	73	79	71	72	71	70
71	68	96	72	79	77	66	71	74	74
79	86	71	70	65	66	72	91	71	95
70	65	72	73	68	84	88	65	73	69
73	66	72	77	65	74	95	87	65	93
92	73	79	68	65	88	97	89	68	74
67	73	81	66	72	71	65	65	78	85
66	74	65	74	65	68	87	70	70	67

These 100 scores do not represent a large data set by industry standards, but a bit more cumbersome to work with when doing manual data entry as we have been doing so far. **Consider the dot plot. Are you able to make interpretations from the dot plot?**

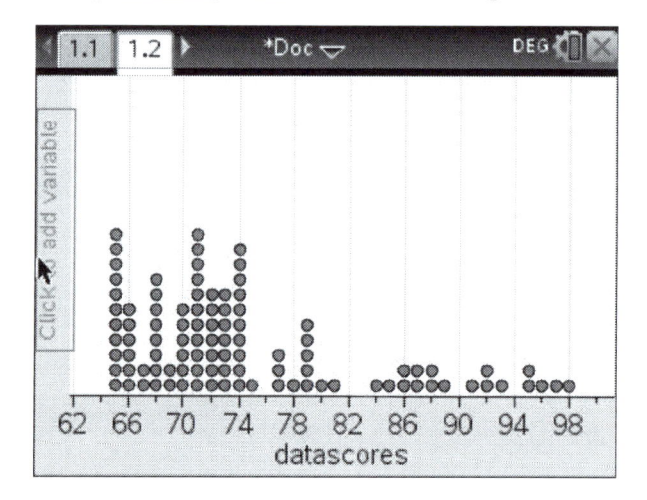

Histogram: **A histogram is used for continuous data, where the bins (horizontal width of bars) represent ranges of data, and the vertical height represents frequency.**

When we consider grades, there is a natural grouping we always desire, especially here at Woodward. Letter grades would be an ideal way to group the data. Consider the histogram of the data, in the graph below.

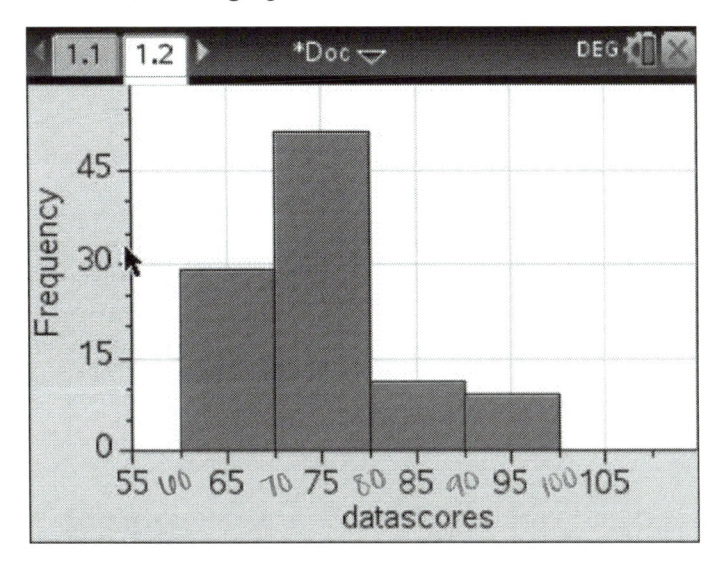

About how many

Ds	Cs	Bs	As
29	52	13	9

What was the median grade on this test?   A   B   (C)   D

How do you know that?

The median is the avg of 50 & 51st elements. (middle #)

Let's consider a dataset of math class sizes. This is a raw dataset. Each individual entry represents a class size.

20	17	14	19
15	18	14	16
13	16	15	12
12	16	13	

Create a ***histogram*** of this dataset. A histogram is a series of rectangles giving a frequency distribution of the values of the dataset. The histogram will allow us to visualize the summary data.

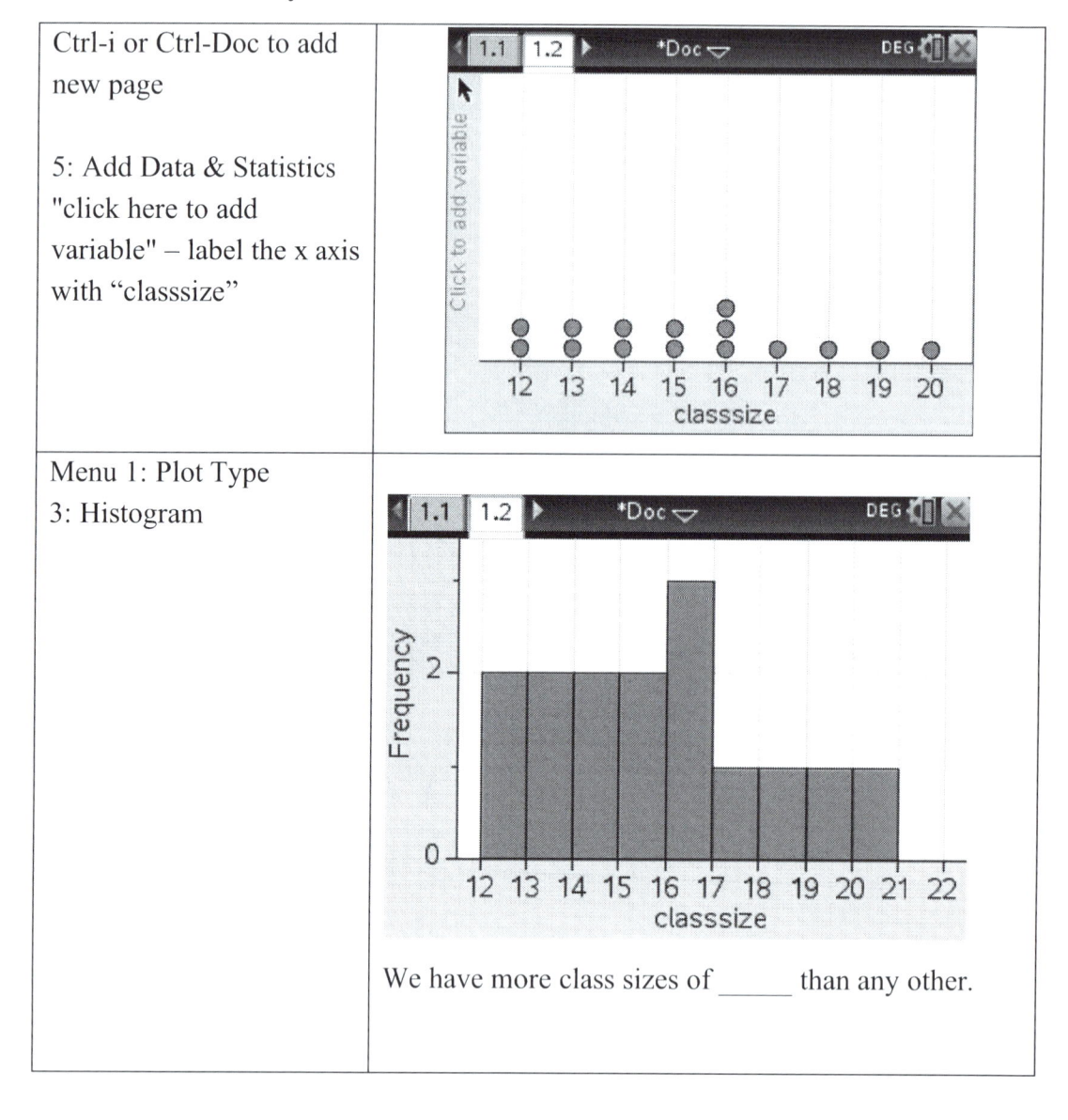

Ctrl-i or Ctrl-Doc to add new page 5: Add Data & Statistics "click here to add variable" – label the x axis with "classsize"	
Menu 1: Plot Type 3: Histogram	 We have more class sizes of _____ than any other.

The bin size refers to the width of each rectangle. Here I have a separate rectangle for each value of class size. The bin width is one. Let's change the bin width to 3. Menu 2: Plot Properties 2: Histogram Properties 2: Bin Settings 1: Equal Bin Width Value of 3	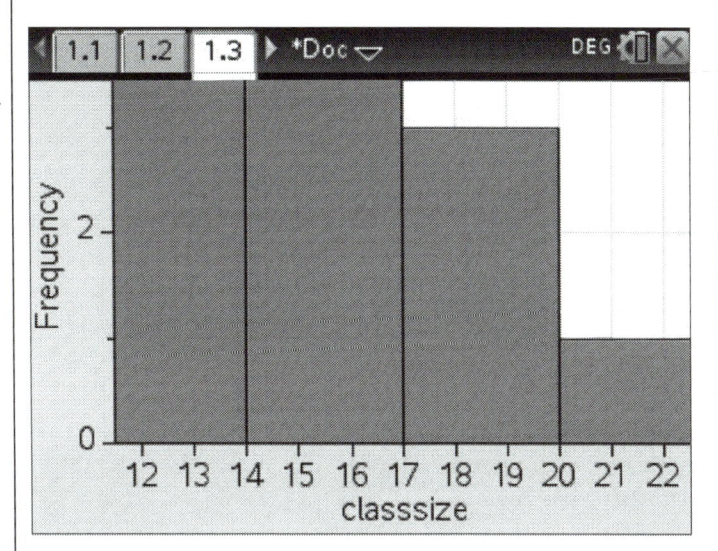 *(histogram screenshot with bins from 12 to 22, classsize)*
Use the window/zoom to better view the histogram. Menu 5: Window/Zoom 2: Zoom Data	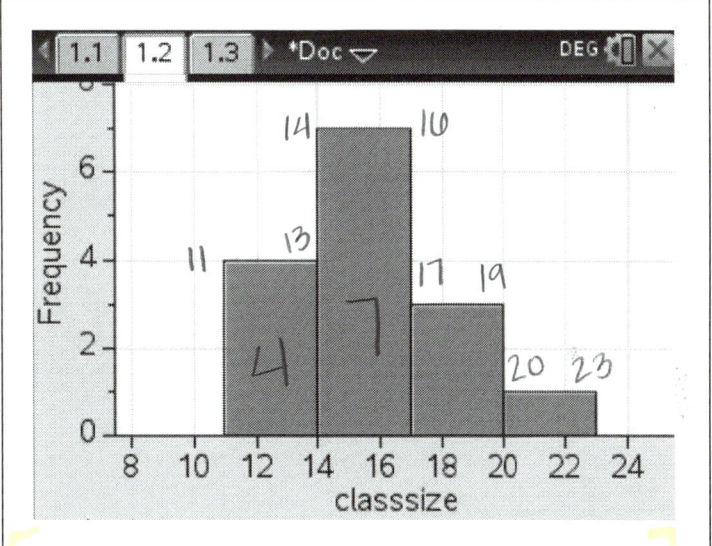 *(histogram screenshot with annotations: 11, 13, 14, 4, 7, 16, 17, 19, 20, 23)* [According to this we have more classes with size in the range __14__ to __17__ than any other range]
Looking at this histogram, what is the range of values for the median class size?	15 datapoints → median = ≈ 8th between 14 & 17
Looking at this histogram which of the following could be the value of the mean? (a v a)	A. 14 B. 15 (circled) C. 16 (circled) D. 18 Smallest = 14.2 largest = 16.2

President Trump Popular Votes in 2016 Election
Percentage Share of Votes among gender and race groups

	Male	Female	Total
White	62%	47%	54%
Black	14%	2%	6%
Hispanic	28%	28%	28%
Total	62%	39%	45%

Source: Pew Research Poll of validated voters, December 2016.

In reading the chart above, we interpret that of the population of white males, ~~62%~~ 10% of them voted for Mr. Trump, etc.

Which two demographics show the largest gap in support for President Trump in the 2016 election? black male & black female

Among white voters, can you determine if there were more male or female voters?

female

Demographics of the Electorate in the 2016 Election

	White	Black	Hispanic	Total
Male	33%	4%	5%	42%
Female	41%	6%	5%	52%
Total	74%	10%	10%	94%

Source: Pew Research Poll of validated voters, December 2016.

According to this chart black voters made up what percentage of the 2016 electorate?

10%

Non-white females make up what percentage of the electorate?

11%

Non-whites made up what percentage of the electorate?

20%

Consider the chart below (Source eMarketer, Feb 2017) detailing the current and projected ages and number of Snapchat users (millions) in the USA.

	2015	2016	2017	2018	2019	2020	2021	
0-11	0.6	0.9	1.2	1.4	1.7	1.8	2.0	3.3
12-17	11.3	13.5	14.8	15.5	15.9	16.2	16.3	1.4
18-24	16.6	19.3	20.9	22.1	22.9	23.5	23.8	1.4
25-34	11.3	16.7	19.1	21.0	22.8	23.7	24.4	2.1
35-44	3.7	5.4	6.6	7.5	8.4	9.2	9.8	2.6
45-54	1.7	3.5	4.5	5.3	6.0	6.4	6.7	3.9
55-64	0.7	1.8	2.4	2.8	3.3	3.7	4.0	5.7
65+	0.2	0.6	0.9	1.3	1.7	2.0	2.2	11 ←
TOTAL	46.2	61.7	70.4	77.1	82.7	86.6	89.2	

What is the projected percentage increase in total number of users from 2015 to 2021?

46.2   89.2 → increase 43

43 is ___% of 46.2

(93.07%)

Which age group shows the greatest percentage gain from 2015 to 2021?

$\frac{2.2}{0.2} = 11$ (65+)

What is the median age for a user in 2015?   (18-24)

overall = 46.2%
median = 23.1% range

What is the projected median age range for a user in 2021?

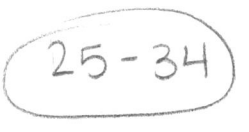

overall = 89.2%
median = 44.6% range   (25-34)

What is the mean age range for users in 2019? (What is the lowest and highest value possible for the mean?)

Scatter Plots

Consider the following data set relating to USA presidential elections

Year	1984	1988	1992	1996	2000	2004	2008	2012	2016
Electoral College Margin of victory	512	315	202	220	5	35	192	126	77

Instead of a data set with one variable, we now have a data set with two variables. Create a scatter plot with the year as the horizontal axis and the margin of victory as the vertical axis.

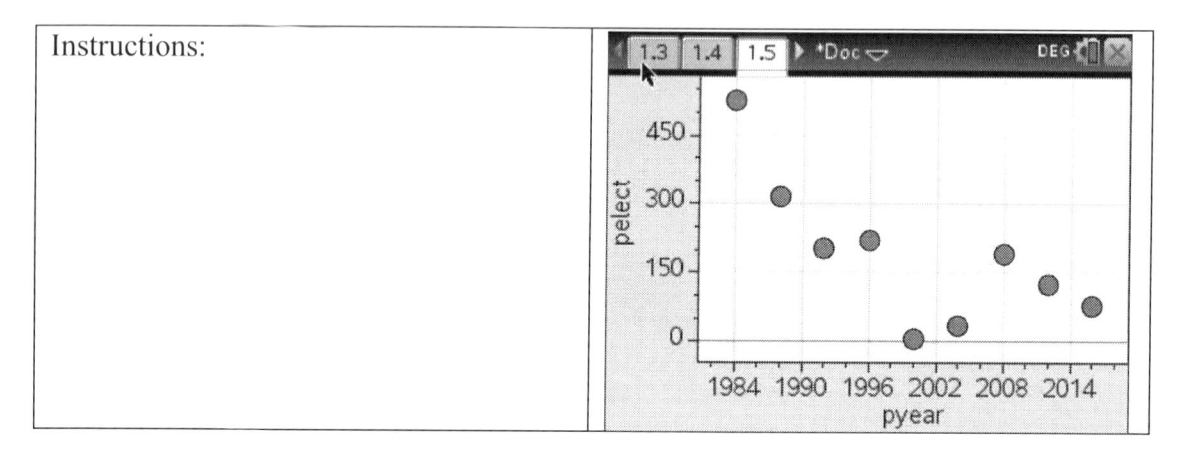

Instructions:

Now that we have two variables, year and electoral college margin of victory how do we interpret the data differently than we did with one variable?

With one variable being time, what kinds of questions can we ask about the data that we couldn't if we just had a list of the margins of victory?

What descriptive language can you use to portray the trend of USA elections since 1984?

In the next lesson we will go further into these datasets and determine models that we can use to project from the data.

Homework 3.2 Histograms, Interpreting Two-variable data, and Scatter Plots

1. Woodward students were surveyed and asked how many hours they spent on Netflix per week. The data is below.

Hours spent watching Netflix					
7	8	6	11	12	7
14	20	17	16	16	11
10	11	14	12	13	15

Use your TI-Nspire to create a histogram of this data.

Modify the histogram to have a bin size of 4. How many students reported watching Netflix between 13 and 17 hours per week? **6 students**

2. Fifteen students were surveyed and asked how many times they attended tutorial that week. The results are below:

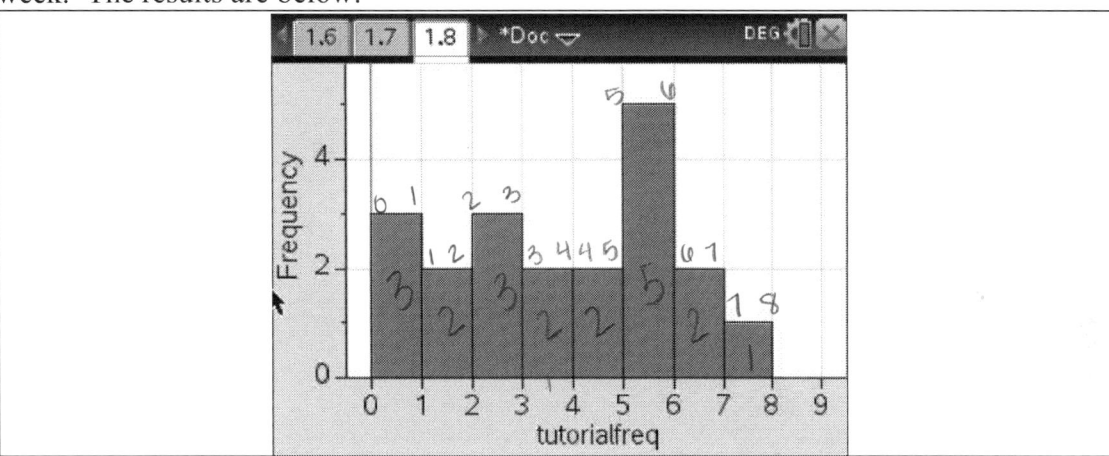

[histogram image with handwritten annotations]

What is the mean value for the number of times a student attended tutorial? (avg)

$3 \cdot 0 + 2 \cdot 1 + 3 \cdot 2 + 2 \cdot 3 + 2 \cdot 4 + 5 \cdot 5 + 2 \cdot 6 + 1 \cdot 7 / 20 = 4.4$

What is the median value for the number of times a student attended tutorial? (middle)

20 data points
median is avg of #10 & #11 = **3.5**

What is the mode value for the number of times a student attended tutorial? (most)

between 5 & 6

Homework 3.2 Histograms, Interpreting Two-variable data, and Scatter Plots

3. Fifty students were surveyed at asked how many times they attended tutorial that week. The results are below:

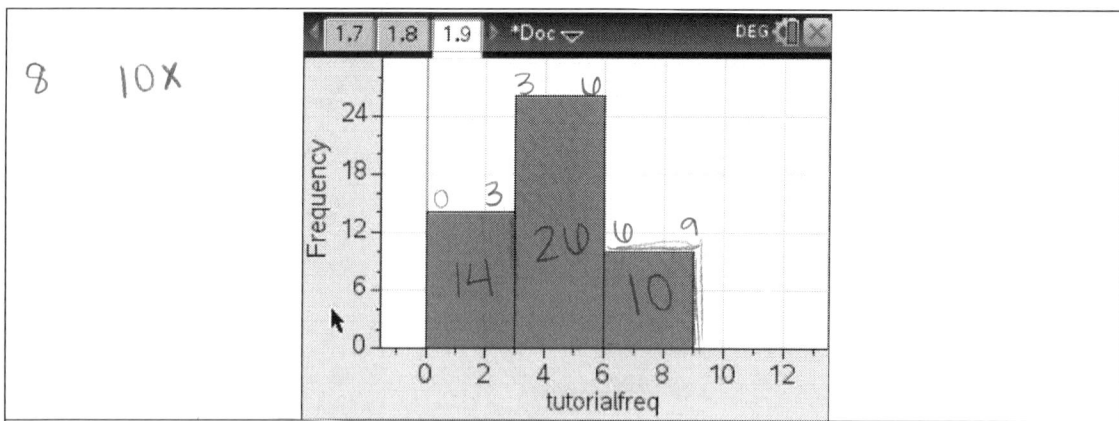

[histogram shown with bars labeled: 0—3: 14, 3—6: 20, 6—9: 10; annotations "8 10✗" and markings "0 3", "3 6", "6 9"]

How many students report attending 0, 1, or 2 times per week?
14

What percentage of the students report attending 6, 7, or 8 times?
$\frac{10}{50} = 20\%$

What are all the possible numbers for the median value?
50 points → median is avg of #25 & #26 3, 4, or 5 → = betwn 3 & 5

Find the minimum and maximum values possible for the mean.
$\frac{0 \cdot 14 + 3 \cdot 20 + 6 \cdot 10}{50} = 2.76$ $\frac{3 \cdot 14 + 6 \cdot 20 + 9 \cdot 10}{50} = 5.76$ betwn 2.76 & 5.76

Can you look at this histogram and determine the mode? Is it possible for the mode to be 8? Explain.
0,1,2 – yes   It is possible. 3,4,5 – yes

Homework 3.2 Histograms, Interpreting Two-variable data, and Scatter Plots

4. Seventy-five high school students were surveyed and asked to identify their favorite genre of music.

Genre	Male	Female	Total
Classic Rock and Roll	4	7	11 ←
Pop	2	7	9
Classical	1	4	5
Rap	20	12	32
Country and Western	2	8	10
R&B	6	1	7
Justin Bieber	0	1	1
	35	40 ←	75

Question	Work	Answer
What proportion of the students prefer rap? What percentage of the students prefer rap music?	$\frac{32}{75}$	42.6%
If there are 1200 students in the school, based on this data, about how many would prefer rap music?	$\frac{32}{1200}$ .426(1200)	(511)
Of the females surveyed, what percentage prefer country and western?	$\frac{8}{40}$	20%
In a class of 40 students who all prefer classic rock and roll, about how many would you expect to be female?	$\frac{7}{11} \cdot 40 = 25.45$	≈ 25 females
What does the survey tell you about the music of Justin Bieber?	not popular → not many people's fave music	

Homework 3.2 Histograms, Interpreting Two-variable data, and Scatter Plots

CHALLENGE PROBLEM

1. A group of tenth graders are surveyed as to which math class they are taking. Complete the table and answer the questions.

	Algebra 1	Geometry	Algebra 2	Total
Female	4	20		74
Male	6		30	46
Total		30	80	120

Which one category of survey respondents accounts for about 25% of the students.

Which two categories of survey respondents accounts for about 50% of the students?

What proportion of the students are not taking Algebra 2?

Homework 3.2 Histograms, Interpreting Two-variable data, and Scatter Plots

2. Consider the histogram of Algebra 2 assessment scores. Assume the only letter grades given are A, B and C. Which of the statements below do we know to be definitively true?

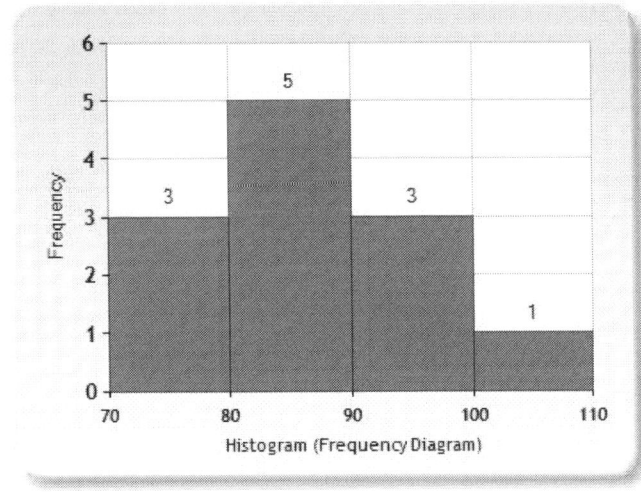

a. The mean score on the test was 85.

b. It is possible that the median score is 90.

c. The median score must be a B.

d. You can add an additional score of 100 and the median will still be greater than or equal to 80 and less than 90.

e. If we knew the mean to be 82, what score must be added to the data set to bring the mean grade down to a C?

3.3. Lines of Best Fit & Linear Regression Models

Linear Regression

Linear regression is a commonly used predictive analysis tool for modeling a line to approximate a dataset. In our examples, we will have one **independent variable**, often *time*. There will also be a **dependent variable**, typically something that can be measured, such as temperature, test scores, or another natural phenomenon.

Let's return to the data set relating to USA presidential elections

Year	1984	1988	1992	1996	2000	2004	2008	2012	2016
Electoral College Margin of victory	512	315	202	220	5	35	192	126	77

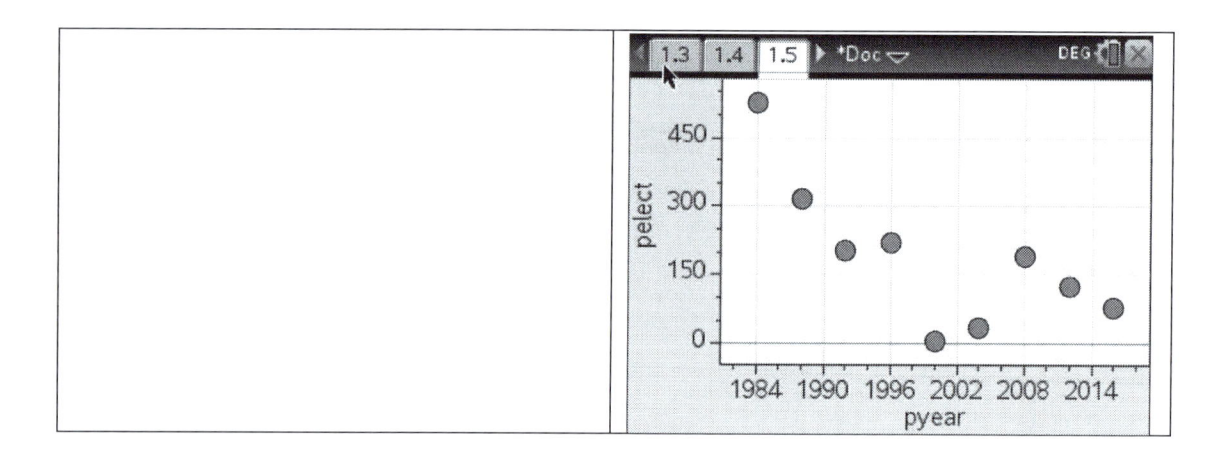

In linear regression, we are considering two variables and asking how well would a line fit the data set? Additionally, we are asking if it is reasonable that the line models a relationship between the two variables.

In the example above, if we can create a model that links year to margin of victory, what would we expect to use that for?

Create a linear model for the data set of election year versus electoral college margin of victory:

To add a line of best fit (linear regression)
 	
 Menu
     4: Analyze
     6: Regression
     1: Show Linear (mx +b)
	$y = -10.6833 \cdot x + 21555.2$

Is linear a good model? $y = -10.4067 + 21120.4$	
Does the model overestimate the actual? 1992, 2000, 2004	
Does the model underestimate the actual? 1984, 2008, 2012, 2016	

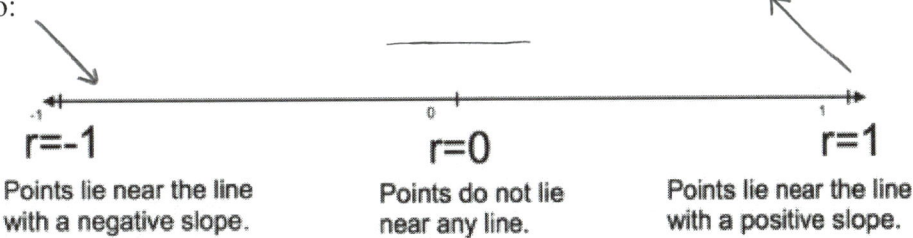

In statistics, the **correlation coefficient**, r, measures the strength and direction of a linear relationship between two variables on a scatterplot. The value of r is always between +1 and −1. To interpret its value, see which of the following values your correlation r is closest to:

$r = -1$ — Points lie near the line with a negative slope.

$r = 0$ — Points do not lie near any line.

$r = 1$ — Points lie near the line with a positive slope.

The correlation coefficient will be a measure that lets us quantitatively assess how accurate the model is.

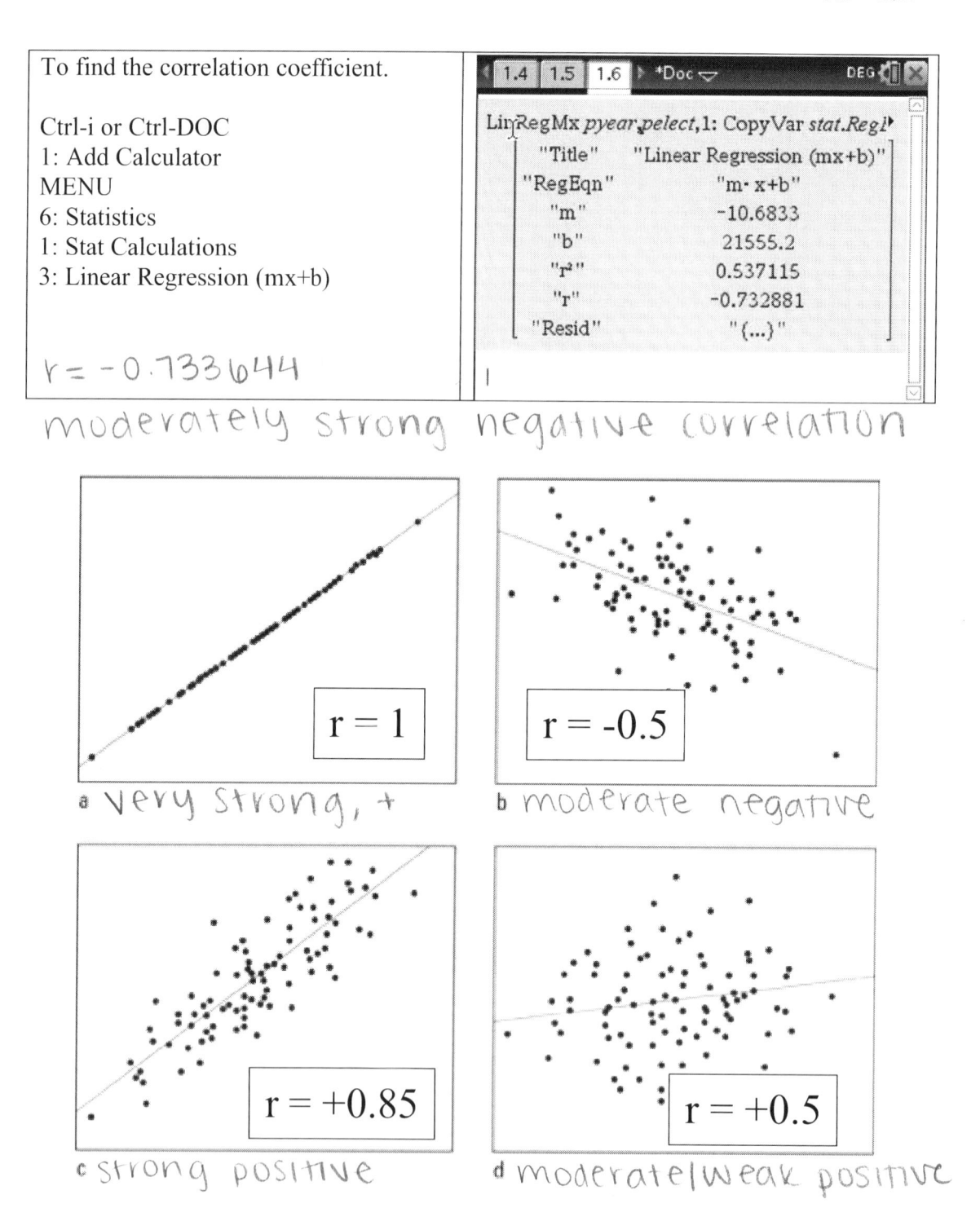

To find the correlation coefficient.

Ctrl-i or Ctrl-DOC
1: Add Calculator
MENU
6: Statistics
1: Stat Calculations
3: Linear Regression (mx+b)

$r = -0.733644$

```
LinRegMx pyear,pelect,1: CopyVar stat.Reg1
"Title"     "Linear Regression (mx+b)"
"RegEqn"    "m·x+b"
"m"         -10.6833
"b"         21555.2
"r²"        0.537115
"r"         -0.732881
"Resid"     "{...}"
```

moderately strong negative correlation

$r = 1$ — a **very strong, +**

$r = -0.5$ — b **moderate negative**

$r = +0.85$ — c **strong positive**

$r = +0.5$ — d **moderate/weak positive**

Similar to slope we see positive correlation for lines of best fit rising from left to right, and negative correlation for lines rising from right to left. The closer the correlation is to +1 or -1, the stronger the dataset fits to a straight line. The closer the correlation is to zero, the weaker the straight-line approximation is of the dataset.

Consider the following data of tenth graders in the USA having reported drinking alcohol in the last 30 days.

Year	%	Year	%	Year	%
2000	41	2007	33	2014	24
2001	39	2008	30	2015	22
2002	35	2009	29	2016	20
2003	35	2010	29		
2004	35	2011	27		
2005	33	2012	28		
2006	34	2013	26		

Let's plot the data on the TI-Nspire with a scatter plot.

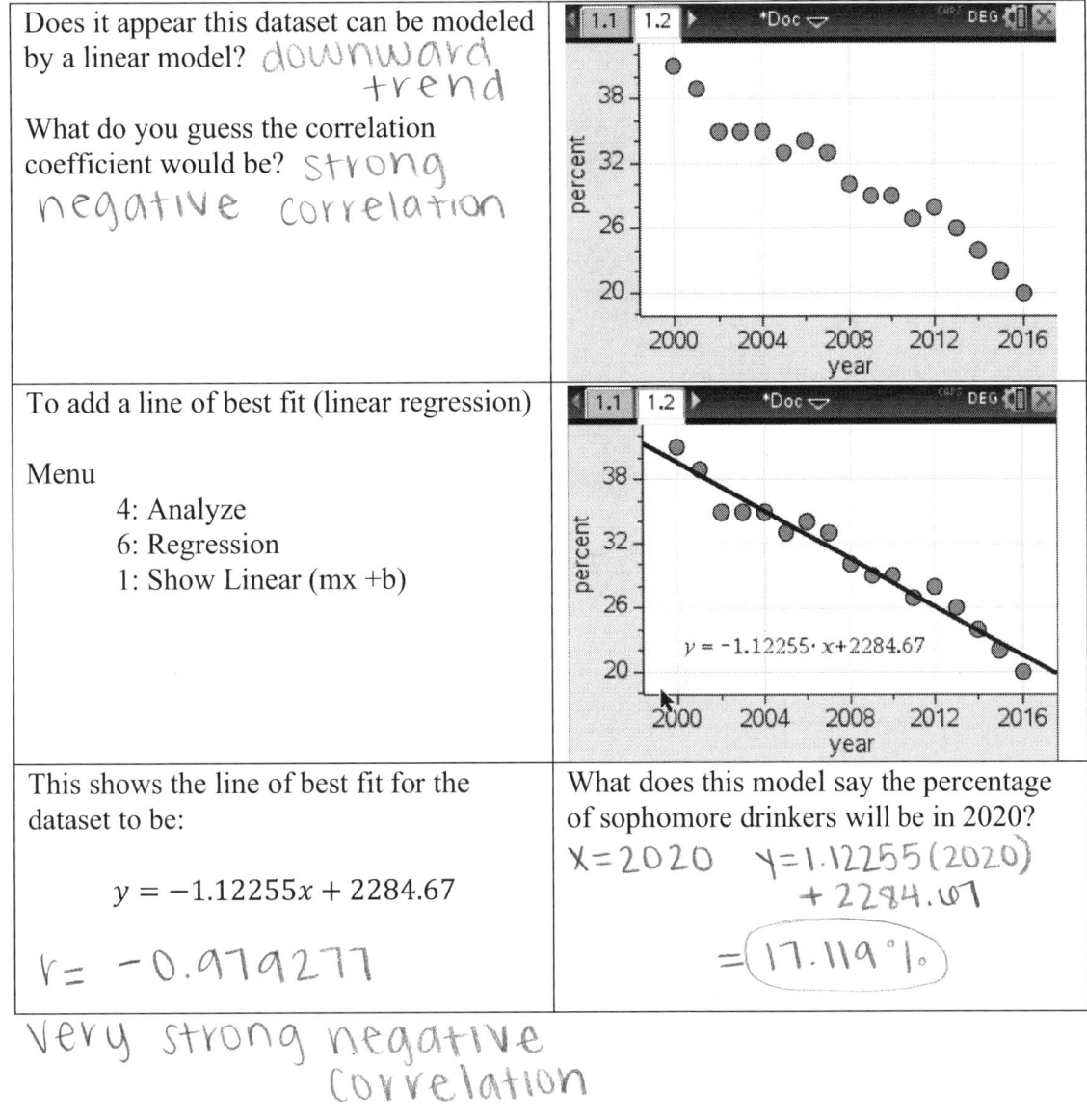

Does it appear this dataset can be modeled by a linear model? downward trend

What do you guess the correlation coefficient would be? strong negative correlation

To add a line of best fit (linear regression)

Menu
    4: Analyze
    6: Regression
    1: Show Linear (mx +b)

This shows the line of best fit for the dataset to be:

$$y = -1.12255x + 2284.67$$

$r = -0.979277$

very strong negative correlation

What does this model say the percentage of sophomore drinkers will be in 2020?

X = 2020   Y = 1.12255(2020) + 2284.67
= 17.119%

| To find the correlation coefficient.

Ctrl-i or Ctrl-DOC
1: Add Calculator
MENU
6: Statistics
1: Stat Calculations
3: Linear Regression (mx+b)

XLIST ->year (or whatever you named the variable)
YLIST-> percent | LinRegMx *year,percent*,1: CopyVar *stat.Reg*.

"Title"   "Linear Regression (mx+b)"<br>"RegEqn"   "m·x+b"
"m"   -1.12255
"b"   2284.67
"r²"   0.958983
"r"   -0.979277
"Resid"   "{...}" |
| You see now that the correlation coefficient, r, is -.979277 | Does this value of r make sense? |

The two main variables in an experiment are the **independent** and **dependent** variable. An independent variable is the variable that is changed or controlled in a scientific experiment to test the effects on the dependent variable. A dependent variable is the variable being tested and measured in a scientific experiment, the variable of most interest. The dependent variable is often thought of as the "response" variable – we are interested in the dependent variable's response to the independent variable in determining if there is correlation.

Example if we are measuring amount of CO_2 emissions by year what are the variables?

Independent variable	Dependent Variable
year	amt of CO_2

What if we are studying nationality and how fast a person can run 1000m?

Independent variable	Dependent Variable
nationality	time 1000m

Consider an experiment where you want to determine if the size of people's feet correlates to their GPA.

Independent variable	Dependent Variable
shoe size	GPA

Consider the dataset with shoe size versus GPA.

Shoe Size	8	9	10	11	10	9	7
GPA	3.5	4.0	2.0	2.9	3.1	3.3	2.75

How do we know if this would be a good dataset to model with a linear line of best fit?

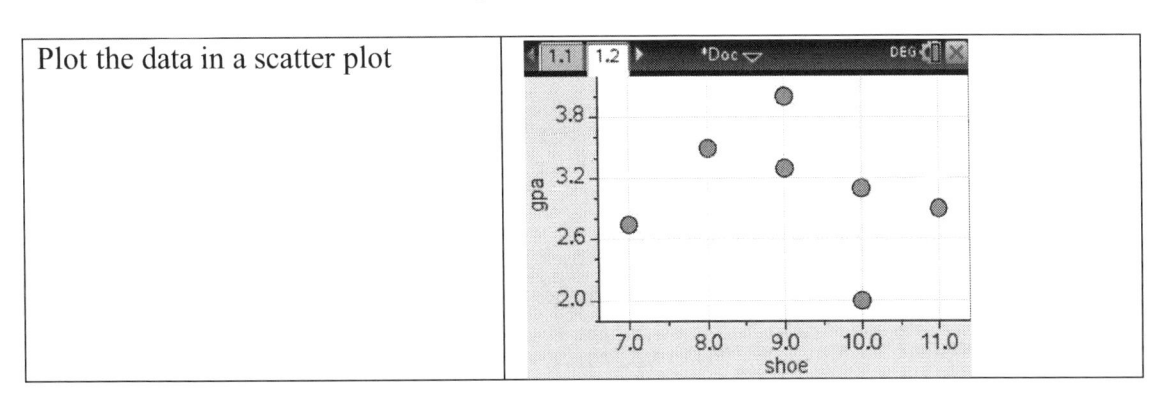

Plot the data in a scatter plot	

Find the Coefficient of Correlation From a calculator page or the spreadsheet MENU 6: Statistics 1: Stat Calculations 3: Linear Regression (mx+b) X List -> shoe size Y List -> gpa	LinRegMx shoe,gpa,1: CopyVar stat.RegEqn "Title" "Linear Regression (mx+b)" "RegEqn" "m· x+b" "m" -0.108553 "b" 4.07105 "r²" 0.053771 ("r" -0.231886) "Resid" "{...}" very weak This shows the line of best fit as: $y = -.108553x + 4.07105$

Find the line of best fit on your graph. MENU 4: Analyze 6: Regression 1: Show Linear mx+b	

Is this a good model for this data?	no, it is weak

3.3—135

Let's look at another model regarding electricity consumption.

Households consume much more electricity when the weather is warmer. The table below show average household energy consumption.

Temperature (°F)	57	62	69	71	81	84
Electricity (kilowatt-hours)	32.3	34.2	36.8	39.6	44.8	43

Independent variable	Dependent Variable
temperature	electricity

Find the linear model expressing the consumption of electricity in kWh as a function of the temperature F°.	$y = 0.455643x + 6.25121$ (slope)
Find the correlation coefficient	$r = 0.971026$
What does this value mean?	It is a very strong positive correlation
What does the slope of the line represent in the context of this problem?	slope = 0.455643 increase in kWh for every degree F increase in temperature.
What does the y-intercept mean in the context of this problem?	y represents at the temp of 0°F the usage is 6.25121 kWh.
What does this model predict the consumption to be at 95°F?	$y = 0.455643(95) + 6.25121$ = 49.5373 consumption in kilowatt hrs at 95°F
What does the model predict the temperature would be for consumption to reach 35 kWh of electricity?	$35 = 0.455643(x) + 6.25121$ = 63.095 °F when 35 kWh are consumed

Homework 3.3 Linear Regression Models

1. In the graphs below, determine if the correlation coefficient, r, is closer to -1, -0.5, 0, 0.5, or 1.

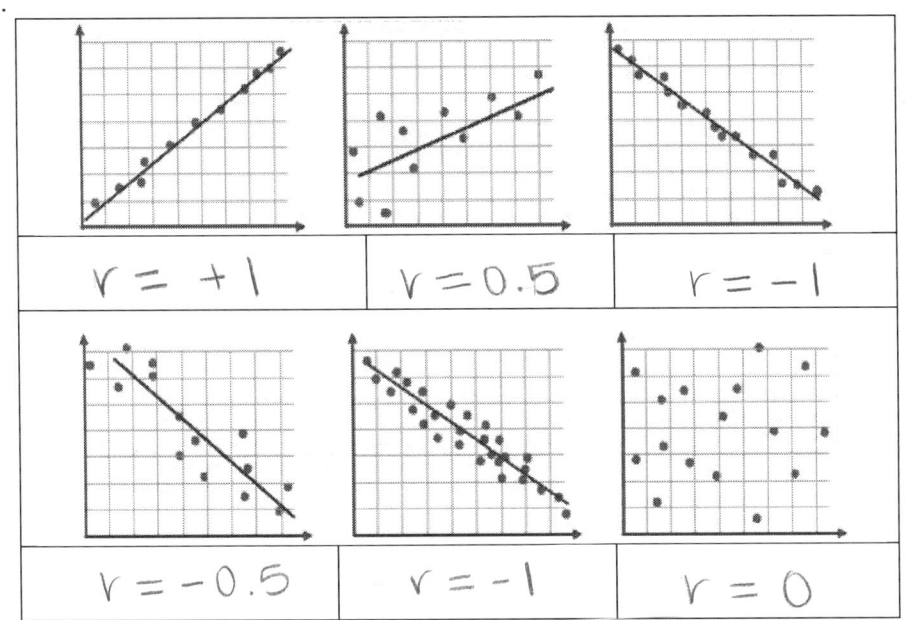

$r = +1$	$r = 0.5$	$r = -1$
$r = -0.5$	$r = -1$	$r = 0$

2. The number of countries participating in the Winter Olympics is listed below.

Year	1998	2002	2006	2010	2014	2018
# of countries	72	77	80	82	89	92

Independent variable	Dependent Variable
year	# of countries

Create a scatter plot on your TI-Nspire and add the line of best fit.

Equation of the line of best fit	$y = 0.985714x - 1897.31$
Coefficient of correlation	$r = 0.989254$
What does the slope tell us about the change in # of countries every 4 years?	there is a 0.985714 increase for every 4 years
How many countries would be expected in 2022? (round to the nearest integer)	$0.985714(2022) - 1897.31$ = 95.8 countries
In what Olympics do we expect to have exceeded 100 countries? (Round to the nearest Olympic year)	$100 = 0.985714x - 1897.31$ = 2026 In 2026, there will be >100 countries.

Homework 3.3 Linear Regression Models

3. Teens were surveyed in 2017, and asked to report their preferred social media platform. https://tinyurl.com/Algebra2SnapChat

The data was recorded every six months, and the number of respondents rating Snapchat as their preferred social media platform is given.

Months since Jan 1 2015	0	6	12	18	24	30
Snapchat percentage	11	17	24	35	39	47

Independent variable	Dependent Variable
months	snapchat

Create a scatter plot on your TI-Nspire and add the line of best fit.

Equation of the line of best fit	$y = 1.22381x + 10.4762$ (slope)
Coefficient of correlation $r = 0.990265$	What intelligence do you obtain from the coefficient of correlation relative to this specific dataset and linear data model? It is a very strong positive trend
What does the slope tell us about the change in teens favoring snapchat every six months?	slope = 1.22381 increase in SC percentage every month. 0 months = 7.34%
If trends hold, how many teens will favor Snapchat on Jan 1, 2019?	48 from 1/1/15 $1.22381(48) + 10.4762$ = 69 teens will favor SC
About how many months from January 1, 2015 was the percentage projected to reach 50%?	$50 = 1.22381x + 10.4762$ = 32 months, 50% SC favorship

Homework 3.3 Linear Regression Models

4. The College Board published data in 2016 regarding the average yearly costs (tuition, room and board, fees) for attending private college in the USA.

Year	Cost in 2016 Thousands of Dollars
1976	16.76
1981	16.63
1986	21.65
1991	25.07
1996	28.14
2001	32.34
2006	36.06
2011	40.45
2016	45.37

Create a scatter plot of the data with years as the horizontal axis and costs as the vertical axis.

Do you think a linear model would work well for this? **yes**

What is the correlation coefficient? Does the value of r lead you to believe this model will work well with a linear model? **$r = 0.986128$ → yes, it's very close to positive one**

Find the equation of the linear line of best fit.

$y = 0.739967x + 1447.81$

What does this model project the costs to be in 2021?

$0.739967(2021) + 1447.81$

$= 2943.27$

In 2021, the cost of school tuition will be $2943.27.

Homework 3.3 Linear Regression Models

CHALLENGE PROBLEM

1. Consider the data below which details the outdoor temperature and the number of announcements that are made at 3:15PM.

Temperature	Observed number of Announcements	Expected number of announcements	Residual (Observed-Expected)
66°	8		
70°	10		
75°	12		
80°	16		

We can model this data with a linear model.

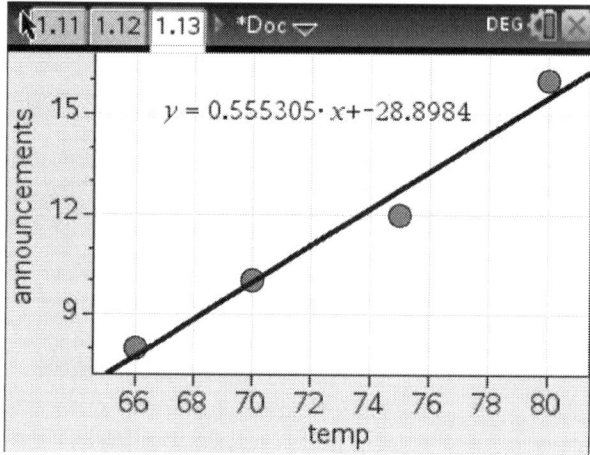

$y = 0.555305 \cdot x + -28.8984$

The residual is the amount by which our model undershoots or overshoots the observed value. Calculate the residuals at each observed point. Use your TI for this plot!

If the plot looks fairly random, then we can assume that linear model is a good fit. Explain.

Sketch a graph of the temperature (horizontal axis) versus the residual (vertical axis). Does this graph indicate that a linear model is appropriate?

Why would the residual plot below indicate a poor fit for a linear model?

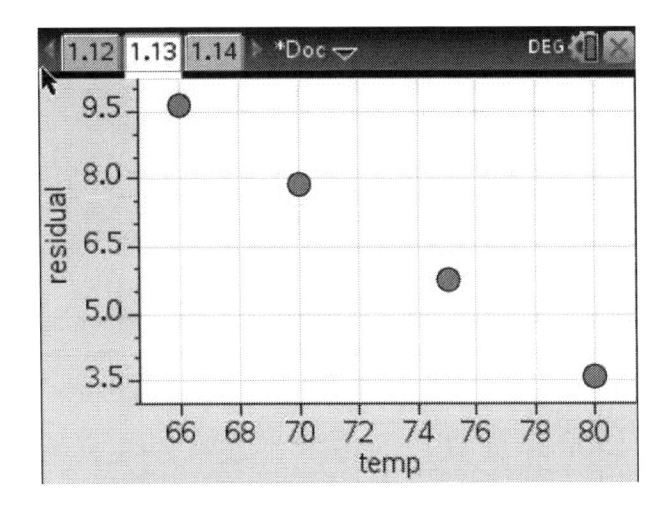

3.4. Monopoly Line of Best Fit

Is there an association between the number of spaces from GO and the property cost in Monopoly?

Property	Spaces from Go	Cost	Property	Spaces from Go	Cost
Mediterranean Avenue	1	60	Kentucky Avenue	21	220
Baltic Avenue	3	60	Indiana Avenue	23	220
Reading Railroad	5	200	Illinois Avenue	24	240
Oriental Avenue	6	100	B&O Railroad	25	200
Vermont Avenue	8	100	Atlantic Avenue	26	260
Connecticut Avenue	9	120	Ventnor Avenue	27	260
St Charles Place	11	140	Water Works	28	150
Electric Company	12	150	Marvin Gardens	29	280
States Avenue	13	140	Pacific Avenue	31	300
Virginia Avenue	14	160	North Carolina Avenue	32	300
Penn Railroad	15	200	Pennsylvania Avenue	34	320
St James Place	16	180	Short Line Railroad	35	200
Tennessee Avenue	18	180	Park Place	37	350
New York Avenue	19	200	Boardwalk	39	400

Using your TI-Nspire, create a scatter plot ($x-axis$ Spaces from go, $y-axis$, Cost)	What is the correlation coefficient? Does there appear to be an association between the two variables?
Find the line of best fit, $y = mx + b$	What does b represent in the context of the problem? What does m represent?
Are there any data points that stand out from the others?	Predict the cost of a new property, Woodward Way, 50 spaces from Go.

3.5. Review

Assessment Checklist. Below are the competencies one should master in preparation for an assessment on data analysis.

- [ ] This unit relies heavily on the TI-Nspire. Be confident in **all** of the methods we have used on the calculator to explore various statistics and data.
- [ ] Find common one variable statistics, mean, median, standard deviation, min, max, sum of all values.
- [ ] Create a box-plot from one-variable data, and know the 5-number summary (min, Q1, Median, Q3, max) and what each value represents relative to the dataset.
- [ ] Know the meaning and how to graphically identify outliers in a dataset with a boxplot.
- [ ] How to create a histogram from a one-variable dataset.
- [ ] Reading information from a histogram, interpreting the mean and median from a histogram.
- [ ] Changing the bin size of a histogram.
- [ ] The coefficient of correlation and what it means.
- [ ] How to graph and calculate the line of best fit for a two-variable dataset
- [ ] Know the context of independent vs dependent variables in data experiments.
- [ ] Answer questions in context of an application problem involving a line of best fit.
- [ ] Look at a chart of two-variable data and answer questions about the data, percentages, proportions, etc.

Review 3.5

1. The Cerro Negro volcano in Nicaragua is an active volcano, part of the Pacific "ring of fire" volcanos. Cerro Negro is not very large compared to a stratovolcano like Mount Rainier in Washington State or Pinatubo in the Philippines. As its name implies, this volcano is made of very dark rocks. The table below shows the last 60 years' worth of eruption data.

Year	Eruption Volume (cubic km) of Ash
1957	0.074
1960	0.095
1961	0.095
1962	0.095
1963	0.096
1968	0.115
1969	0.115
1971	0.142
1992	0.152
1995	0.152
1996	0.160

What is the average of all eruptions over this time period?
0.117364

What is the median amount of eruptions?
0.115

What is the total volume of eruptions in cubic km over the entire period? (Σx)
1.291 km³

What is the standard deviation?
0.029449

What is the value of r and what does it tell us relative to this dataset?
$r = 0.934921$; the correlation is a strong problem

What is the equation of the line of best fit?
$y = 0.001850 x + -3.54214$

What does the slope of the line tell us in context of the problem?
slope = 0.001850
There is a 0.001850 increase in the eruption volume per year.

What is the expected ash volume in 2018?
1990 - 2018
$0.001850(2018) + -3.54214 = 0.203268$ km³ by 2018.

In what year do we expect the ash volume to exceed $.207\ km^3$?
$0.207 = 0.001850 x - 3.54214$
In 2020, the ash volume would exceed .207 km³.

Review 3.5

2. Besides the *coefficient of correlation*, r, we have used in these exercises, statisticians square the coefficient of correlation, and use that measure. Consider this *coefficient of determination*, r^2, as a measure to describe correlation. Which dataset would have the **weakest** linear correlation? (Circle the best answer) *unsquare the "r"*

(A.) $r = .987$	B. $r^2 = .980$	C. $r = -0.992$

$r = 0.989949$

3. All the visualizations below show data for the age of USA presidents at the time of their inauguration.

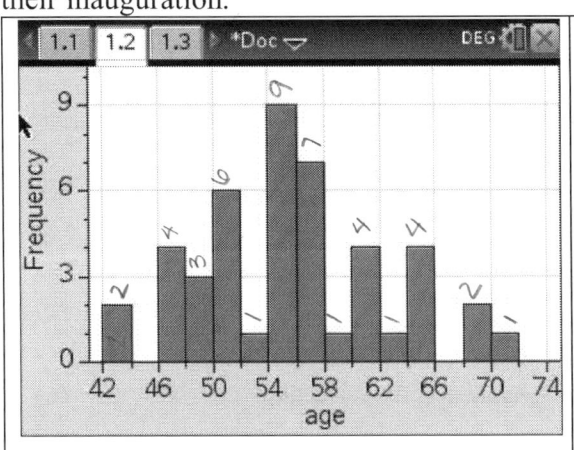

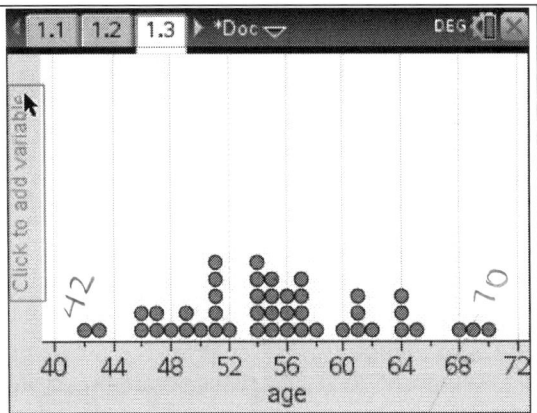

median → middle

mode → most

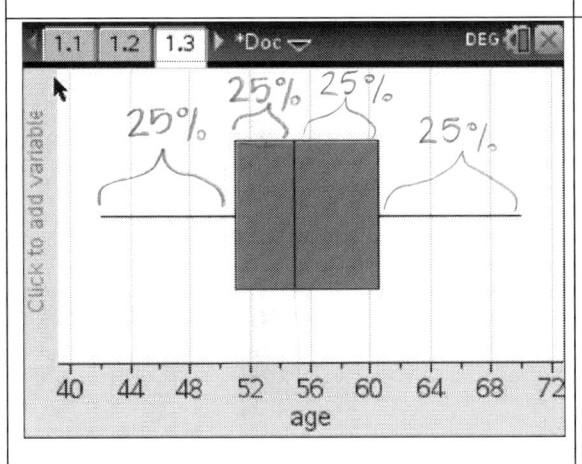

mean → avg

What is the median age of the president at the time of inauguration? **55 yrs**
#=45
#22 & #23

What is the difference between the oldest and youngest president? **28 yrs**
oldest = 70
youngest = 42

What percentage of presidents were between 51 and 55 at the time of their inauguration?
(look at box plot) **25%**

How many presidents were older than 68? **2 presidents**

The youngest 25% of the presidents were aged **42** to **51**.

Review 3.5

4. The second worst collapse in Atlanta in 2017 was the bridge over I-85. The worst collapse was the Atlanta Falcons in Super Bowl LI. For that season, the offensive scoring totals of the team were:

| 24 | 35 | 45 | 48 | 23 | 24 | 30 | 33 | 43 | 15 | 38 | 28 | 42 | 41 | 33 | 38 | 36 | 44 | 28 |

Create a box plot and give the values of:

Minimum	Q1	Median	Q3	Maximum
15	28	35	42	48

In their top 25% highest scoring games, the team scored between __42__ and __48__ points. between Q3 and maximum

5. Consider the following dataset. You've been told by one of your parents that the scatter plot below and the coefficient of correlation indicate this dataset is not modelled well by a linear line of best fit. Your other parent suggests you put your cell phone away and think about breaking the data into two datasets. You decide to take you parent's advice, on both accounts. Could you see a way to model the dataset linearly if you considered it two separate sets? Draw your modeled lines of best fit and discuss. Would one part of the model have a stronger correlation than the other? What are your best guesses for the coefficients of correlation for the two datasets, and what reasoning backs up your guesses?

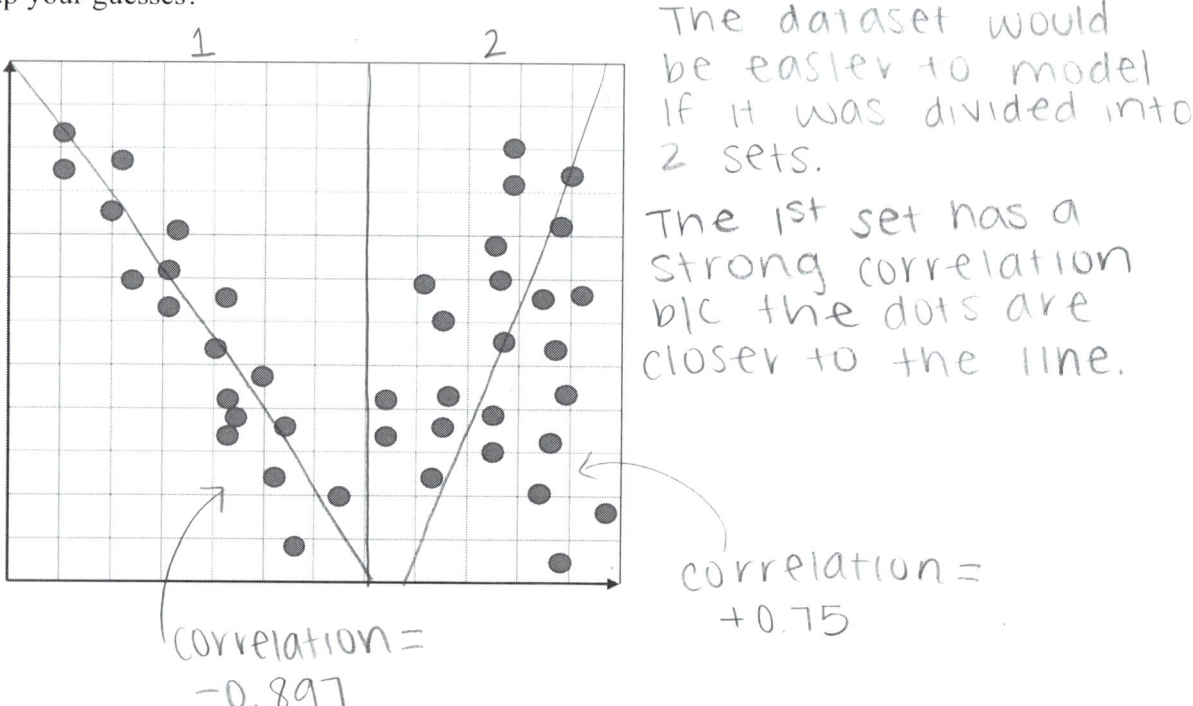

The dataset would be easier to model if it was divided into 2 sets.

The 1st set has a strong correlation b/c the dots are closer to the line.

correlation = −0.897

correlation = +0.75

Review 3.5

6. The table below shows the results of a survey of car owners. The owners were asked if they received a ticket last year and the color of their car.

	Speeding Ticket	No Speeding Ticket	Total
Red Car	19	132	151
Non-red Car	43	465	508
Total	62	597	659

What proportion of those surveyed were red car owners who received a ticket?

$$\frac{19}{659} = 0.028832$$

Among those who reported receiving a ticket, what percentage did not own a red car?

$$\frac{43}{62} = 69.3548\%$$

If a red car owner is selected at random, what is the probability that he/she has received a ticket?

$$\frac{19}{151} = 0.125828$$

If a non-red car owner is selected at random, what is the probability that he/she received a ticket?

$$\frac{43}{508} = 0.084646$$

If the results of this survey were extended to a town of 10,000 people, about how many would report receiving no tickets?

$$\frac{597}{659} \cdot 10000 = 9059$$

If a person were selected at random, what would be the probability that the person either drove a red car OR received a speeding ticket?

red car = 151

speeding ticket = 62

$$\frac{19+132+43}{659} = 0.294385$$

$$\frac{62+151-19}{659} = 0.294385$$

you have to subtract 19 because it applies to both red & ticket.

4. PIECEWISE LINEAR FUNCTIONS

4.1. Relations/Function/Domain/Range

In this unit we will study functions that are broken up into multiple sections or units, piecewise functions. Quite often in modeling real life scenarios, there is not one function that describes the entire behavior, so being able to define, interact with, and graph piecewise functions are all helpful.

As a review, we need to be solid with ==expressing intervals in interval notation,== and the meaning of relations, functions, domains and ranges.

Let a and b be real numbers, $a < b$.

set of all #s x, such that x is → >a & <b

Set of real numbers in inequality notation	Interval notation	Region on the real number line
$\{x \mid a < x < b\}$	(a, b)	open circle at a, open circle at b
$\{x \mid a \leq x < b\}$	$[a, b)$	closed at a, open at b
$\{x \mid a < x \leq b\}$	$(a, b]$	
$\{x \mid a \leq x \leq b\}$	$[a, b]$	closed at a, closed at b
$\{x \mid x < b\}$	$(-\infty, b)$	arrow left, open at b
$\{x \mid x \leq b\}$	$(-\infty, b]$	
$\{x \mid x > a\}$	$(-\infty, a)$	closed at a, arrow right
$\{x \mid x \geq a\}$	$(-\infty, a]$	closed at a, arrow right
$\{x \mid -\infty < x < \infty\}$	$(-\infty, \infty)$	

Write the following in interval notation

$x \leq 5$	$(-\infty, 5]$
$-2 \leq x < 10$	$[-2, 10)$
$x \geq 0$	$[0, \infty)$
$x \leq -2$ or $x > 10$	$(-\infty, -2]$ or $(10, \infty)$
$x < 0$	$(-\infty, 0)$
$x > 0$ and $x \leq 5$	$(0, 5]$

A **relation** is an association between two datasets.

Your name	Favorite Foods
Sydney	chocolate
Gavin	ice
Wyatt	chicken
Austin	strawberries

DOMAIN   RANGE

The table above is a relation that associates your name with a favorite food.

For any relation the **domain** is the set of input values. The **range** is the set of output values.

Suppose we define a relation as follows:

Your first name → # of years you have attended Woodward

Domain	Range
Amanda → Lily → Aidan →	5 10 3

==A *function* is a relation with the property that each element of the domain is only paired with one element of the range.==

A relation will NOT be a function if we can find at least one element of the domain paired with more than one element in the range.

Look at the relations defined below:

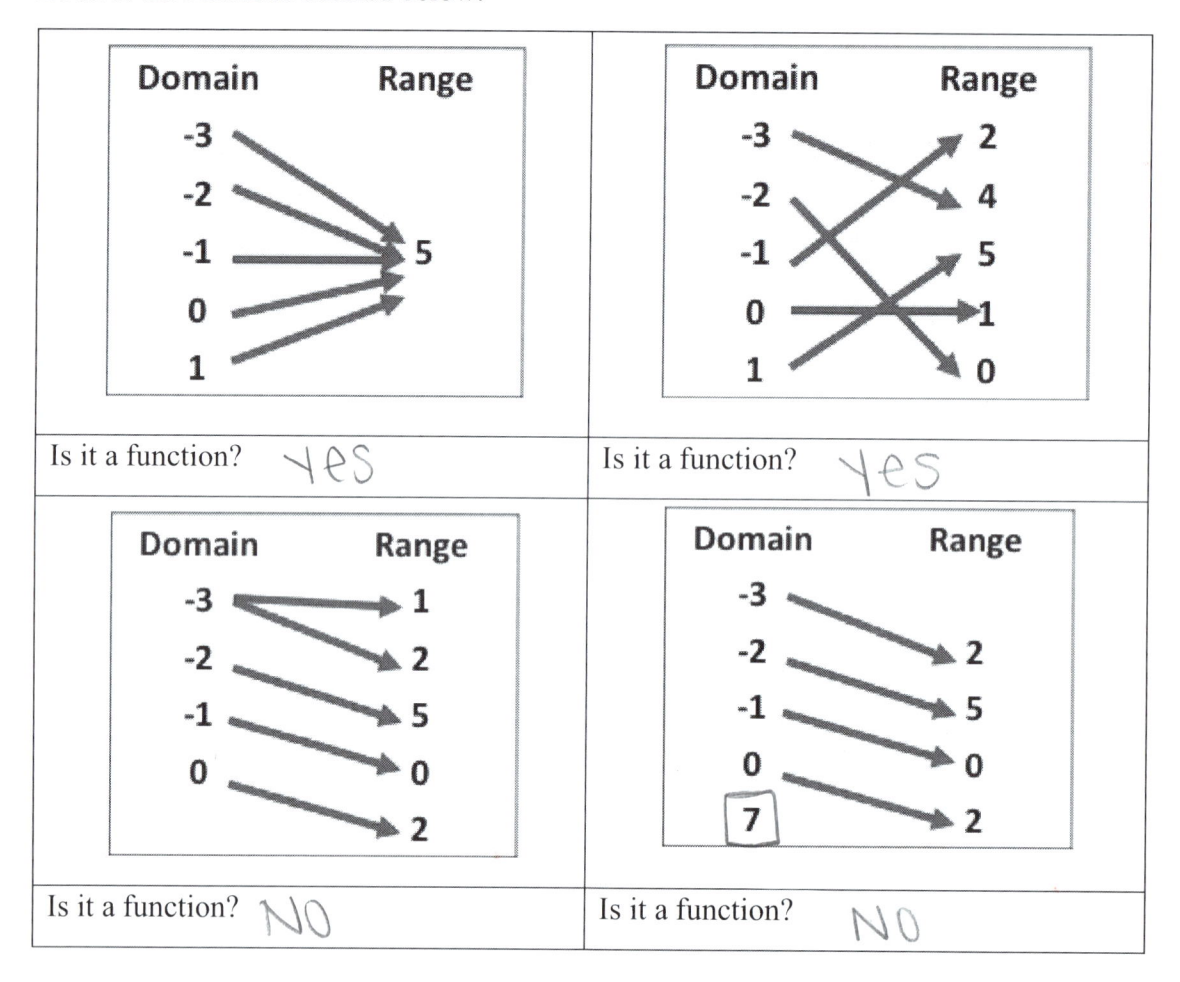

Domain → Range -3, -2, -1, 0, 1 → 5	Domain → Range -3 → 1, -2 → 0, -1 → 5, 0 → 2, 1 → 4 (crossed)
Is it a function? Yes	Is it a function? Yes
Domain → Range -3 → 1, -3 → 2, -2 → 2, -1 → 5, 0 → 0, 0 → 2	Domain → Range -3 → 2, -2 → 5, -1 → 0, 0 → 2, 7 → 2
Is it a function? NO	Is it a function? NO

From the table above, we see that if we have repeated domain values, mapped to different range values, the relation is NOT a function.

Consider the relation, F, defined by the order pairs:

$$F = \{(0,1), (1,1), (2,1), (3,2)\}$$

What is the domain of F?	0, 1, 2, 3
What is the range of F?	1, 1, 1, 2
Is F a function?	Yes

$$G = \{(-5,2), (-4,3), (-2,0), (0,5), (2,1), (3,2)\}$$

What is the domain of G?	-5, -4, -2, 0, 2, 3
What is the range of G?	2, 3, 0, 5, 1, 2
Is G a function?	Yes

$$H = \{(-5,2), (-4,3), (-2,0), (0,5), (-2,1), (3,4)\}$$

What is the domain of H?	-5, -4, -2, 0, -2, 3
What is the range of H?	2, 3, 0, 5, 1, 4
Is H a function?	NO, there are 2 values for "-2"

In addition to inspecting a relation given in set notation, we can also look at a graph and determine if the graph represents a function or not.

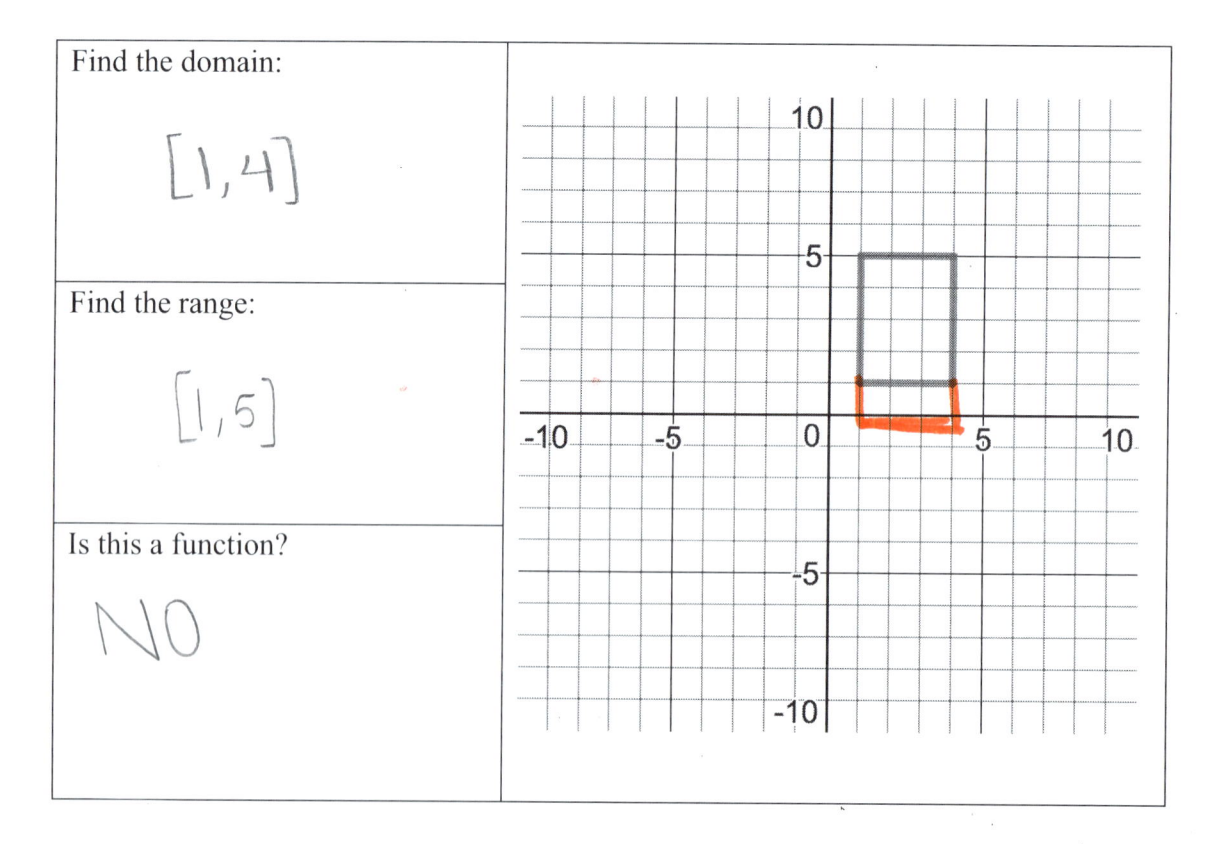

Find the domain: (x axis)

(-6, 4]

Find the range: (y axis)

[-2, 6)

Is this a function?

Yes

Find the domain:

[1, 4]

Find the range:

[1, 5]

Is this a function?

No

Find the domain:

$(-\infty, \infty)$

Find the range:

$(-\infty, 4]$

Is this a function?

Yes

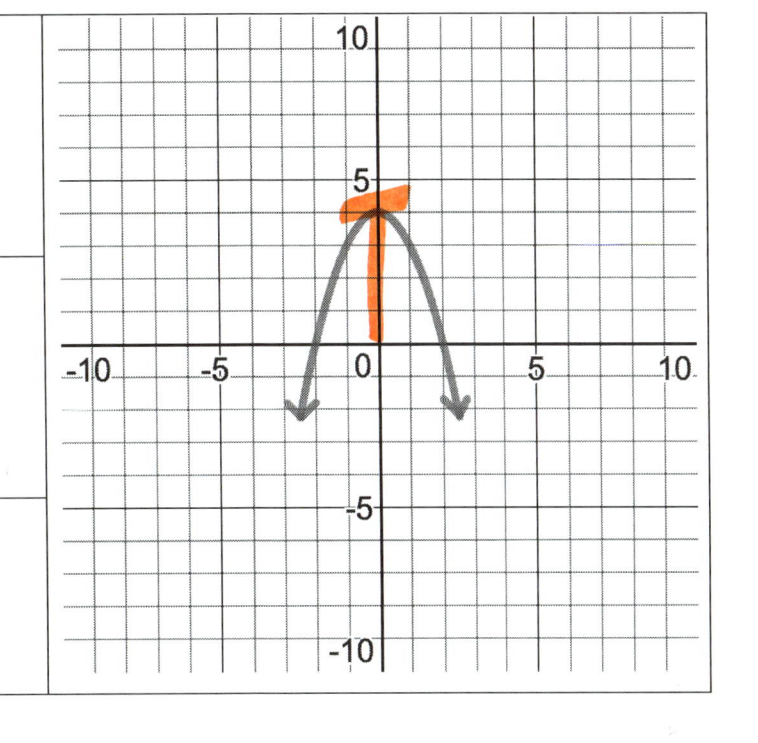

Find the domain:

$(-5, 4] \cup [6]$

Find the range:

$(-3, 2] \cup [5] \cup [3]$

Is this a function?

NO

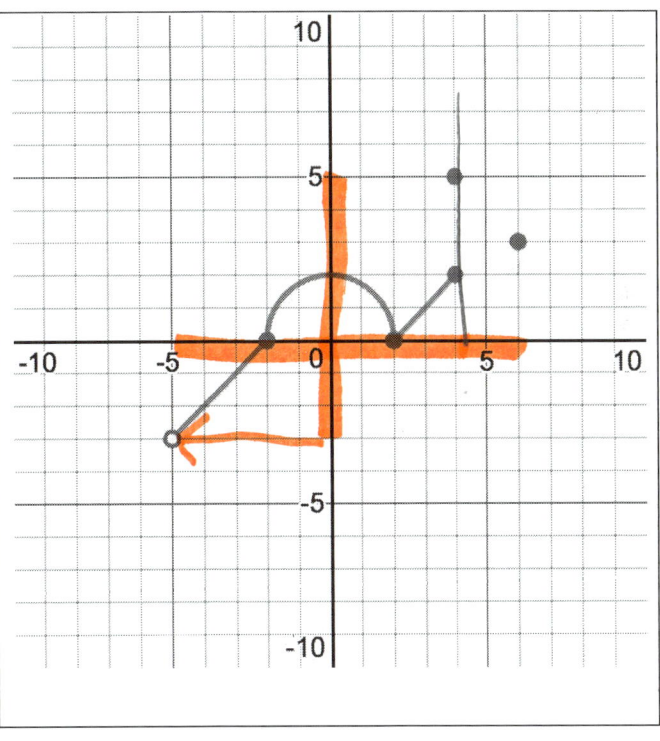

(4, 5)
(4, 2)

Homework 4.1 Relations/Functions/Domain/Range

1. Write the following in interval notation.

$x > -7$	$(-7, \infty)$
$-\pi \leq x < 2\pi$	$[-\pi, 2\pi)$
$x \geq 14$	$[14, \infty)$
$x \geq 2$ or $x < -10$	$[2, \infty) \cup (-\infty, -10)$
$x < 0.1$	$(-\infty, 0.1)$
$x > 0$ and $x \leq 14$	$(0, 14]$

2. For each relation given, specify the domain and range, and tell whether the relation is a function or not.

$F = \{(-1,2), (2,-1), (3,0), (4,0)\}$		
Domain	Range	Function?
-1, 2, 3, 4	2, -1, 0, 0	yes

$G = \{(-1,2), (2,-1), (3,-1), (1,2)\}$		
Domain	Range	Function?
-1, 2, 3, 1	2, -1, -1, 2	yes

$F = \{(-1,2), (2,-1), (-1,0), (4,9)\}$		
Domain	Range	Function?
-1, 2, -1, 4	2, -1, 0, 9	NO

Homework 4.1 Relations/Functions/Domain/Range

For each graph displayed give the domain and range, and state whether the graph represents a function or not.

3.

Find the domain:

$(-3, \infty)$

Find the range: NO DOTS

$(-\infty, \infty)$

Is this a function?

NO

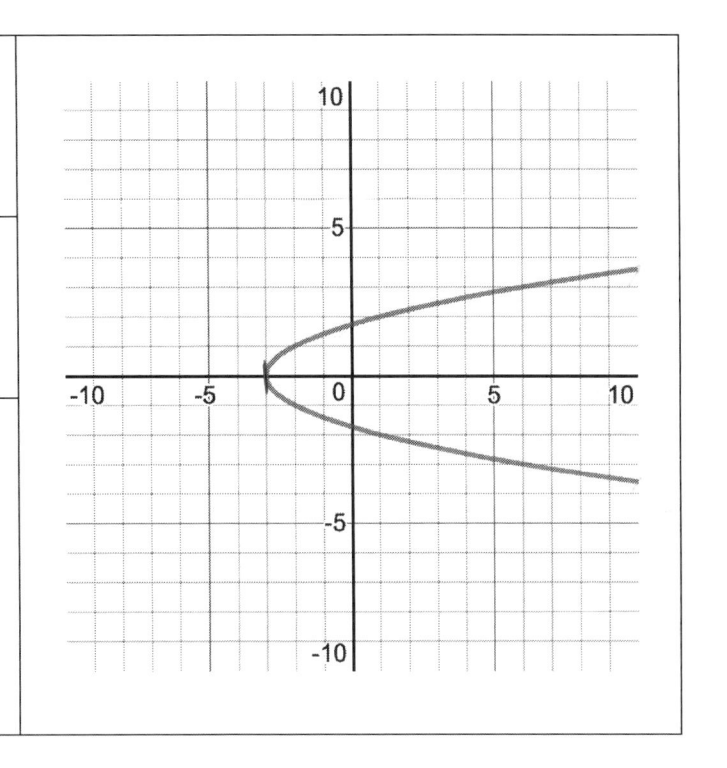

4.

Find the domain:

$[-7, 8)$

Find the range:

$[-3, 3] \cup [-6, -4)$

Is this a function?

Yes

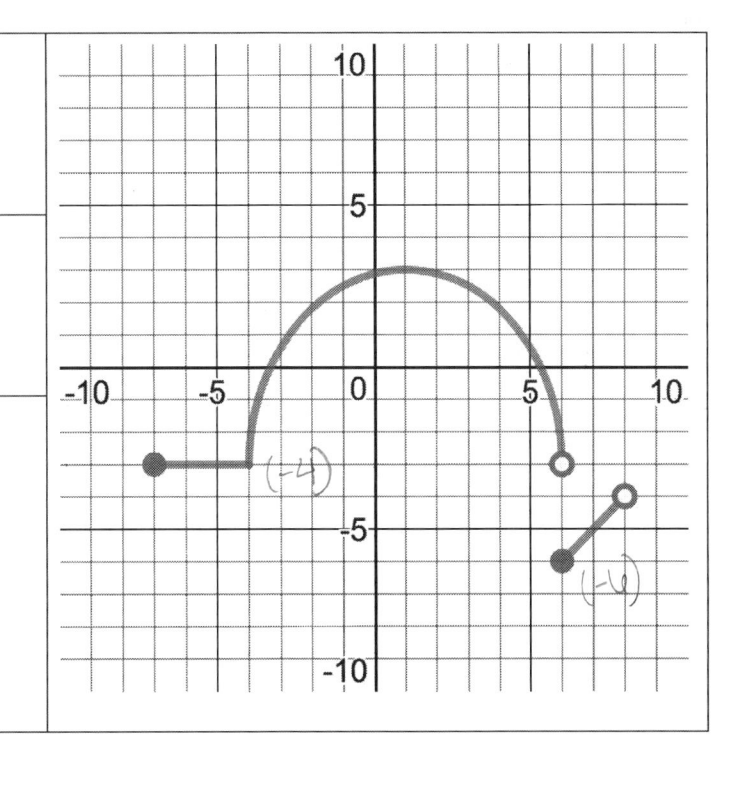

Homework 4.1 Relations/Functions/Domain/Range

5.

Find the domain: [-4, 4)
Find the range: [0, 6]
Is this a function? Yes

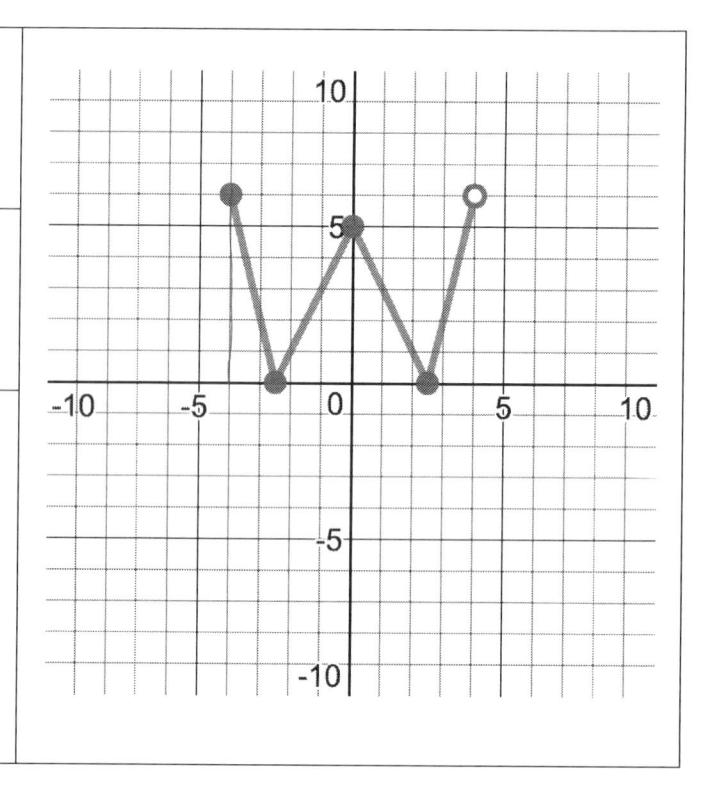

6.

Find the domain: $(-\infty, 1]$
Find the range: $(-\infty, \infty)$
Is this a function? Yes

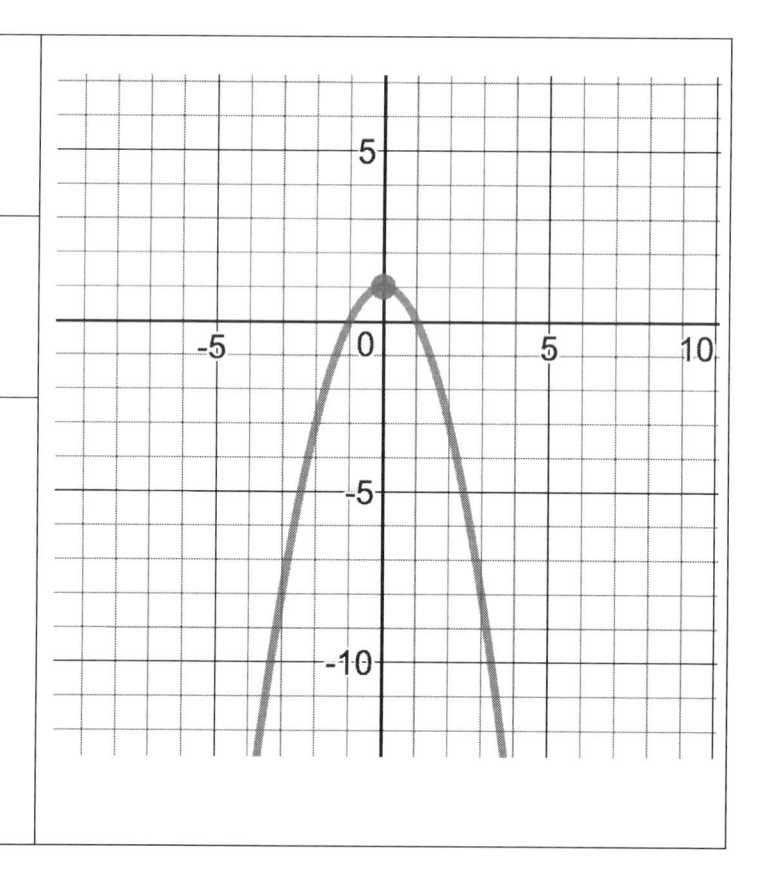

Homework 4.1 Relations/Functions/Domain/Range

CHALLENGE PROBLEMS

1. Write the single interval that satisfies the statement below.

The set of all real numbers, x, such that
$$x \in (-12,14] \cap [-3,10) \cap (0,14) \cap [1,20] \cap [-17,4)$$

2. Consider the functions below:
$$F = \{(0,1), (1,2), (2,3), (3,4)\}$$
$$G = \{(0,3), (1,3), (2,2), (3,0)\}$$

Domain of F	Range of F
Domain of G	Range of G

Consider the function $H = \dfrac{F}{G}$

Find $H(0)$	$H(1)$	$H(2)$	$H(3)$
Domain of H		Range of H	

3. Consider the function graphed below. Give the domain and range.

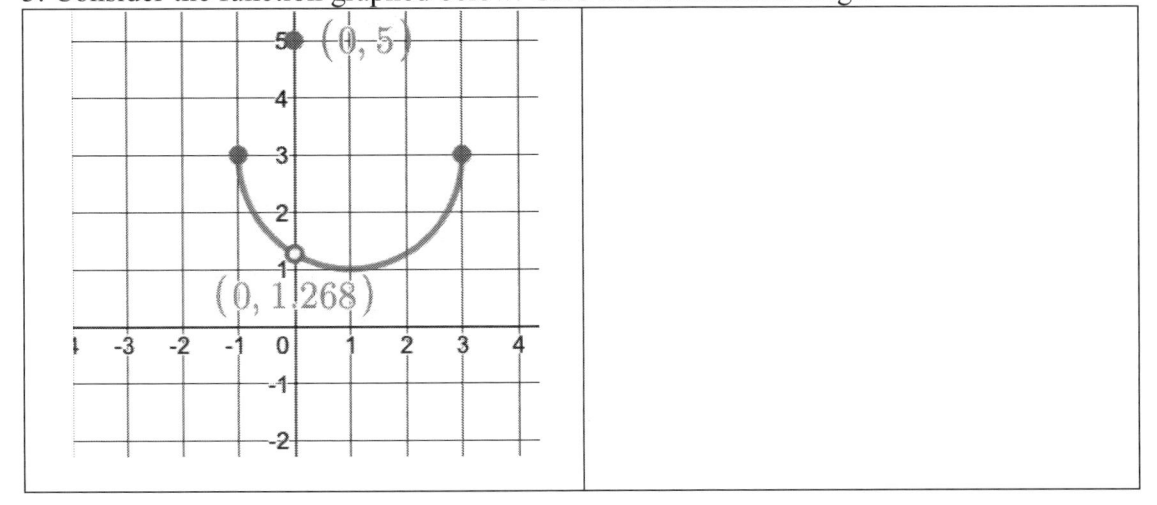

Points marked: $(0,5)$ and $(0, 1.268)$

4.2. Introduction to piecewise functions

Definition
In mathematics, **a *piecewise-defined*** function (also called a *piecewise* function) is a function which is **defined** by multiple sub-functions, each sub-function applying to a certain interval of the main function's domain (a sub-domain).

Think about an airplane flight...
You are flying to Disneyland. For the first 30 minutes, your plane climbs from an altitude of 1000 feet (ground level at ATL) to 40,000 feet. For the next 3 hours, the plane flies at this "cruising altitude". The plane then spends the last 30 minutes descending to 100 feet.

Draw a picture that uses time in flight along the x-axis and altitude along the y-axis.

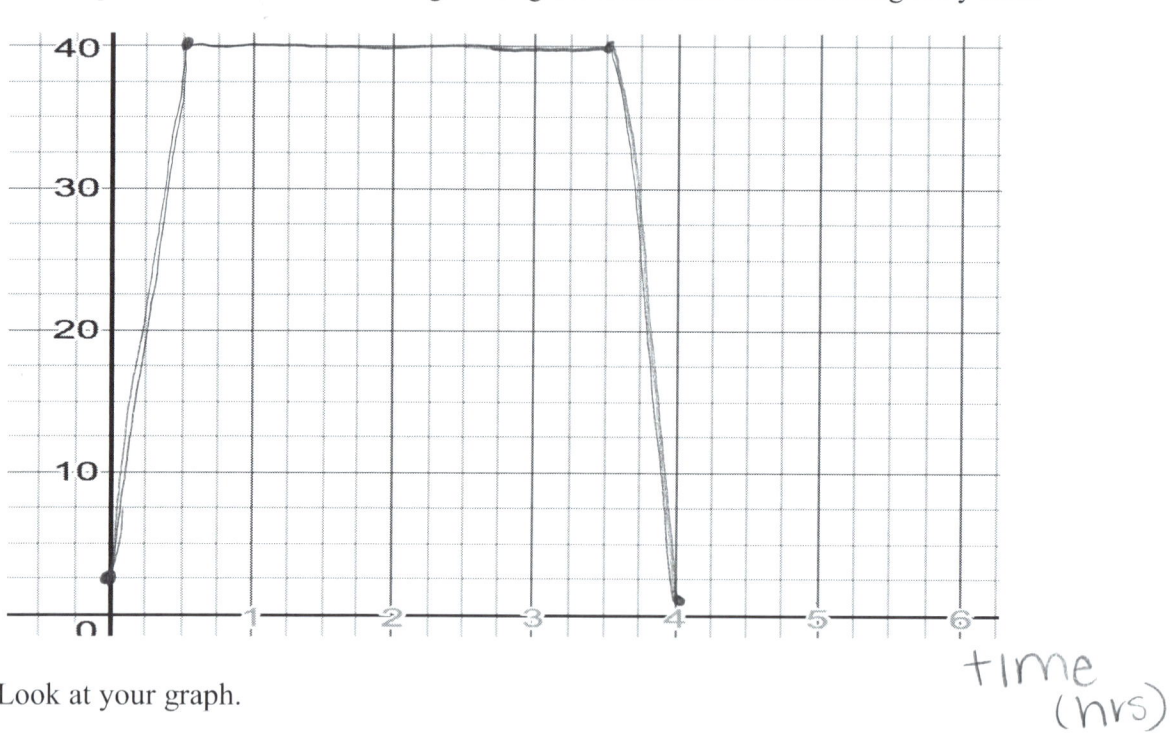

height [graph showing altitude vs time (hrs), climbing from ~0 to 40 between 0 and 0.5, level at 40 from 0.5 to 3.5, descending from 40 to ~0 between 3.5 and 4]

Look at your graph.

Can one equation represent the entire flight? **no**

If, yes, what would that equation look like?

If, no, how many equations? **3**

What type of equations would model this situation? **linear**

We will need to gain expertise in how piecewise functions are defined and graphed.

Consider the table below. You see the two ways we will interact with piecewise functions, either via the mathematical definition of the function, or its graph.

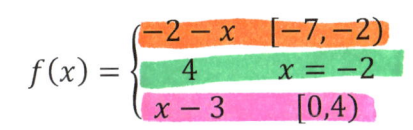

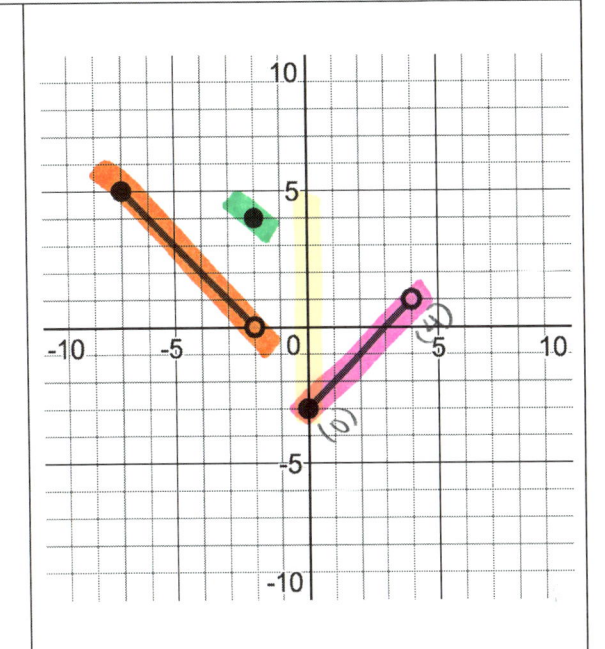

$$f(x) = \begin{cases} -2-x & [-7,-2) \\ 4 & x=-2 \\ x-3 & [0,4) \end{cases}$$	*(graph showing three pieces: line from (-7,5) to (-2,0) open, point at (-2,4), and line from (0,-3) to (4,1) open)*
Mathematical notation for a piecewise function, the function is defined in 3 separate pieces, with the domain for each piece or section specified. You can see that this function has 3 different **sub-functions**, each with its own **sub-domain**.	Graphically, the function with three distinct pieces, 2 lines and a point, and note the open/closed cirles on the line segments. Again, just as in the mathematical notation to the left, we see 3 distinct **sub-functions**, and each with its **sub-domain**.
What is the range of f? range = $[-3, 5]$	

4.2—159

Evaluating a piecewise function
Consider the following piecewise function.

$f(x) = \begin{cases} 4x - 3, & x > 3 \\ 5x + 2, & x \leq 3 \end{cases}$	Rewrite the function but expressing the domains in interval notation $f(x) = \begin{cases} 4x - 3 & (3, \infty) \\ 5x + 2 & (-\infty, 3] \end{cases}$

Evaluate the following:

$f(-2)$ $5x+2$ $5(-2)+2$ $-10+2$ $\boxed{-8}$	$f(4)$ $4x-3$ $4(4)-3$ $\boxed{13}$
$f(3)$ $5x+2$ $5(3)+2$ $\boxed{17}$	$f(0)$ $5x+2$ $5(0)+2$ $\boxed{2}$

You see that you must pay close attention to the function definition, especially the sub-domains to know how to evaluate piecewise functions.

Graphing functions with restricted domains.

Graph $f(x) = 2x - 3$, $x \in [-2, 4]$

x	y
-2	-7
4	5

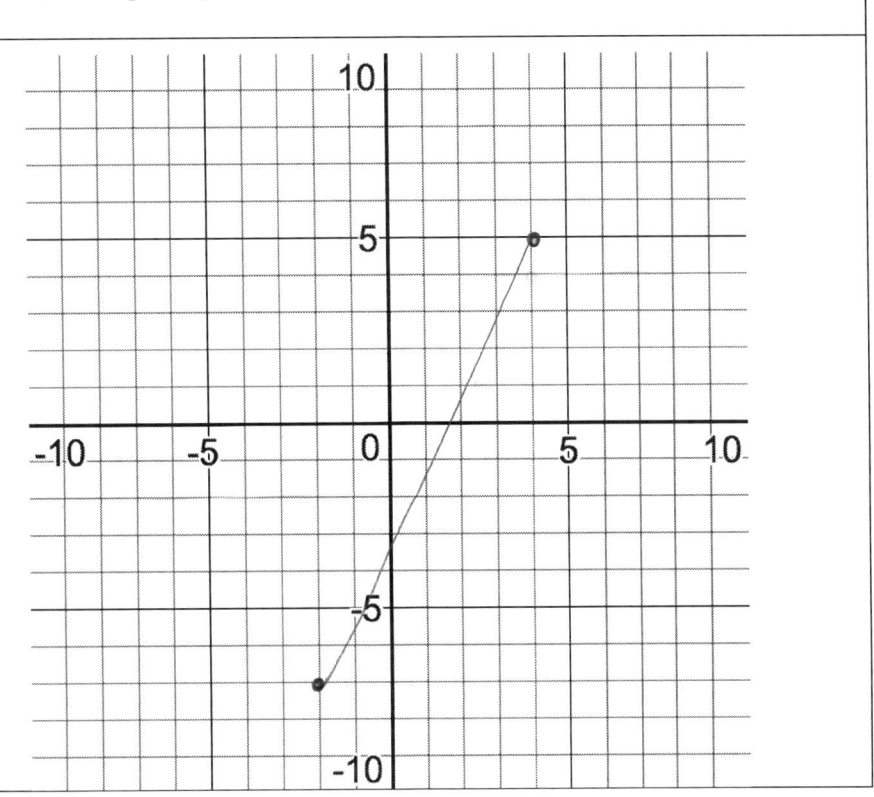

Graph $f(x) = 3x - 5$, $x \in [1, \infty)$

x	y
1	-2
2	1
4	7

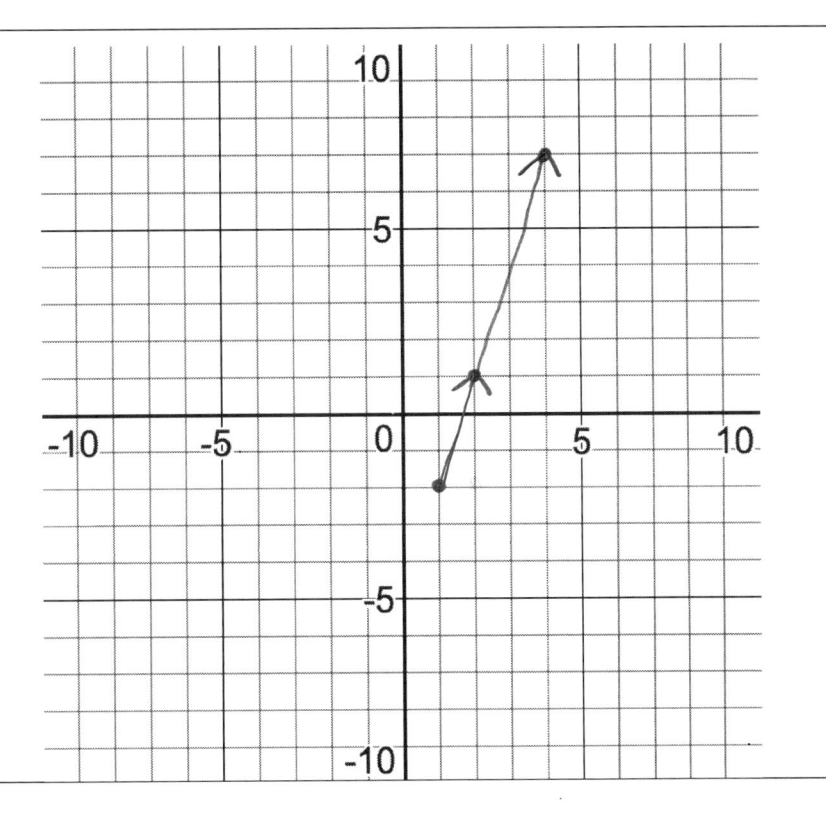

Graphing piecewise functions

Graph the following piecewise linear function.

$$f(x) = \begin{cases} x & x < -1 \\ x+1 & -1 \leq x \leq 2 \\ -1 & x > 2 \end{cases}$$

Rewrite with interval notation:

$$f(x) = \begin{cases} x & (-\infty, -1) \\ x+1 & [-1, 2] \\ -1 & (2, \infty) \end{cases}$$

x	f(x)=x
-1	-1
-5	-5

x+1	
-1	0
2	3

Then proceed to graph each piece of the function, paying close attention to the subdomains.

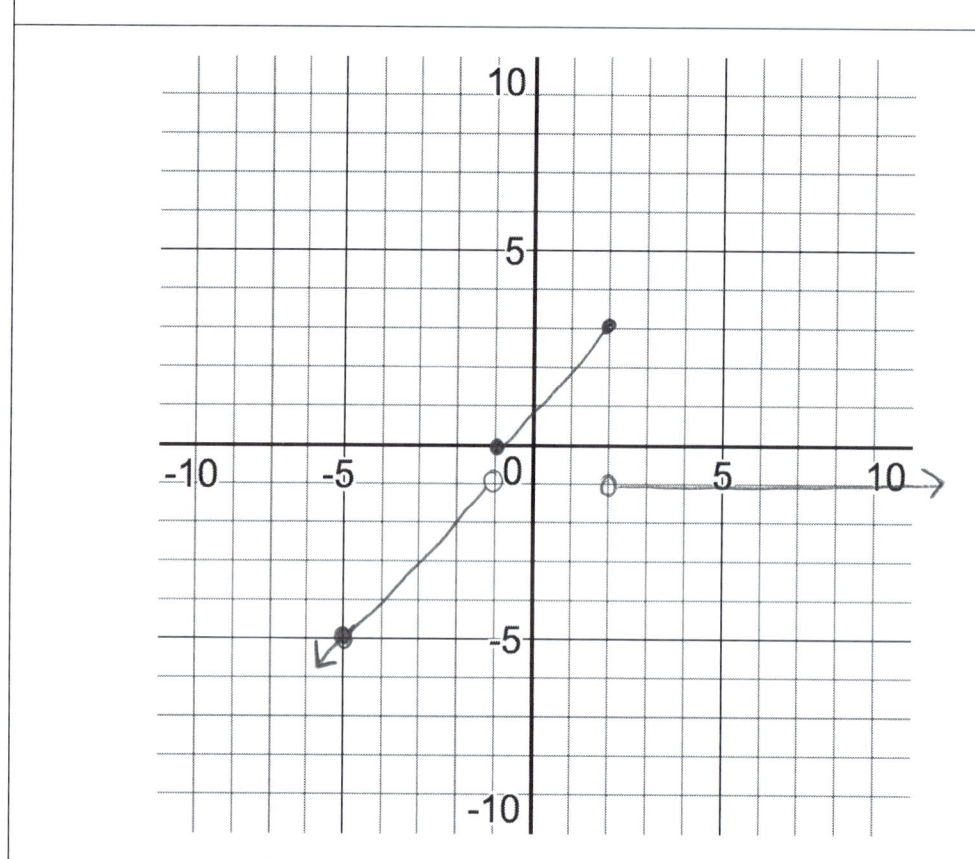

Is f a function? **YES** NO

$f(3) = -1$

$f(2) = x+1 = 3$

Graph the following piecewise linear function.

$$f(x) = \begin{cases} x+2 & x > 2 \\ 2x & -3 \leq x \leq 2 \\ x-2 & x < -3 \end{cases}$$

Rewrite with interval notation:

$$f(x) = \begin{cases} x+2 & (2, \infty) \\ 2x & [-3, 2] \\ x-2 & (-\infty, -3) \end{cases}$$

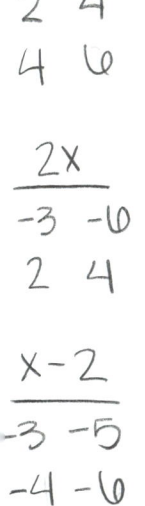

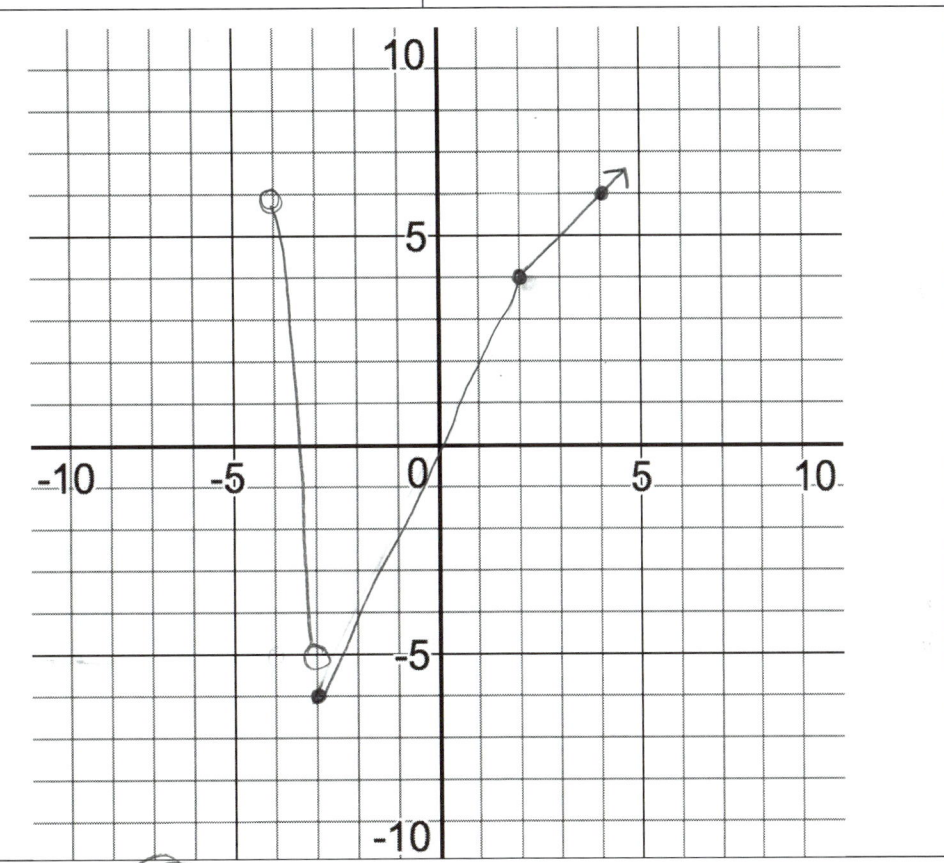

$x+2$
2	4
4	6

$2x$
-3	-6
2	4

$x-2$
-3	-5
-4	-6

Is f a function? (YES)   NO

$f(-3) = -6$

$f(2) = 4$

Homework 4.2 Introduction to Piecewise Functions

1. Consider the function below. Evaluate the function at the requested points.

$$f(x) = \begin{cases} x - 2 & x \in (-2, 1] \\ -2x & x \in (1, 3] \\ x + 5 & x \in (5, \infty) \end{cases}$$

$f(2)$ (1, 3] $-2(2) = -4$	$f(7)$ (5, ∞) $7 + 5 = 12$	$f(1)$ (-2, 1] $-2(1) = -2$
$f(0)$ (-2, 1] $0 - 2 = -2$	$f(-2)$ undefined	$f(4)$ undefined

2. Graph the following function on the specified domain.

$$f(x) = \frac{1}{2}x \quad x \in (-4, 2]$$

x	y
-4	-2
2	1

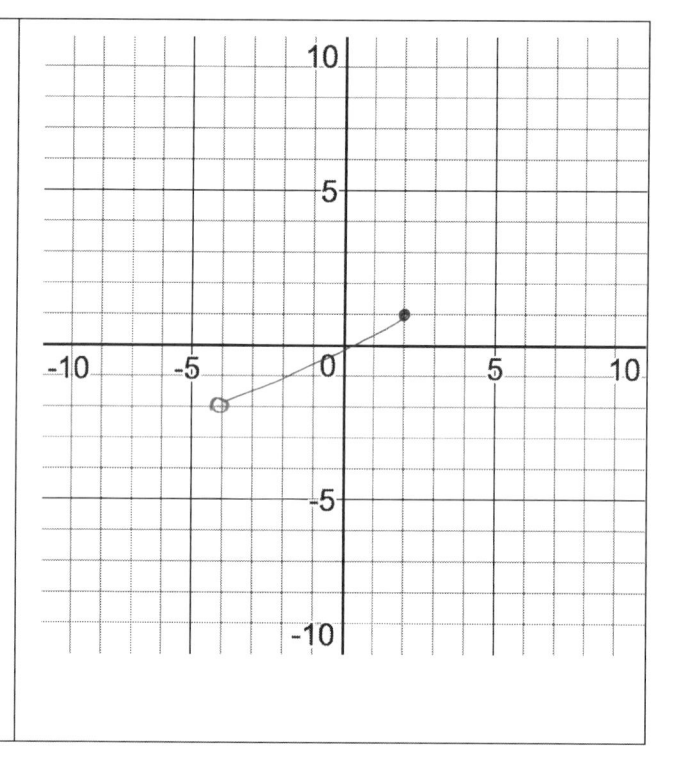

4.2—165

Homework 4.2 Introduction to Piecewise Functions

3. Graph the following function on the specified domain.

$f(x) = 2x + 3, x \in (1, \infty)$

2x+3	
1	5
2	7

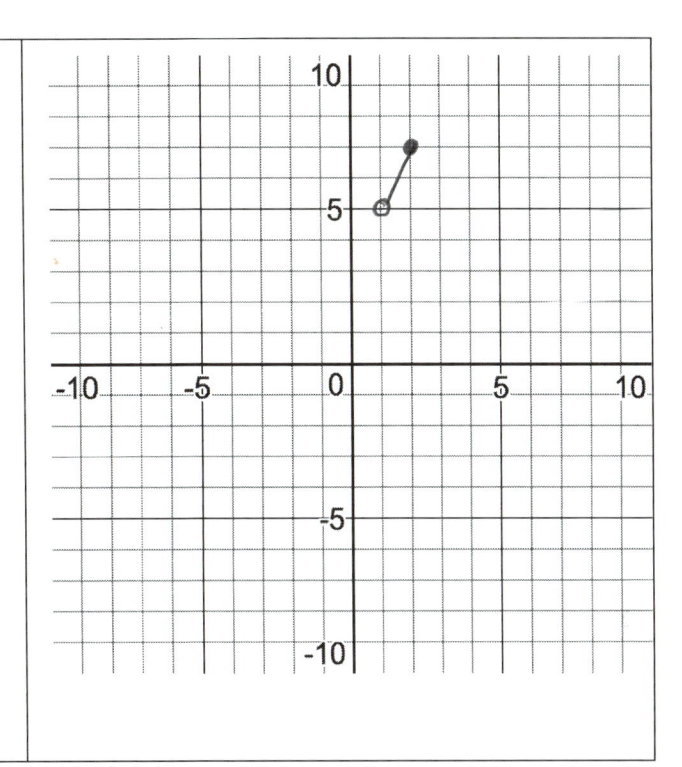

4. Graph the following piecewise function.

$f(x) = \begin{cases} -\dfrac{3}{2}x - 2 & x \leq -2 \\ -3 & -2 < x < 2 \\ 2x + 1 & x \geq 2 \end{cases}$

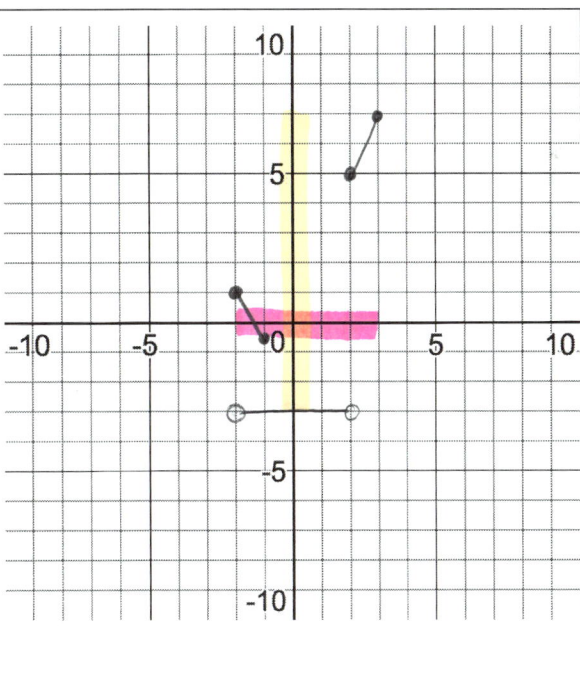

Is f a function?

(YES)   NO

$f(-2) = -\dfrac{3}{2}(-2) - 2 = 1$

$f(2) = -3$

Domain: $[-2, 3]$

Range: $[-3, 7]$

$(-\infty, -2]$

-2	1
-1	$-\dfrac{1}{2}$

$(-2, 2)$

-2	-3
2	-3

$[2, \infty)$

2	5
3	7

Homework 4.2 Introduction to Piecewise Functions

Graph the following piecewise functions.

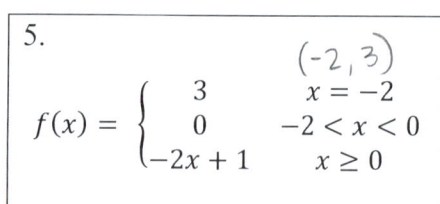

5.

$$f(x) = \begin{cases} 3 & x = -2 \\ 0 & -2 < x < 0 \\ -2x + 1 & x \geq 0 \end{cases}$$

(-2, 3)

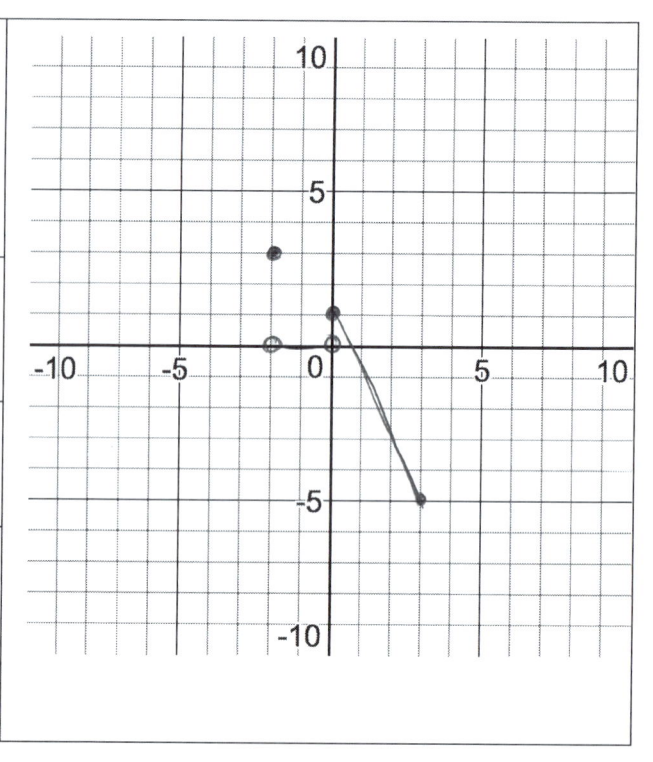

Is f a function? **YES** NO

$f(0) = -2(0) + 1 = 1$

$f(-2) = 3$

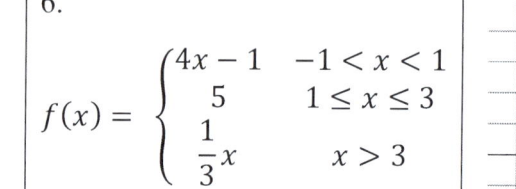

Side notes:
(-2, 0)
| -2 | 0 |
| 0 | 0 |

[0, ∞)
| 0 | 1 |
| 3 | -5 |

6.

$$f(x) = \begin{cases} 4x - 1 & -1 < x < 1 \\ 5 & 1 \leq x \leq 3 \\ \frac{1}{3}x & x > 3 \end{cases}$$

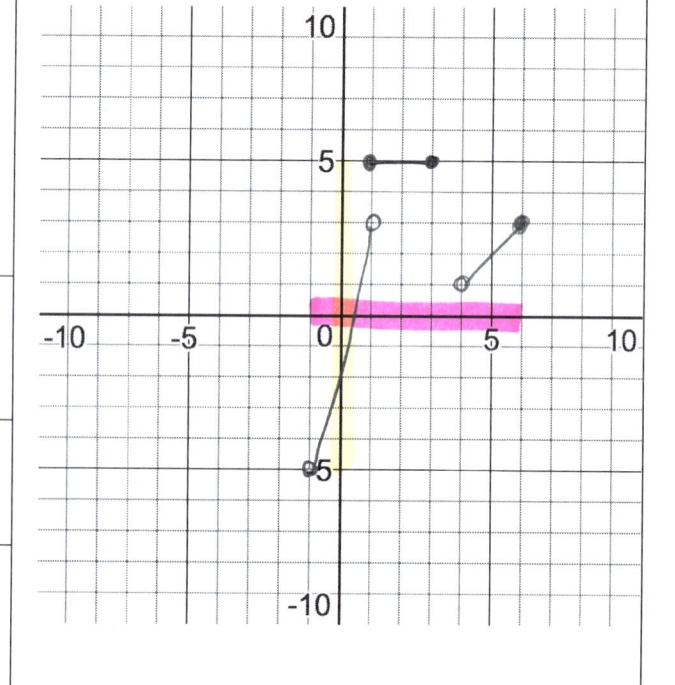

Is f a function? **YES** NO

$f(-5) = \frac{1}{3}(-5) = \frac{-5}{3}$

$f(3) = 5$

Domain: $[-1, 4]$

Range: $[-5, 5]$

Side notes:
(-1, 1)
| -1 | -5 |
| 1 | 3 |

[1, 3]
| 1 | 5 |
| 3 | 5 |

(3, ∞)
| 3 | 1 |
| 6 | 2 |

Homework 4.2 Introduction to Piecewise Functions

CHALLENGE PROBLEMS
1. Graph the piecewise function.

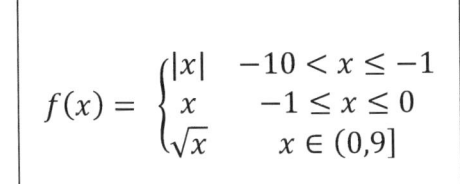

$$f(x) = \begin{cases} |x| & -10 < x \leq -1 \\ x & -1 \leq x \leq 0 \\ \sqrt{x} & x \in (0,9] \end{cases}$$

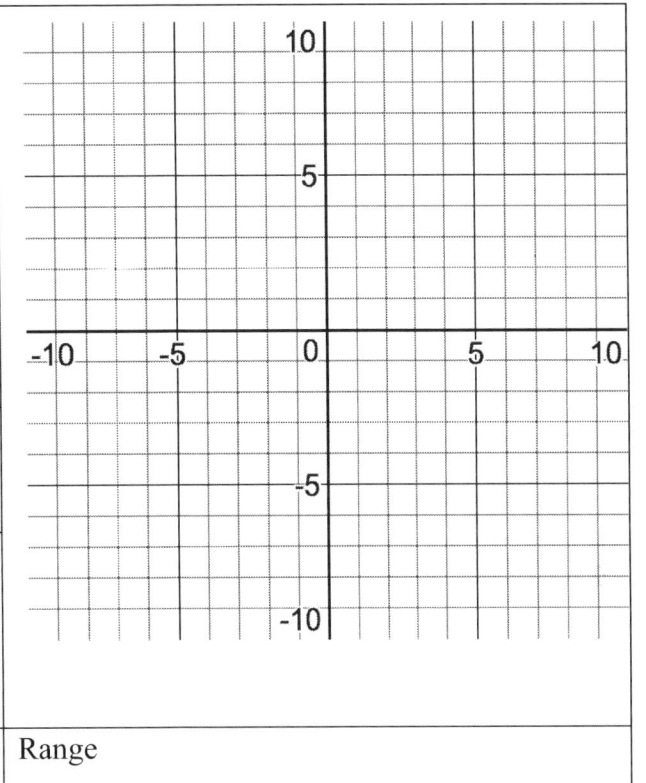

Is f a function?

      YES   NO

$f(10) =$

$f(-10) =$

Domain

Range

2. Draw the graph of a piecewise linear function with a domain [-5,5] that has graphical elements on the following points. (These points appear on the graph as either shaded or unshaded.) Give the definition of the function you graphed.

$$(-5,0), (-5,5), (0,5), (5,0), (5,-3)$$

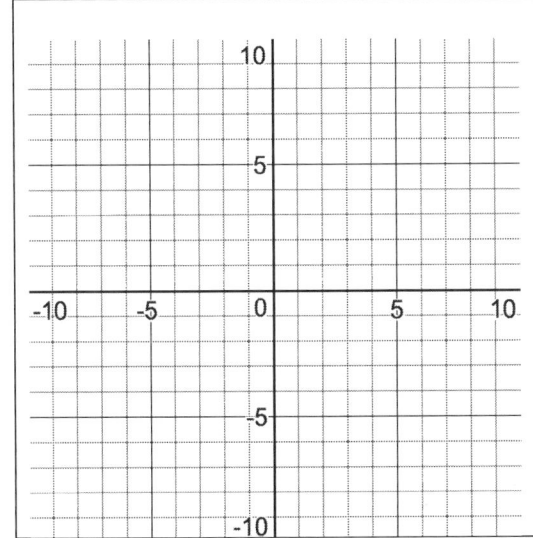

$f(x) = \{$

4.3. Graphing Piecewise Functions

In this lesson we'll continue graphing piecewise functions, and also refresh our skills on writing equations of lines, given two points, or a point and the slope.

Graph the following piecewise linear function.

$$f(x) = \begin{cases} x+5 & (-\infty, -1) \\ 2x-1 & (0, 3] \\ -x+2 & (3, 7) \end{cases}$$

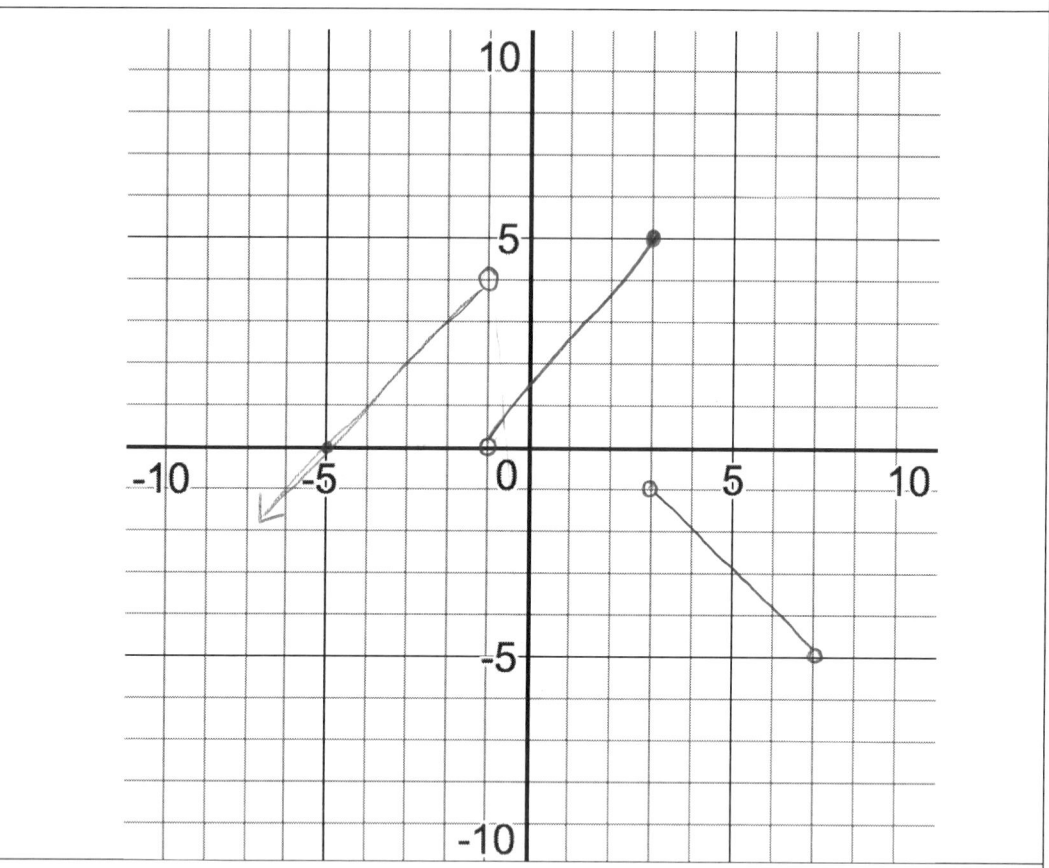

x	x+5
-1	4
-5	0

x	2x-1
0	-1
3	5

x	-x+2
3	-1
7	-5

$f(3) = 5$

$f(8) =$ undefined

Domain	Range
$(-\infty, -1) \cup (0, 7)$	$(-\infty, 5]$

The following is a piecewise function you are familiar with, but probably have never graphed.

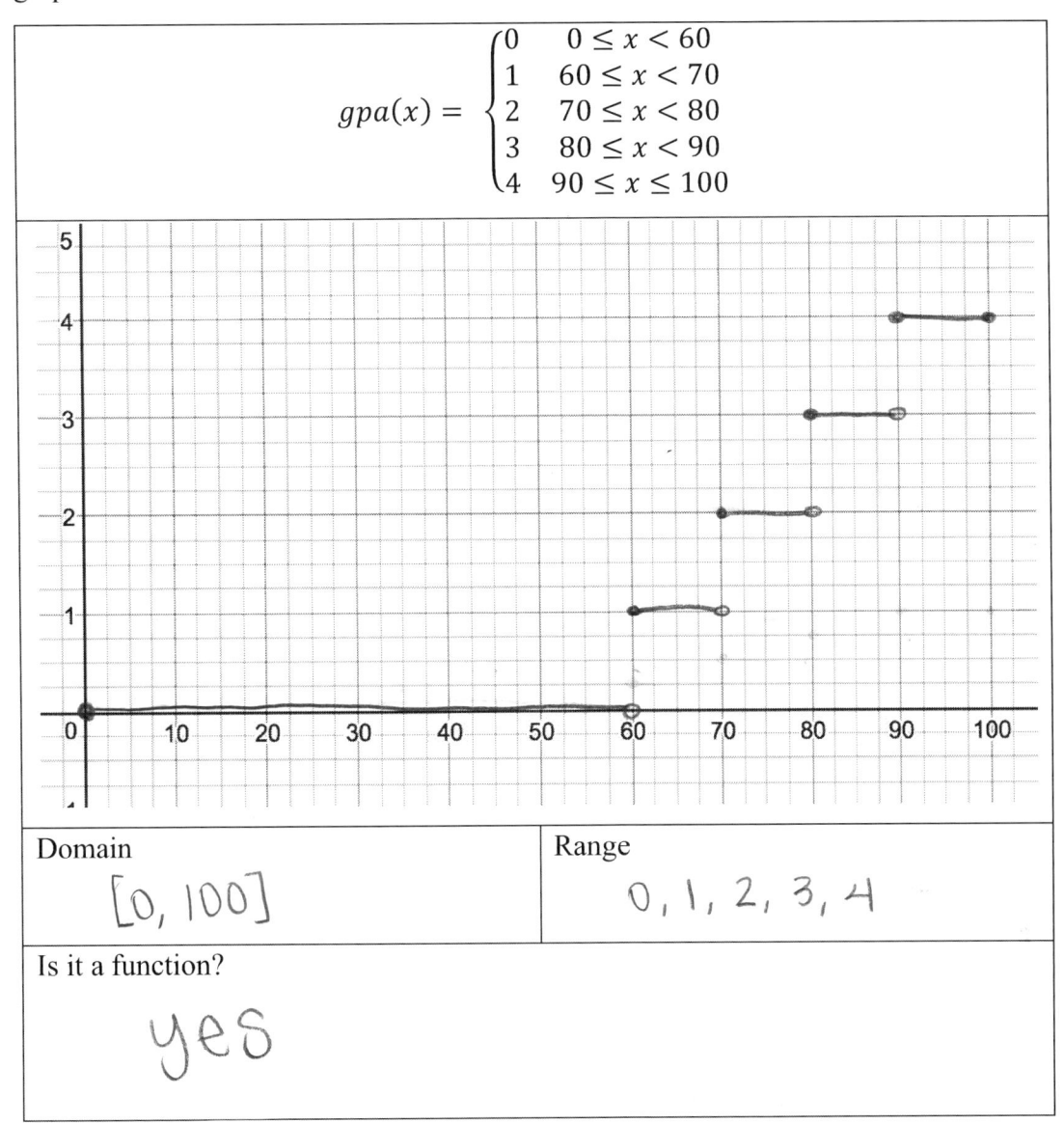

$$gpa(x) = \begin{cases} 0 & 0 \leq x < 60 \\ 1 & 60 \leq x < 70 \\ 2 & 70 \leq x < 80 \\ 3 & 80 \leq x < 90 \\ 4 & 90 \leq x \leq 100 \end{cases}$$

Domain

[0, 100]

Range

0, 1, 2, 3, 4

Is it a function?

yes

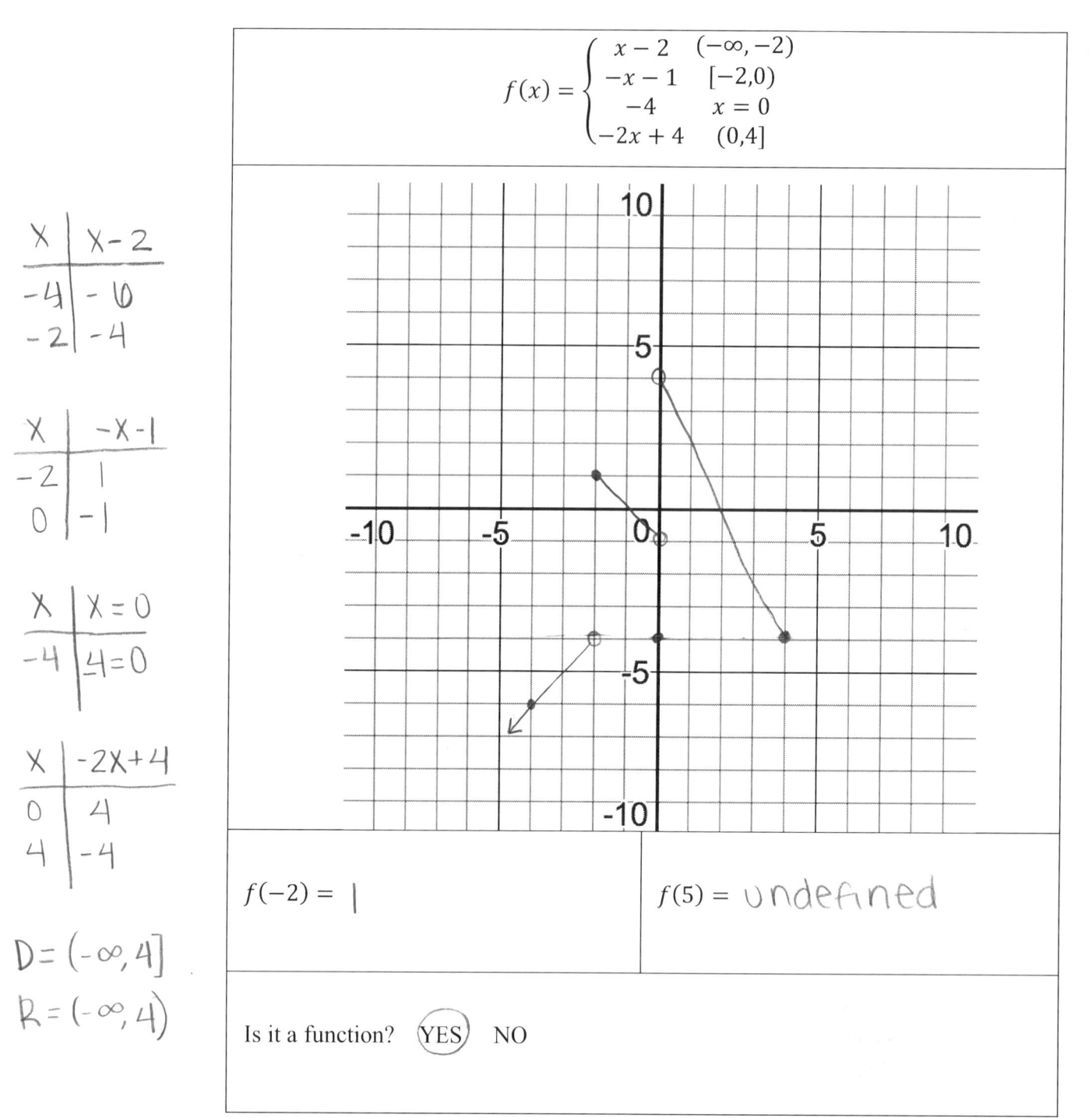

$$f(x) = \begin{cases} x-2 & (-\infty,-2) \\ -x-1 & [-2,0) \\ -4 & x=0 \\ -2x+4 & (0,4] \end{cases}$$

x	x-2
-4	-6
-2	-4

x	-x-1
-2	1
0	-1

x	x=0
-4	4=0

x	-2x+4
0	4
4	-4

$D = (-\infty, 4]$

$R = (-\infty, 4)$

$f(-2) = 1$

$f(5) = $ undefined

Is it a function? **YES** NO

In the next lesson we will start by viewing the graph of a piecewise function, and then write its equation.

Solid skills on writing the equations of lines are needed for this, so we will review this material here.

Equations of lines

Point Slope Form	Slope Intercept Form
$y - y_1 = m(x - x_1)$	$y = mx + b$
$m = slope$	$m = slope$
(x_1, y_1) is a point on the line	$(0, b)$ is the y-intercept

Find the equation of the line passing through $(-2,4)$ & $(3,6)$. Write the equation in function notation, e.g. $f(x) = \cdots$.

$m = \dfrac{6-4}{3+2} = \dfrac{2}{5}$

$y - 6 = \dfrac{2}{5}(x-3)$

$y = \dfrac{2}{5}(x-3) + 6$

Find the equation of the line passing through $(0,6)$ with slope -2. Write the equation in function notation. $m = -2 \quad yint = 6$

$y = -2x + 6$

Find the equation passing through $(3,7)$ with slope -1.

$y - 7 = -1(x-3)$
$y = -1(x-3) + 7$
$y = -x + 10$

By inspecting the graph below, write the equation of the line.

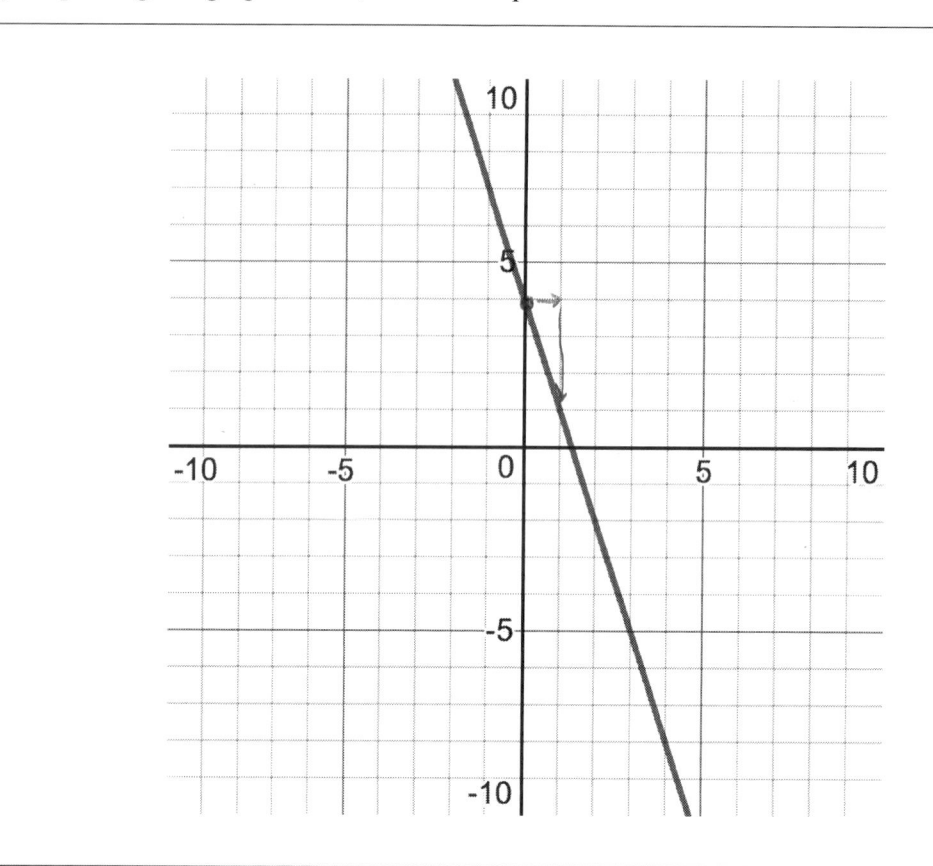

Find the y-intercept
$(0, 4)$
Find the slope
$m = \dfrac{-3}{1}$
Equation of the line
$y = -3x + 4$

By inspecting the graph below, write the equation of the line.

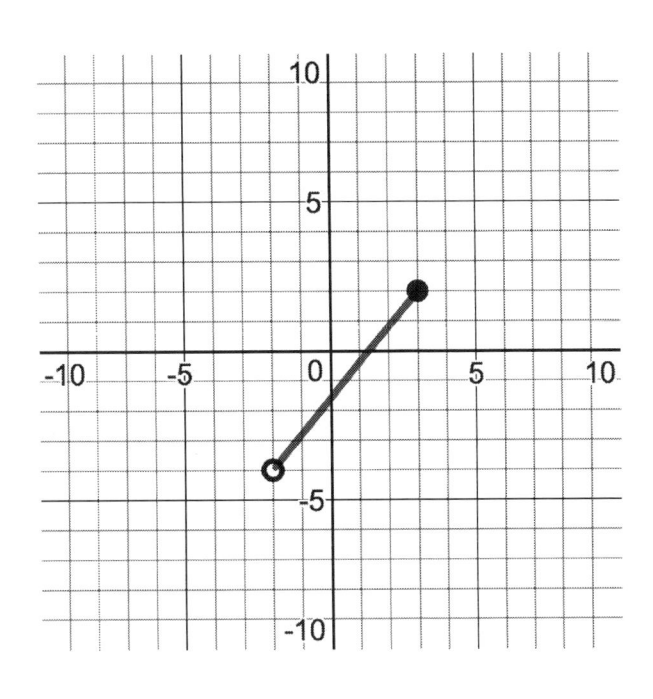

Find a point on the line
$(3, 2)$ $(-2, -4)$

Find the slope
$m = \dfrac{2+4}{3+2} = \boxed{\dfrac{6}{5}}$

Write the equation of the line
$y - 2 = \dfrac{6}{5}(x - 3)$
$y = \dfrac{6}{5}(x - 3) + 2$
$y = \dfrac{6}{5}x - 1$

4.3—174

Homework 4.3 Graphing Piecewise Functions

Graph the following piecewise functions.

1.

$$f(x) = \begin{cases} x-1 & x \leq -3 \\ -x+1 & -3 < x < 3 \\ 2 & x \geq 3 \end{cases}$$

$(-\infty, -3]$
$[3, \infty)$

x	x−1
−3	−4
−4	−5

x	−x+1
−3	4
3	−2

x	2
3	2
6	2

Is f a function? **YES** NO

$f(-3) = -4$

$f(3) = 2$

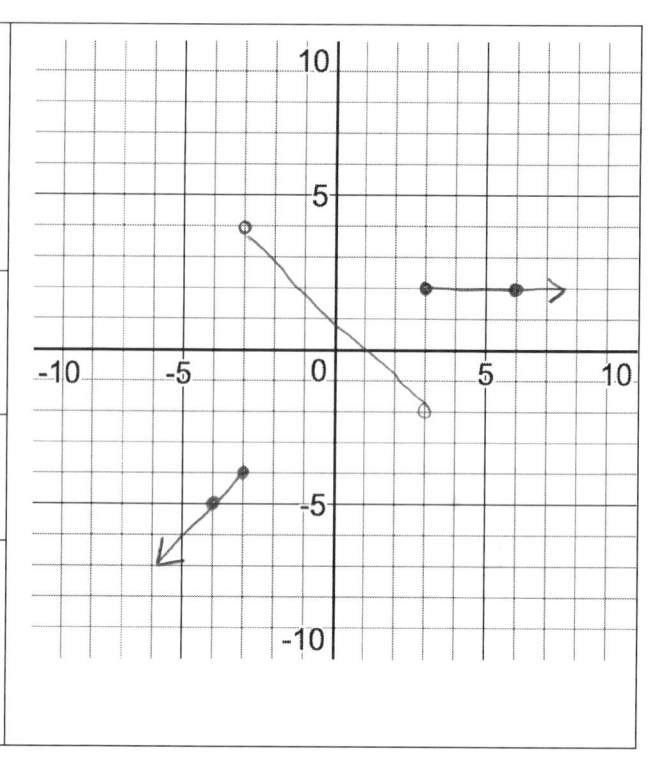

2.

$$f(x) = \begin{cases} 3 & x \leq -1 \\ 0 & -1 < x < 3 \\ x & x \geq 3 \end{cases}$$

$(-\infty, -1]$ $(-1, 3)$
$[3, \infty)$

x	3
−1	3
−4	3

x	0
−1	0
3	0

x	x
3	3
7	7

Is f a function? **YES** NO

$f(-1) = 3$

$f(3) = 3$

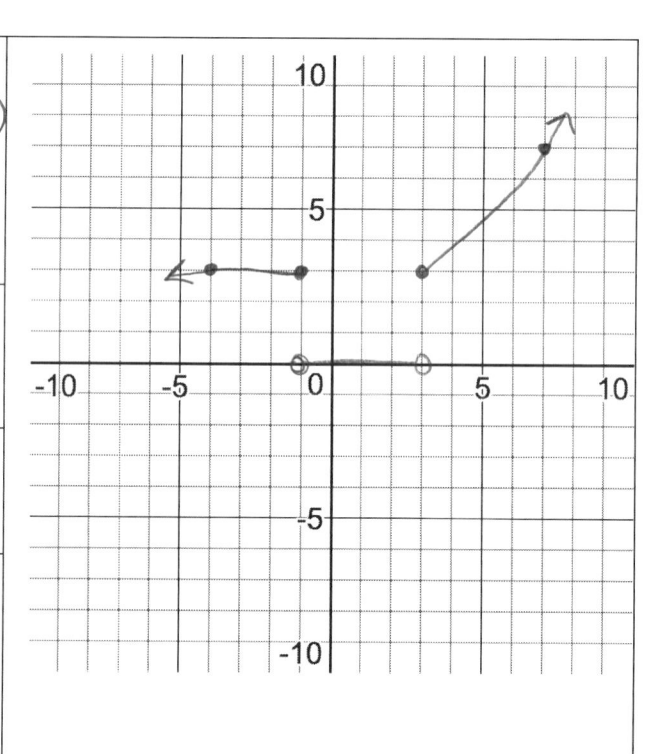

Domain: $(-\infty, \infty)$

Range: $(0) \cup [3, \infty)$

Homework 4.3 Graphing Piecewise Functions

Graph the following piecewise functions.

3.
$$f(x) = \begin{cases} \frac{1}{2}x + 1 & x \in [-8, 0] \\ -x + 4 & x \in [0, 4] \\ 2 & x \in (4, \infty) \end{cases}$$

Is f a function?

YES (NO)

$f(-3) = \frac{1}{2}(-3) + 1$
$= \frac{-3}{2} + 1 = \left(\frac{-1}{2}\right)$

$f(4) = -4 + 4 = 0$

Domain: $[-8, \infty)$

Range: $(-3, 4]$

x	$\frac{1}{2}x+1$
-8	-3
0	1

x	$-x+4$
0	4
4	0

x	2
4	2
6	2

4.
$$f(x) = \begin{cases} 4 & x \leq -2 \quad (-\infty, -2] \\ -x - 1 & -2 < x \leq 3 \\ x - 2 & x > 3 \quad (3, \infty) \end{cases}$$

Is f a function?

(YES) NO

$f(-3) = 4$

$f(3) = -4$

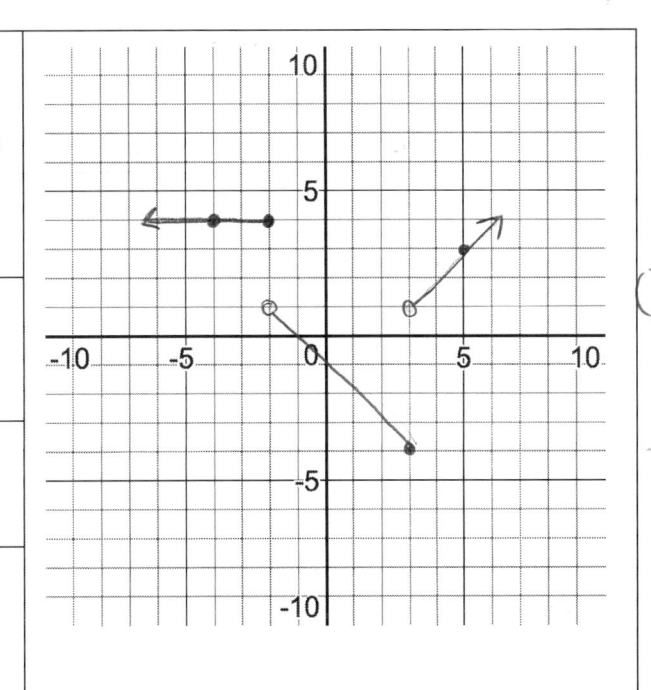

x	4
-2	4
-4	4

x	$-x-1$
(-2	1
[3	-4

x	$x-2$
(3	1
5	3

Homework 4.3 Graphing Piecewise Functions

5. Give the equation of the line passing through (4,1) and (6,2)

$m = \frac{1-2}{4-6} = \frac{-1}{-2}$ $\left(\frac{1}{2}\right)$

$y - 1 = \frac{1}{2}(x-4)$ $\boxed{y = \frac{1}{2}x - 3}$

$f(x) = \frac{1}{2}x - 3$

6. Give the equation of the line passing through (0,7) with slope of -2.

$y = -2x + 7$

$f(x) = -2x + 7$

7. Give the equation of the line, and the domain restriction for the graph below.

$(-5, 2)\ (5, -2)$

$m = \frac{2 - -2}{-5 - 5} = \frac{4}{-10}$

$y - 2 = \frac{4}{-10}(x + 5)$

$y = \frac{4}{-10}x + 7$

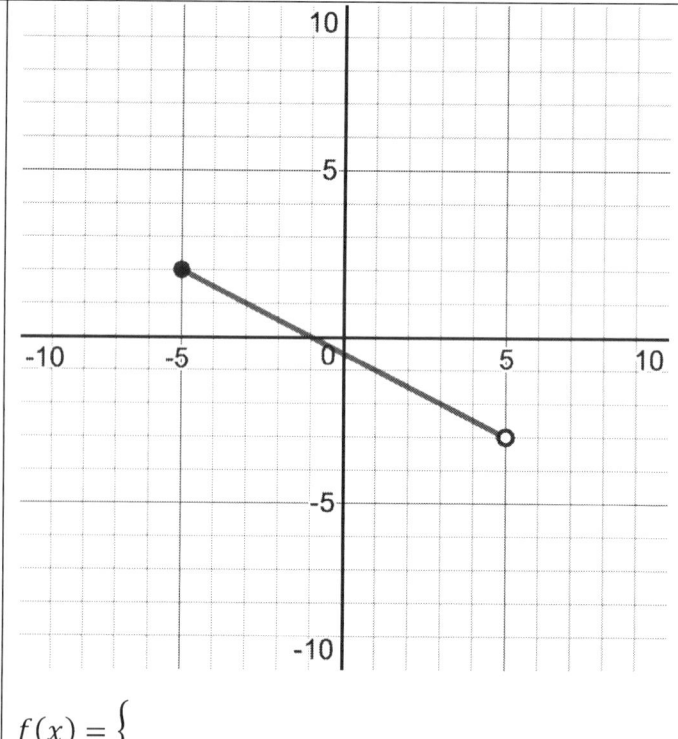

$f(x) = \{$

Homework 4.3 Graphing Piecewise Functions

CHALLENGE PROBLEMS

1. The Eagle Electric Scooter company now provides electric scooters for all students. This will help the students travel the campus quickly. By using the app on your smartphone, you can reserve a scooter, or receive hours for a cell phone violation, whichever comes first. The pricing of the scooter is a flat fee of $1.00 plus $0.50 per minute for the first 5 minutes and $0.25 per minute for the minutes of use after 5. The scooter locks after 10 minutes of use. Graph a function that shows the TOTAL cost of the rental as a function of time in minutes. What is the cost of a 7 minute rental? A 10 minute rental?

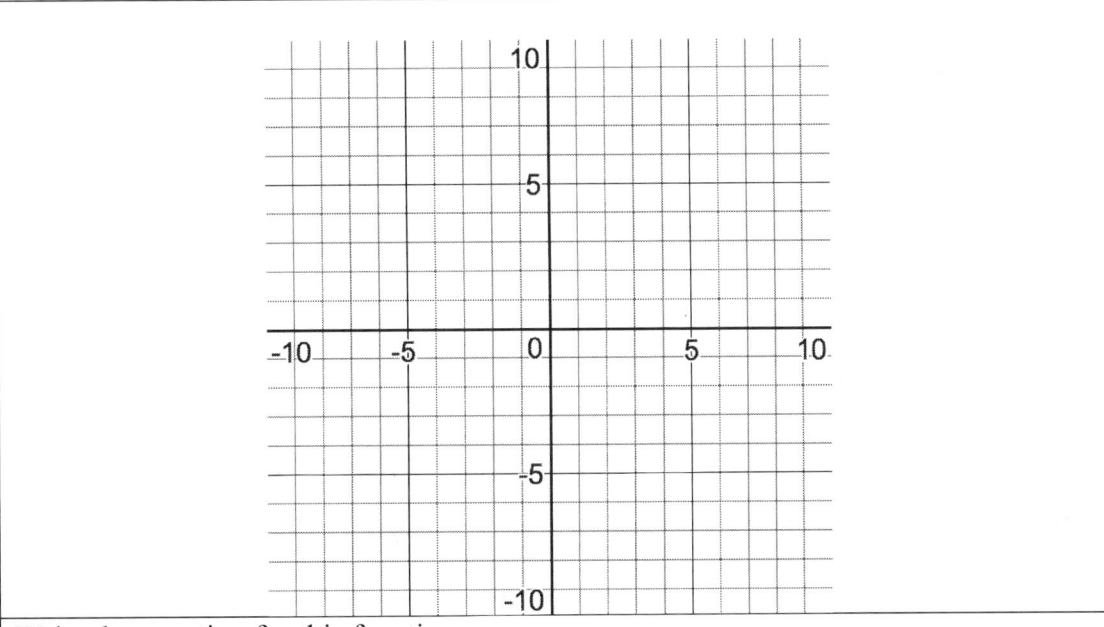

Write the equation for this function

$$f(x) = \begin{cases} \end{cases}$$

4.4. Writing Equations of Piecewise Functions

In this lesson we will be writing equations of piecewise functions from the graph.

Point Slope Form	Slope Intercept Form
$y - y_1 = m(x - x_1)$	$y = mx + b$
$m = slope$	$m = slope$
(x_1, y_1) is a point on the line	$(0, b)$ is the y-intercept

If you know the slope and the y-intercept, use the Slope Intercept Form of the equation. If you know two points, or the slope and a point other than the y-intercept, use the Point Slope Form then convert to Slope Intercept Form.

equation of a horizontal line = x

Write the equation of the function below:

How many pieces make up this function?

We can see that the function will have subdomains of:

$[-7, -3)$
$[-3, 1)$
$x = 1$
$(1, 5]$

We need to find the value of the function for each of these intervals.

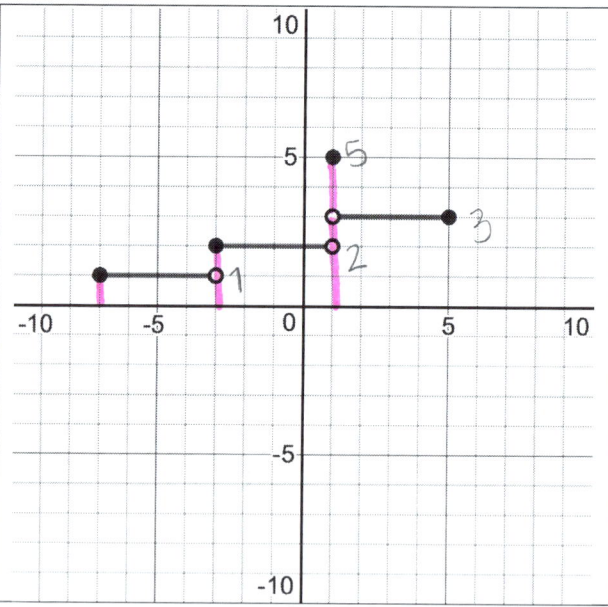

Domain
$[-7, 5]$

$$f(x) = \begin{cases} 1 & [-7, -3) \\ 2 & [-3, 1) \\ 5 & x = 1 \\ 3 & (1, 5] \end{cases}$$

Range
$\{1, 2, 3, 5\}$

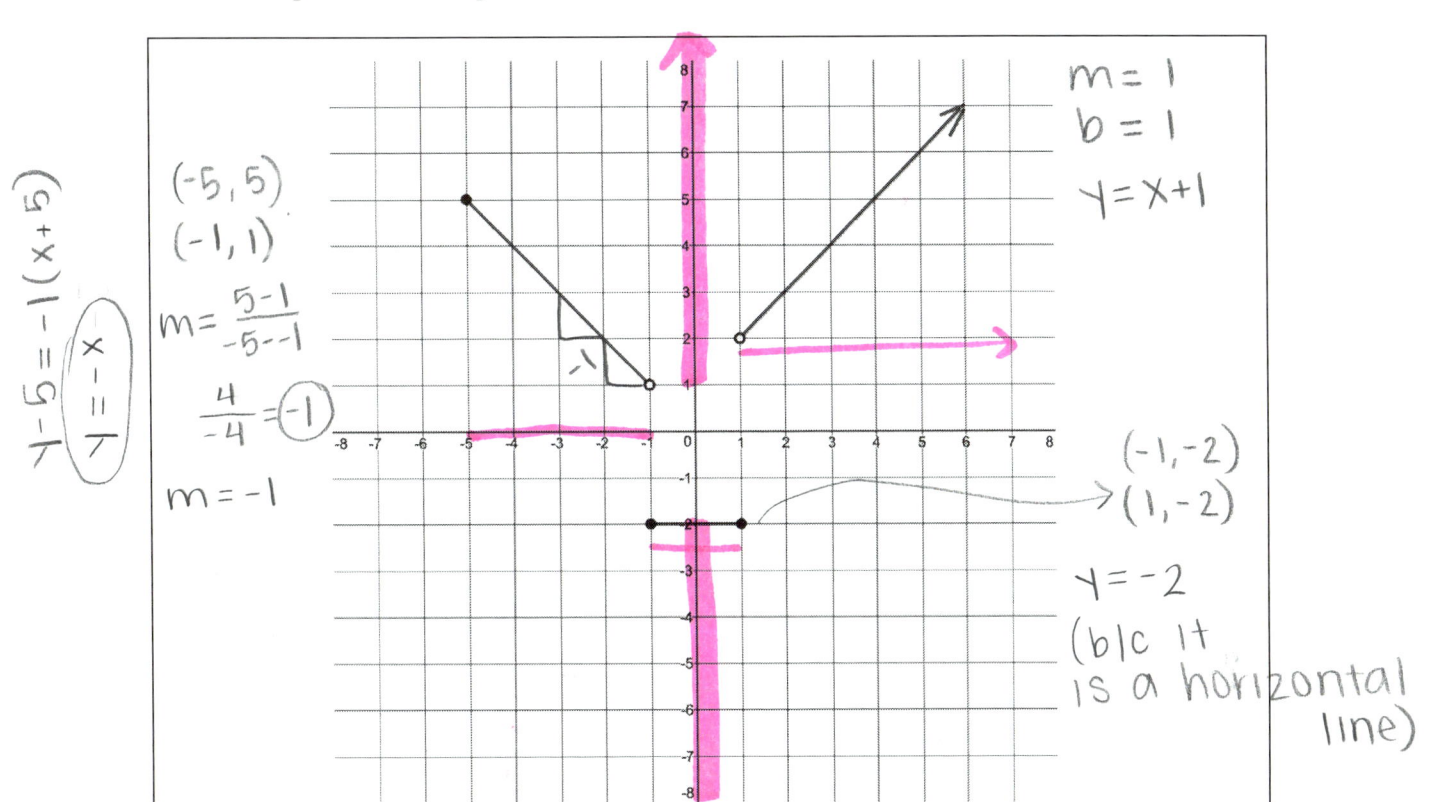

Write the equation for the piecewise function below.

$y-5 = -1(x+5)$
$y = -x$

$(-5, 5)$
$(-1, 1)$
$m = \dfrac{5-1}{-5--1}$
$\dfrac{4}{-4} = -1$
$m = -1$

$m = 1$
$b = 1$
$y = x+1$

$(-1, -2)$
$(1, -2)$
$y = -2$
(b/c it is a horizontal line)

$$f(x) = \begin{cases} -x & [-5, -1) \\ -2 & [-1, 1] \\ x+1 & (1, \infty) \end{cases}$$

Domain
$[-5, \infty)$

Range
$[-2] \cup (1, \infty)$

4.4—180

Write the equation for the piecewise function below.

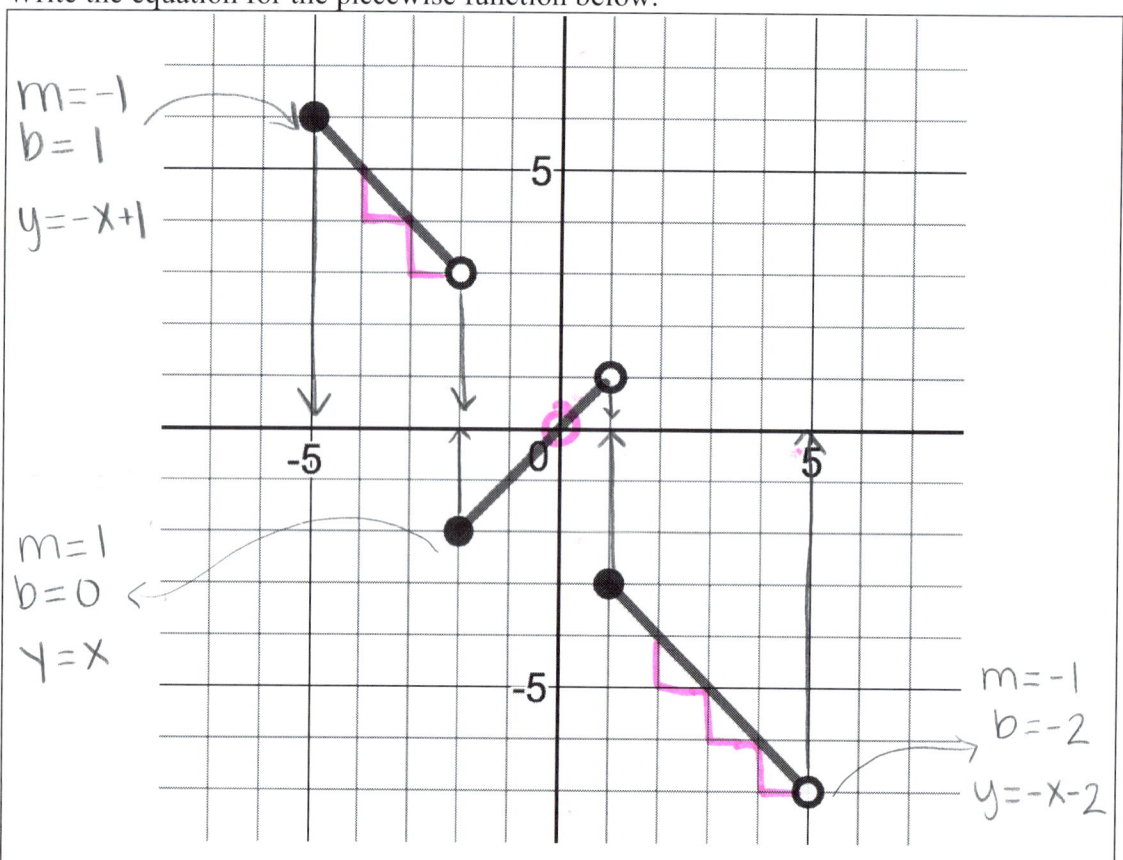

m = -1
b = 1
y = -x+1

m = 1
b = 0
y = x

m = -1
b = -2
y = -x-2

$$f(x) = \begin{cases} -x+1 & [-5,-2) \\ x & [-2,1) \\ -x-2 & [1,5) \end{cases}$$

Domain X

[-5, 5)

Range Y

(-7, -3] ∪ [-2, 1) ∪ (3, 6]

Write the equation for the piecewise function below.

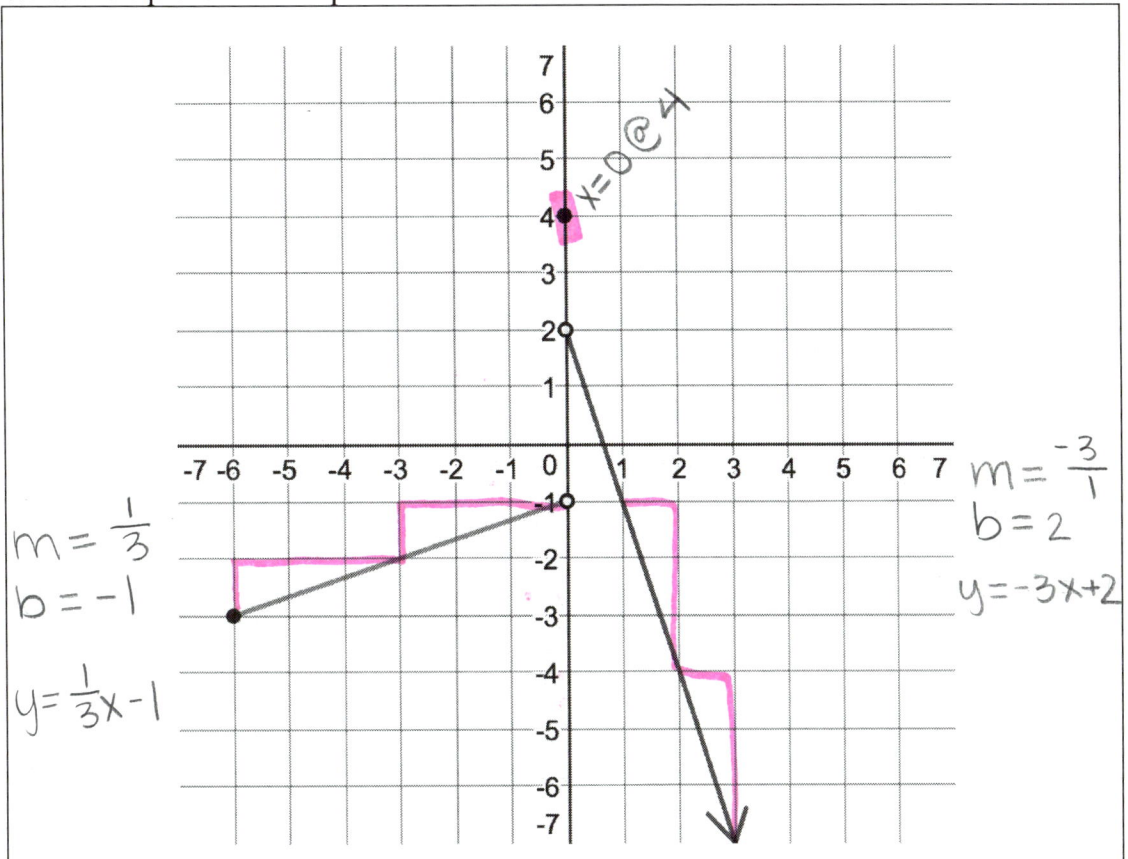

$m = \frac{1}{3}$
$b = -1$
$y = \frac{1}{3}x - 1$

x=0 @ 4

$m = \frac{-3}{1}$
$b = 2$
$y = -3x + 2$

$$f(x) = \begin{cases} \frac{1}{3}x - 1 & [-6, 0) \\ 4 & x = 0 \\ -3x + 2 & (0, \infty) \end{cases}$$

Domain
$[-6, \infty)$

Range
$(-\infty, 2) \cup [4]$

Write the equation for the piecewise function below.

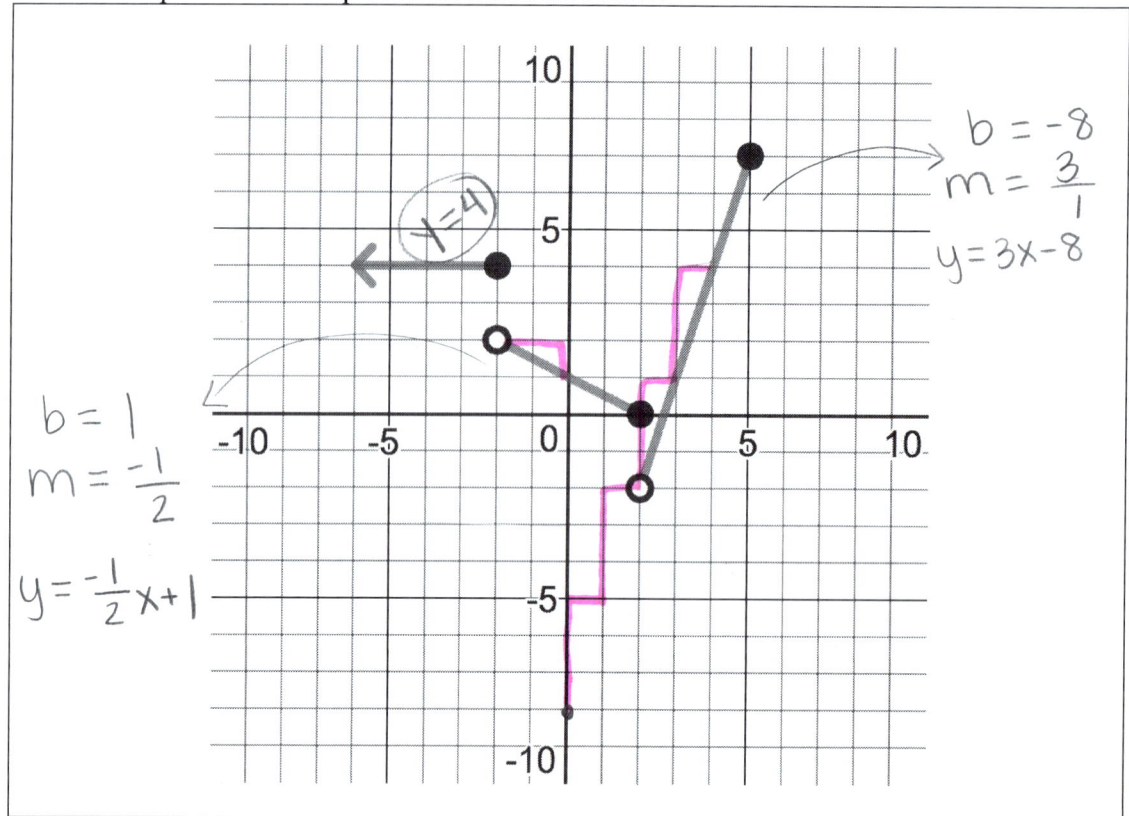

$$f(x) = \begin{cases} 4 & (-\infty, -2] \\ -\frac{1}{2}x+1 & (-2, 2] \\ 3x-8 & (2, 5] \end{cases}$$

Domain: $(-\infty, 5]$

Range: $(-2, 7]$

4.4—183

Homework 4.4 Writing Equations of Piecewise Functions

Write the equation of the piecewise function.

1.

1 $m = -1$, $b = -2$ $y = -x - 2$

2 $m = 2$, $b = -3$ $y = 2x - 3$

3 $m = -1$, $b = -1$

$$f(x) = \begin{cases} -x-2 & (-7, -1] \\ 2x-3 & (-1, 4) \\ -x-1 & [4, 7] \end{cases}$$

$f(-1) = -1 - 2 = \boxed{-3}$

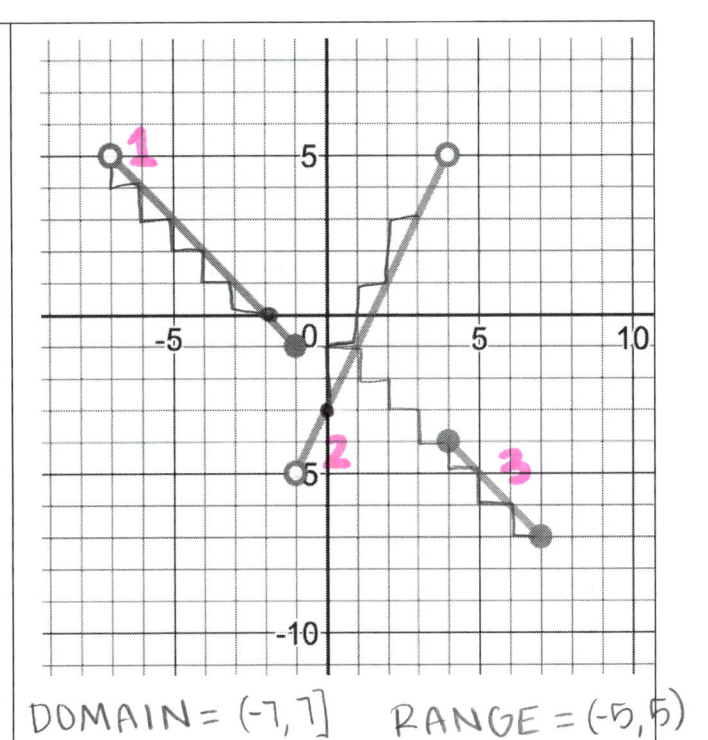

DOMAIN = $(-7, 7]$ RANGE = $(-5, 5)$

2.

1 $= -2$

2 $m = -1$, $b = 2$

3 $m = 1$, $b = 0$

$$f(x) = \begin{cases} -2 & (-9, -2] \\ -x+2 & (-2, 3) \\ x-1 & [3, 8] \end{cases}$$

$f(3) = 2$
$f(-2) = -2$

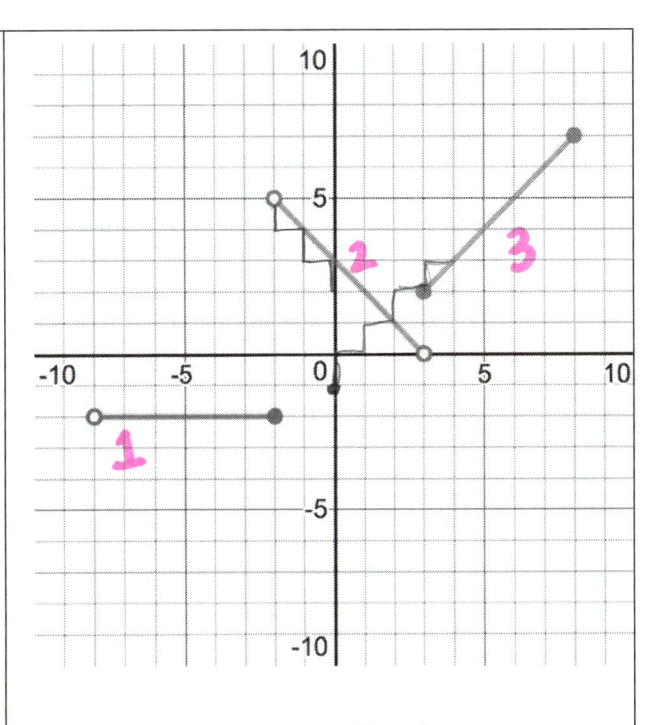

DOMAIN = $(-9, 8]$
RANGE = $(0, 5)$

Homework 4.4 Writing Equations of Piecewise Functions

Write the equation of the piecewise function.

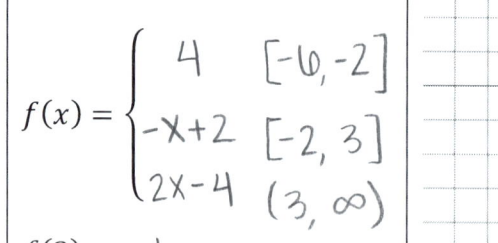

3.

$$f(x) = \begin{cases} 4 & [-6, -2] \\ -x+2 & [-2, 3] \\ 2x-4 & (3, \infty) \end{cases}$$

$f(3) = -1$

1 $= 4$

2 $m = -1$
$b = 2$ $y = -x+2$

3 $m = 2$
$b = -4$

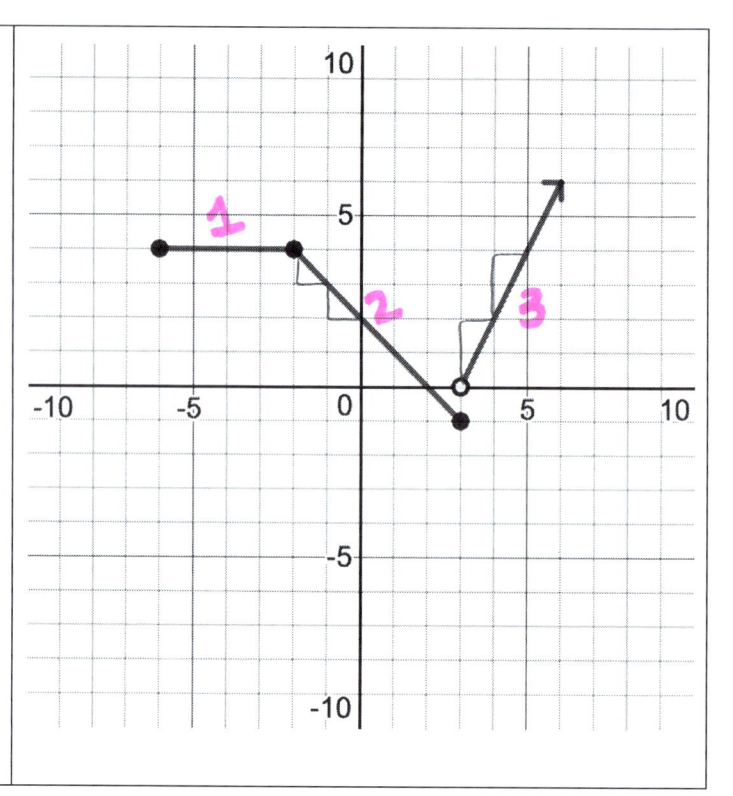

4.

1 $m = -2$
$b = 2$

2 $= 2$

3 $= 4$

4 $= 4$

$$f(x) = \begin{cases} -2x+2 & (-\infty, 0] \\ 4 & (0, 2) \\ 4 & (2, 7] \\ 2 & x = 2 \end{cases}$$

$f(0) = 2$
$f(2) = 2$

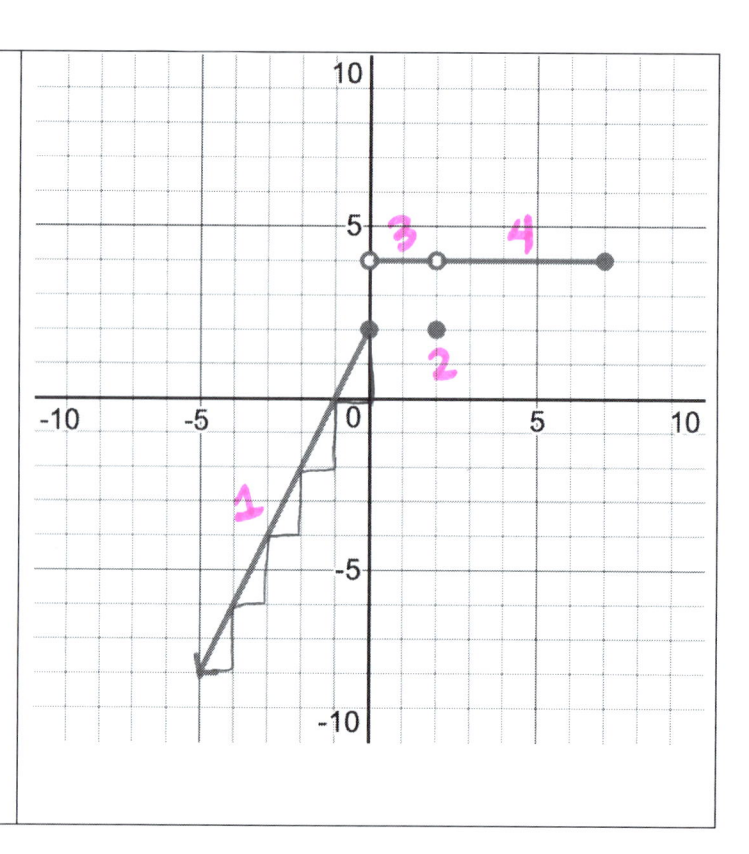

Homework 4.4 Writing Equations of Piecewise Functions

CHALLENGE PROBLEMS

Give the equation for the function graphed below.

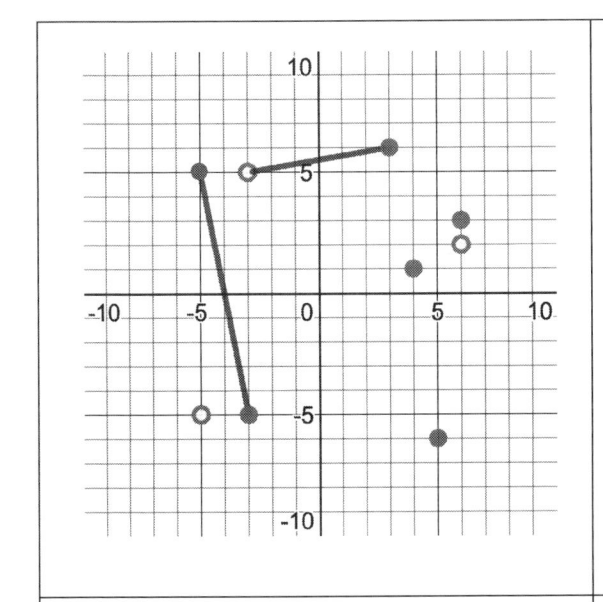

	Domain
[graph]	
	Range
$y-$ intercept	$x-$ intercept
$f(x) = \begin{cases} & \\ & \\ & \\ & \end{cases}$	
Find the value of $\dfrac{f(-5)}{f(5)}$	

4.5. Review

Assessment Checklist. Below are the competencies one should master in preparation for an assessment on piecewise linear functions.

- [ ] Proficient with interval notation, going from inequality to interval notation, graphing on a number line
- [ ] Identifying whether a relation is a function or not, given either set notation or a graph
- [ ] Writing the domain and range from the set definition of a relation/function.
- [ ] Identifying the domain and range from a graph of a relation/function.
- [ ] Given the equation of a piecewise function, evaluate the function at given points
- [ ] Graphing a line with a restricted domain
- [ ] Graphing of piecewise functions
- [ ] Writing equation of lines passing through two points
- [ ] Writing equation of lines given the slope and a point
- [ ] Writing the equation of lines given the slope and y-intercept
- [ ] Writing the equation of a line from its graph
- [ ] Writing the equation of a piecewise function from its graph
- [ ] Specify the sub-domains of a piecewise function when writing its equation from a graph.

Review 4.5

1. Graph each inequality and convert to interval notation.

   a. $x \geq 5$

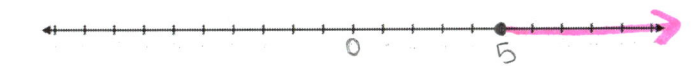

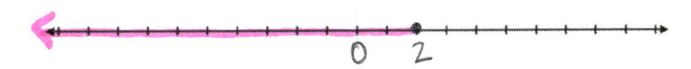

   [number line with closed dot at 5, shaded right; labeled 0, 5]

   b. $x \leq 2$

   [number line with closed dot at 2, shaded left; labeled 0, 2]

   c. $6 \geq x > -4$

   [number line with open dot at -4, closed dot at 6, shaded between; labeled -4, 0, 6]

   d. $x \leq -1$ or $x > 9$

   [number line with closed dot at -1 shaded left, open dot at 9 shaded right; labeled -1, 0, 9]

2. Evaluate the function for the given value.

$$f(x) = \begin{cases} -2x + 3, & -4 < x \leq 4 \quad (-4, 4] \\ \frac{1}{2}x - 2, & 4 < x < 10 \quad (4, 10) \end{cases}$$

   a. $f(2) = -2(2) + 3 = \boxed{-1}$

   b. $f(4) = -2(4) + 3 = \boxed{-5}$

   c. $f(10) = $ undefined

4.5—187

Review 4.5

3. Graph the equation for the stated interval.

$f(x) = -3x + 2$ if $x \in [-2, 3]$

x	-3x+2
-2	-3(-2)+2 = 8
3	-3(3)+2 = -7

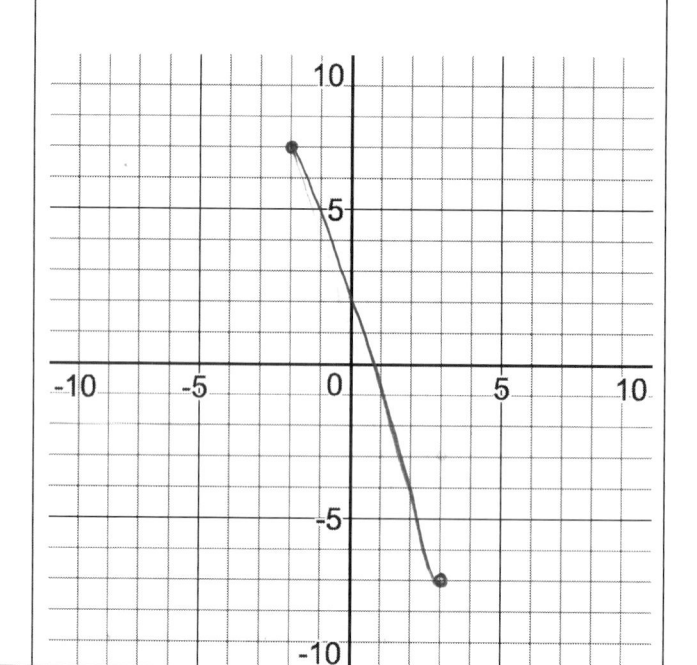

[-2, 8]
[3, -7]

$f(x) = 5x - 4$ if $x \in [-1, \infty)$

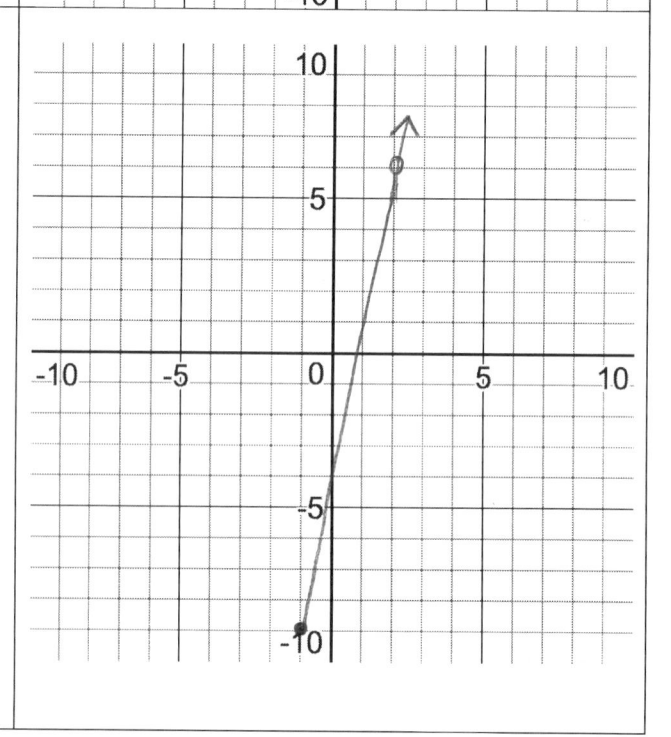

x	5x-4
-1	-9
2	6

4.5—189

−∞/∞ points are a closed circle!

Review 4.5

4. Graph the piecewise function with the given information.

$$f(x) = \begin{cases} -2x + 5, & x \in (-\infty, 2] \\ x + 2, & x \in (2, \infty) \end{cases}$$

x	-2x+5
2	1
-2	9

x	x+2
2	4
4	6

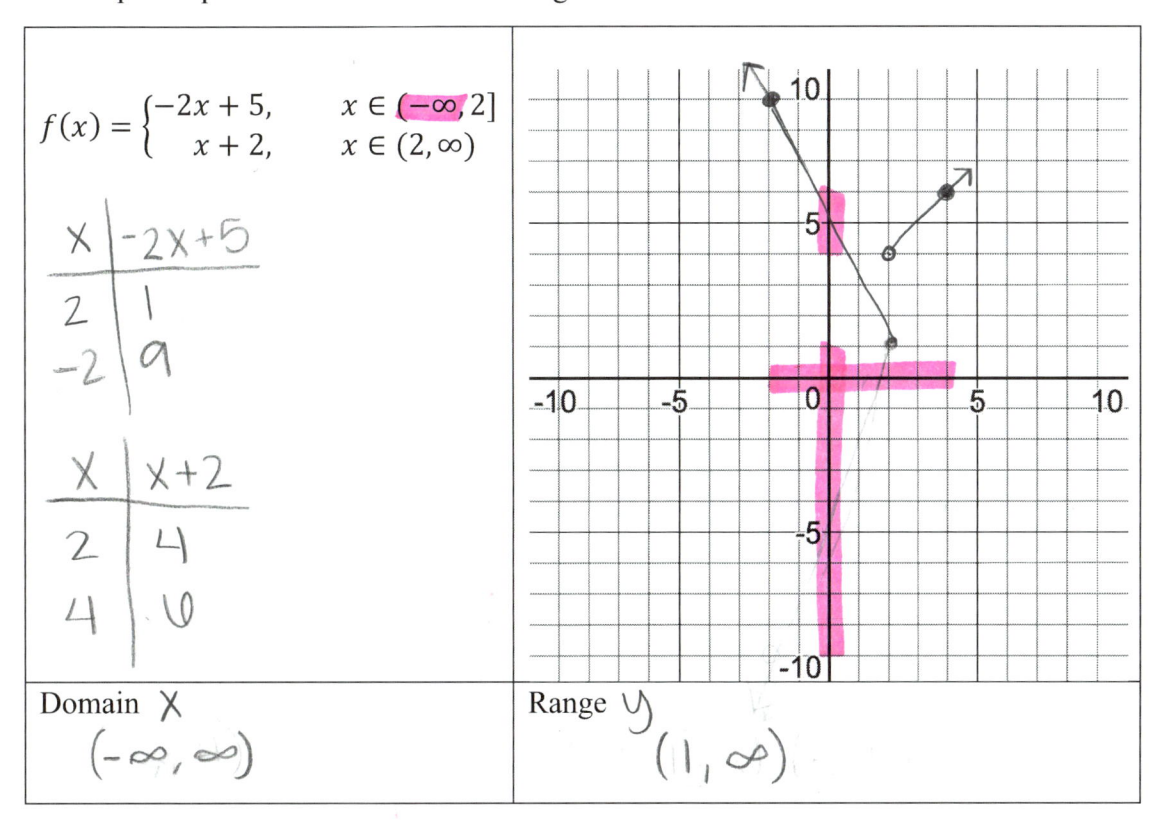

Domain x
$(-\infty, \infty)$

Range y
$(1, \infty)$

$$f(x) = \begin{cases} -x + 2 & x \in (-\infty, -4) \\ -2 & x \in [-4, 1] \\ x + 3 & x \in (1, 5] \end{cases}$$

x	-x+2
-4	6
-3	5

x	x+3
1	4
5	8

x	-2
-4	-2
1	-2

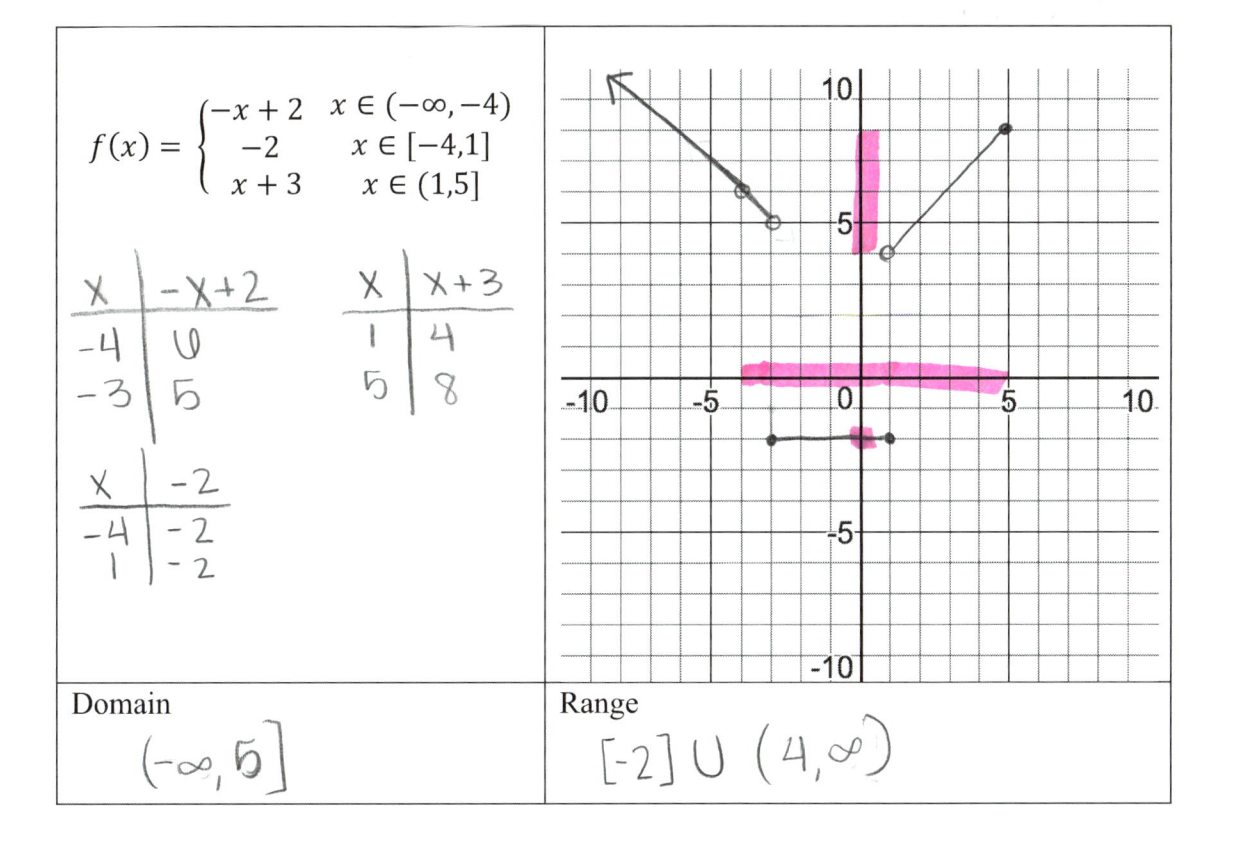

Domain
$(-\infty, 5]$

Range
$[-2] \cup (4, \infty)$

Review 4.5

5. Write the equation of the piecewise function shown:

1 $m = \frac{1}{2}$
   $b = 1$

2 $m = \frac{2}{1}$
   $b = -2$

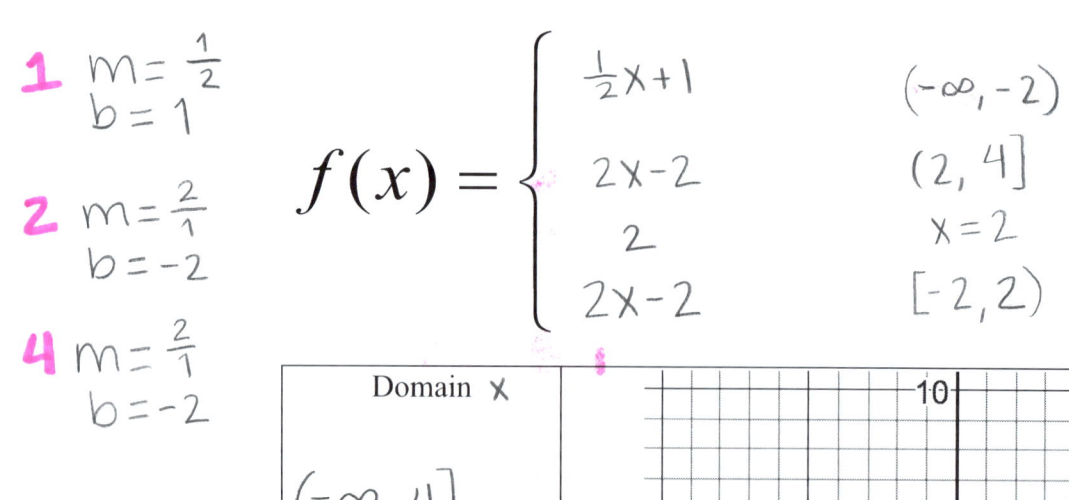

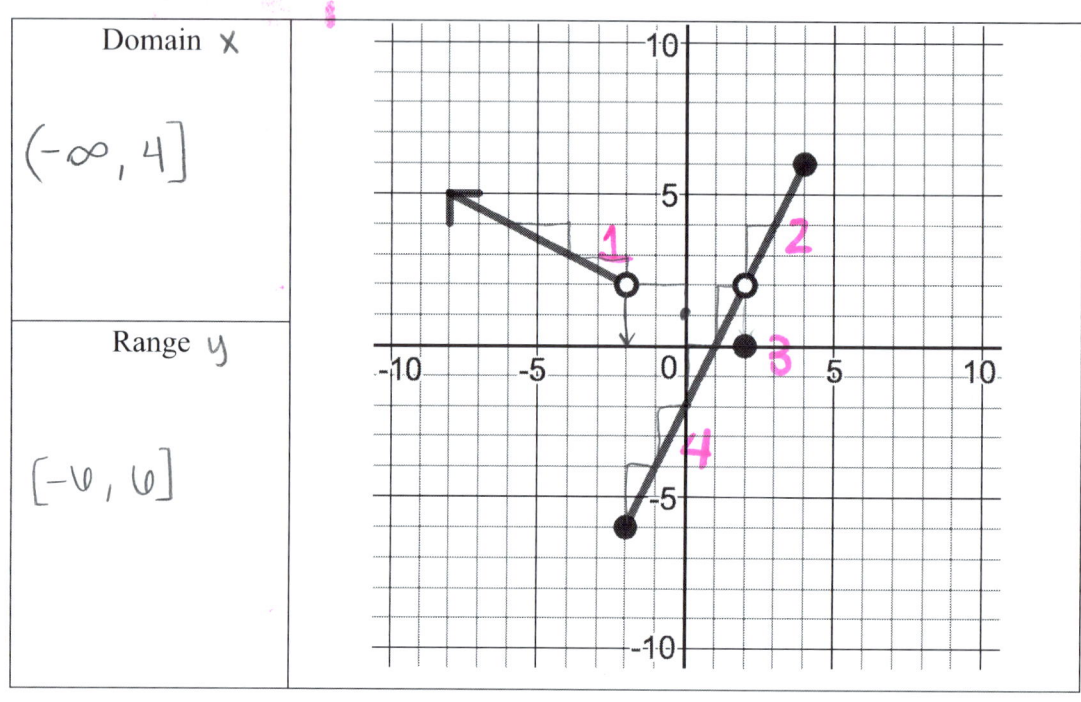

4 $m = \frac{2}{1}$
   $b = -2$

$$f(x) = \begin{cases} \frac{1}{2}x + 1 & (-\infty, -2) \\ 2x - 2 & (2, 4] \\ 2 & x = 2 \\ 2x - 2 & [-2, 2) \end{cases}$$

Domain x	
$(-\infty, 4]$	
Range y	
$[-6, 6]$	

6. Consider the relation defined by
   $F = \{(0,1), (1,0), (2,0), (3,1), (-1,0), (2,-1)\}$

What is the domain of F?	What is the range of F?	Is F a function? Why? Why not?
0, 1, 2, 3, -1, 2	1, 0, 0, 1, 0, -1	no, repeated domain

Review 4.5

7. Consider the graph below and provide the requested information.

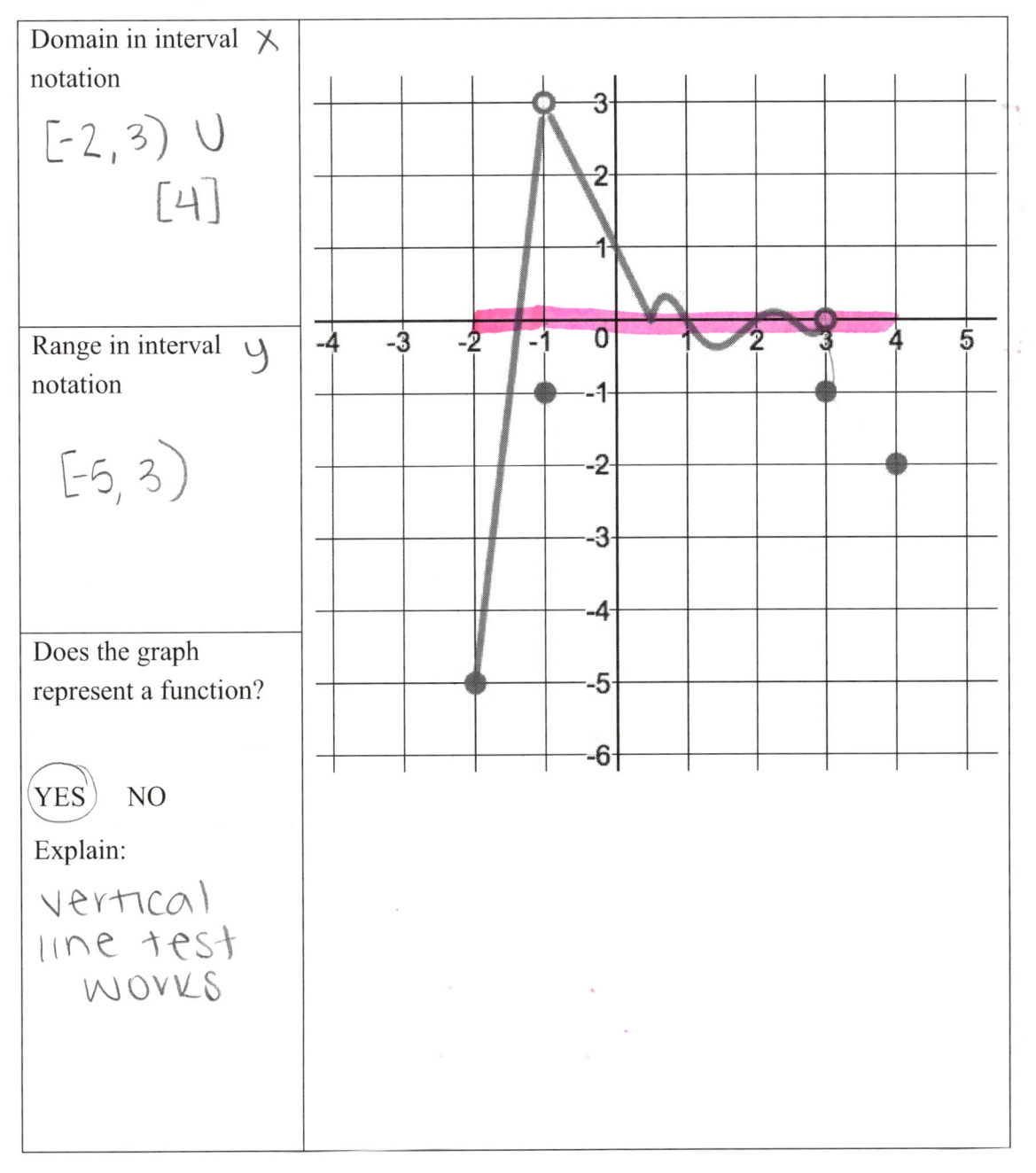

Domain in interval notation x $[-2, 3) \cup$ $[4]$	
Range in interval notation y $[-5, 3)$	
Does the graph represent a function? (YES) NO Explain: vertical line test works	

5. QUADRATIC FUNCTIONS

5.1. Complex Numbers, Addition, Subtraction, Multiplication & Powers

Before we journey too far with our study of quadratic equations, we need to spend a little amount of time solidifying our understanding of complex numbers.

Some key questions:
What is a real number? *number that can be expressed as a distance on a # line*
What are imaginary and complex numbers?
How do we add/subtract complex numbers?
How do we multiply complex numbers?
How do we quickly evaluate powers of imaginary numbers?
How do we divide complex numbers?

Everything we do with complex numbers is based on the solution to

$$\sqrt{-1}$$

Can we evaluate this using real numbers, what we know about square roots, or even our calculators?

The square root of -1 is called imaginary because it did not "make sense" to the people that first discovered them. (It might not make sense to you either...for now!)

Italian mathematicians, Gerolamo Cardano and Rafael Bombelli developed the rudimentary rules of complex numbers during the Renaissance in the 1500s.

Complex numbers are used in real life to explain the mechanics of alternating current (AC) circuits, and the physics of electromagnetism.

The base definition of complex numbers starts with assigning a value to the square root of negative 1.

$$i = \sqrt{-1}$$

Evaluate the following:

$\sqrt{4}$	2
$\sqrt{-4}$	$\sqrt{(-1)4} = 2i$

Write the numbers below as square roots of real numbers.

5	$\sqrt{25}$
$5i$	$\sqrt{-25}$
$-5i$	$-\sqrt{-25}$

Definition:
Complex numbers may be written in the form $a + bi$, where a and b are real numbers, and $i = \sqrt{-1}$.

For any complex number $z = a + bi$, we will refer to the real part of z as a and the imaginary part of z as bi.

Examples of complex numbers

Complex Number	Real part	Imaginary part
$5 + 7i$	5	$7i$
$4 - 3i$	4	$-3i$
$8i$	0	$8i$
16	16	0
$\pi - \pi i$	π	$-\pi i$

Adding and Subtracting Complex Numbers

In adding and subtracting complex numbers, you will notice that i works just like a variable.

For example, $2x + 3x = 5x$, and $2i + 3i = 5i$. But remember that i is NOT a variable. Variables can represent any value, but i is always $\sqrt{-1}$.

5.1—194

To add complex numbers, $a + bi$, and $c + di$, ==you simply add the real parts,== and the ==complex parts.==

$$(a + bi) + (c + di) = (a + c) + (b + d)i$$

Perform the following addition and subtractions of complex numbers:

$(3 + 2i) + (6 - 5i)$	$3 + 6 + 2i - 5i = 9 - 3i$
$(3 + 2i) - (6 - 5i)$	$3 - 6 + 2i + 5i = -3 + 7i$
$(2 + 3i) + (-5 + 4i) + 1$	$2 - 5 + 1 + 3i + 4i = -2 + 7i$
$(1 + i) + 2(3 - i)$	$(1+i) + (6 - 2i) = 1 + 6 + i - 2i = 7 - i$

Powers of i

Recall that when you see x^3, that is a variable expression. When you see 2^3, that is a numerical value that we can evaluate, $2^3 = 8$.

What about i^3 or i^{12}? We will be able to evaluate these?

Think back to our basic definition, $i = \sqrt{-1}$.

Using this base definition, we can start to evaluate power of i.

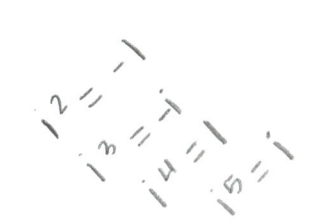

Power of i	Calculation	Answer
i	By definition, $i = \sqrt{-1}$	i
i^2	$i^2 = \sqrt{-1}^2 = -1$	-1
i^3	$i^3 = \sqrt{-1} \cdot \sqrt{-1} \cdot \sqrt{-1} = i^2 \cdot i = (-1)(i)$	$-i$
i^4	$i^4 = i^2 \cdot i^2 = -1 \cdot -1 = 1$	1
i^5	$i^5 = i^4 \cdot i = 1 \cdot i$	i
i^6	$i^6 = i^5 \cdot i = i \cdot i = i^2 = -1$	-1
i^7	$i^7 = i^6 \cdot i = -1 \cdot i = -i$	$-i$
i^8	$i^8 = i^4 \cdot i^4 = (1)(1) = 1$	1
i^9	$i^9 = i^4 \cdot i^4 \cdot i = i$	i
i^{10}	$i^{10} = i^4 \cdot i^4 \cdot i^2 = 1 \cdot 1 \cdot -1$	-1

You see the pattern repeating above. We find that for all integers, $n \geq 0$, i^n will either be $i, -1, -i,$ or 1. The pattern repeats for every 4th integer power.

Find the value of i^{17}.

First divide 17 by 4. 4)17, 4, 16, 1	
Rewrite i^{17}, grouping the powers or 4	$i^{17} = (i^4)^4 i$
Simplify	$i^{17} = \boxed{i}$

Find the value of i^{234}

$$i^{234} = \underbrace{(i^4)^{58}}_{(-1)} \cdot i^2 \qquad i^2 \cdot -1 = \boxed{-1}$$

So, given a problem, i^n, $n \geq 0$, When dividing by 4, your remainder will be either 0, 1, 2, or 3. Your final answer will be i raised to the power of the remainder.

Power of i	Remainder when dividing by 4	Answer
i^{83}	3	$i^3 = -i$
i^{50}	2	$i^2 = -1$
i^{121}	1	i
i^{180}	0	$i^0 = 1$

5.1—197

Multiplying Complex Numbers
The multiplication of complex numbers is a natural extension of what you have done previously when you multiply binomial expressions.   a+bi

FOIL
(2+3x)(4−5x)

$(2 + 3i)(4 - 5i)$	$8 - 10i + 12i - 15i^2$ $8 + 2i - 15(-1)$ $\boxed{23 + 2i}$
$(2 + i)(2 - i)$	$4 - 2i + 2i - i^2$ $4 - i^2 = 4 - (-1) = \boxed{5}$
$i(3 + 2i)$	$3i + 2i^2$ $3i + 2(-1)$ $\boxed{-2 + 3i}$
$(4 + 3i)(2 - 6i)$	$8 - 24i + 6i - 18i^2$ $8 - 18i - 18(-1)$ $\boxed{26 - 18i}$

Homework 5.1 Complex Numbers, Addition, Subtraction, Multiplication

Simplify.

1. $(3i) - (4i)$	2. $(-7 - 2i) - (-8 + 2i)$
$-i$	$-7 + 8 - 2i - 2i$
	$1 - 4i$

3. $(1 + 11i) + (-16 - 20i)$	4. $-17 + (8 - 17i) + 20$
$1 - 16 + 11i - 20i$	$-17 + 8 + 20 - 17i$
$-15 - 9i$	$-9 + 20 - 17i$
	$11 - 17i$

5. $(-7 - 11i) - 4 - (16i)$	6. $(2i) - (6 - 9i) - (-6 + 7i)$
$-7 - 4 - 11i - 16i$	$-6 + 6 + 2i - 9i + 7i$
$-11 - 27i$	$4i$

7. $(50 - 22i) - (-3 - 39i) - (5 + 99i)$	8. $-2i + (78 - 75i) - (-77 - 3i)$
$50 + 3 - 5 - 22i + 39i - 99i$	$78 + 77 - 2i - 75i + 3i$
$48 - 82i$	$155 - 74i$

5.1—199

Homework 5.1 Complex Numbers, Addition, Subtraction, Multiplication

Simplify.

9. $(7i)(3i)$	10. $-5(7i)(-3i)$
$21i^2 = 21(-1) = -21$	$-5(-21i^2)$ $-5(-21(-1))$ $-5(21) = -105$

11. $(1+2i)^2$	12. $(6+9i)(1-2i)$
$1^2 + 2i^2$ $1 - 2$ -1	$6 - 12i + 9i + 18i^2$ $6 - 3i + (18 \cdot -1)$ $6 - 3i - 18$ $-12 - 3i$ **24-3i**

13. $(1-i)(2+3i)$	14. $(2i)^5$
$2 + 3i - 2i - 3i^2$ $2 + i - 3(-1)$ $5 + i$	$32i^5 = 32i^4 i$ $= 32(1)(i)$ $= 32i$

15. i^{34}	16. i^{71}	17. i^{1200}
$4\overline{)34}$ quotient 8 rem 2 $\underline{32}$ 2 $(i^4)^8 i^2$ $(1)(-1)$ -1	$4\overline{)71}$ quotient 17 rem 3 $\underline{4}$ 31 $\underline{28}$ 3 $(i^4)^{17} i^3$ $-i$	$i^0 = 1$

Homework 5.1 Complex Numbers, Addition, Subtraction, Multiplication

CHALLENGE PROBLEMS

1. For a complex number $z = a + bi$, the absolute value of the complex number, as with real numbers, is the distance from the origin. To plot a complex number, we let the horizontal axis represent the real part, and the vertical axis the imaginary part of the number. Plot the point $z = 6 + 7i$, and find the value of $|z|$ (i.e. how far is z from $0 + 0i$)

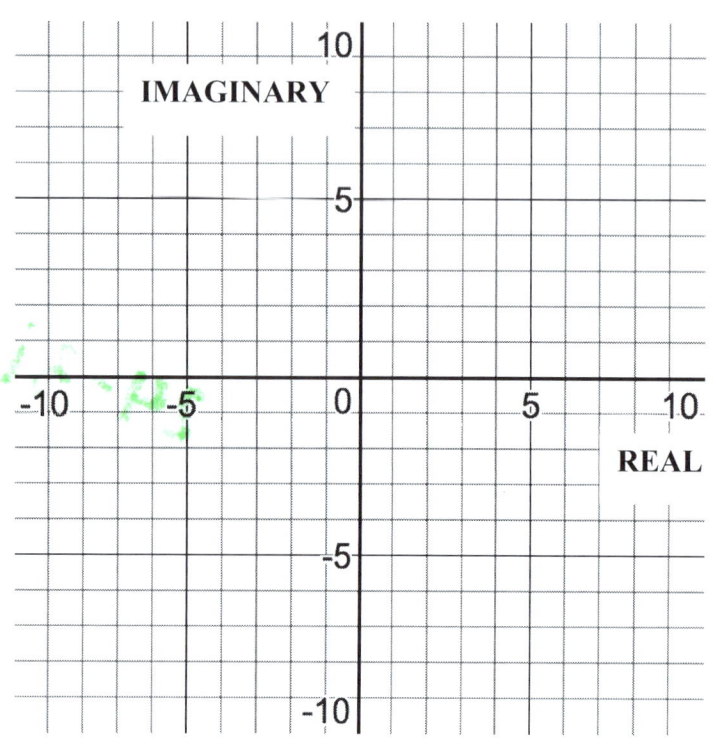

2. Simplify $\sqrt{(i^5 - 1)(i^5 + 1)}$

5.2. Complex Numbers, Division

In the last lesson, we introduced complex numbers and discussed the addition, subtraction, powers and multiplication of complex numbers.

As a review, you should be able to simplify:

---

i^{211}

$4\overline{)211}^{52}$
$\underline{208}$
3

$(i^4)^{52} \cdot i^3 = -i$

$(1-i)(2+i)$

$2 + i - 2i - i^2$
$2 - i - i^2$
$3 - i$

$i^3(1+i)$

$i^3 + i^4 = i^2 i + i^2 i^2$
$ -1\cdot i + -1\cdot-1$
$ -i + 1$
$ 1 - i$

$\sqrt{-25}\, i^{-3i^2}$

$\sqrt{-25}\; i^{(-3)(-1)}$

$\sqrt{-25}\; i^3$

$\sqrt{-25}\,(-i)$

$5i(-i) = -5i^2 = \boxed{5}$
$\downarrow$
-1

Today, we will turn our attention to what it means to divide by complex numbers.

First let's look at the ***definition*** of the conjugate of a complex number.

For any complex number, $a + bi$, the **complex conjugate** is simply $a - bi$. You notice that the real part remains the same, and the sign is changed for the imaginary part.

Complex Number	Complex Conjugate $a-bi$
$5 - 3i$	$5+3i$
$-6 - 2i$	$-6+2i$
$2i$	$-2i$
$i + 21$ $21+i$	$-i+21$ OR $21-i$

What happens when you multiply a complex number by its complex conjugate

Complex Number	Complex Conjugate
$a + bi$	$a - bi$
Multiplication $(a + bi)(a - bi)$ $a^2 - abi + abi - b^2 i^2$ $a^2 - b^2(-1) = a^2 + b^2$	
Final Answer is what kind of number? real	

Multiplying any complex number by its conjugate will yield a real number.

5.2—203

So, in order to divide by a complex number, we must turn the denominator into a real number.

$$\frac{1+2i}{2-3i}$$

What can we multiply by and get a real number in the denominator?

Division of Complex numbers
For any complex number, $a + bi$, $b \neq 0$, to **divide** any expression by $a + bi$, you multiply the expression by one in the form $\frac{a-bi}{a-bi}$, where $a - bi$ is the complex conjugate of $a + bi$.

Perform the division.

$\dfrac{1+2i}{2-3i}$

$\dfrac{1+2i}{2-3i}\left(\dfrac{2+3i}{2+3i}\right)$ ← complex conjugate

$\dfrac{(1+2i)(2+3i)}{(2-3i)(2+3i)} = \dfrac{2+4i+3i+6i^2}{4+6i-6i-9i^2}$

$\dfrac{2+7i-6}{4+9} = \dfrac{-4+7i}{13} \quad \boxed{\dfrac{-4}{13}+\dfrac{7}{13}i}$

$\dfrac{3}{i}$

$\dfrac{3}{i}\left(\dfrac{-i}{-i}\right) = \dfrac{-3i}{-(-1)} = \boxed{-3i}$

$\dfrac{1+i}{1-i}$

$\dfrac{1+i}{1-i}\left(\dfrac{1+i}{1+i}\right) = \dfrac{(1+i)(1+i)}{(1-i)(1+i)} = \dfrac{1+i+i+i^2}{1+i-i-i^2}$

$\dfrac{1+2i+(-1)}{1-(-1)} \quad \boxed{\dfrac{2i}{2}} \to \boxed{i}$

Perform the division.

$\dfrac{2i}{1+i}$

$\dfrac{2i}{1+i}\left(\dfrac{1-i}{1-i}\right) = \dfrac{2i-2i^2}{1-i+i-i^2} = \dfrac{2i-2(-1)}{1-(-1)}$

$\dfrac{2+2i}{2} = \boxed{1+i}$

$\dfrac{5}{i+7}$

$\dfrac{5}{i+7} = \dfrac{5}{7+i}\left(\dfrac{7-i}{7-i}\right) = \dfrac{35-5i}{49-7i+7i-i^2} = \dfrac{35-5i}{49+1}$

$\dfrac{35}{50} - \dfrac{5}{50}i = \boxed{\dfrac{7}{10} - \dfrac{1}{10}i}$

$\dfrac{3}{i^2} = -3$

↳ $\dfrac{3}{-1}$

Homework 5.2 Complex Numbers, Division

Simplify.

1. $\dfrac{9}{9i}\left(\dfrac{i}{i}\right) = \dfrac{9i}{9i^2} = \dfrac{i}{i^2} = \dfrac{i}{-1}$
   $= \boxed{-i}$

2. $\dfrac{4i}{-6+10i} = \dfrac{2i}{-3+5i}\left(\dfrac{-3-5i}{-3-5i}\right) =$
   $\dfrac{-6i-10\boxed{i^2}=-1}{9-25\boxed{i^2}=-1}$
   $\dfrac{10(-1)-6i}{9+25} \quad \dfrac{-10-6i}{34} \quad \boxed{\dfrac{5}{17}-\dfrac{3}{17}i}$ *(note: should be $-\dfrac{5}{17}$)*

3. $(3+9i) \div (9+2i)$
   $\dfrac{3+9i}{9+2i}\left(\dfrac{9-2i}{9-2i}\right) = \dfrac{27+81i-6i-18i^2}{81-4i^2}$
   $\dfrac{27+75i-18(-1)}{81-4(-1)} = \dfrac{45+75i}{85}$
   $\boxed{\dfrac{9}{17}+\dfrac{15}{17}i}$

4. $\dfrac{-2+5i}{-7-6i}\left(\dfrac{-7+6i}{-7+6i}\right)$
   $\dfrac{14-12i-35i+30i^2}{49-42i+42i-36i^2} = \dfrac{-16-47i}{49+36}$
   $\dfrac{-16-47i}{85} \quad \boxed{\dfrac{-16}{85}-\dfrac{47}{85}i}$

5. $\dfrac{5}{5+i}\left(\dfrac{5-i}{5-i}\right) \quad \dfrac{25-5i}{25-5i+5i-i^2}$
   $\dfrac{25-5i}{26}$
   $\boxed{\dfrac{25}{26}-\dfrac{5}{26}i}$

6. $-\dfrac{1+i}{1-i}\left(\dfrac{1+i}{1+i}\right) = -\dfrac{1+i+i+i^2}{1+i-i-i^2} = \dfrac{-2i}{2}$
   $= \boxed{-i}$

7. $\dfrac{2+3i}{i^3} = \dfrac{2+3i}{i^2 i} = \dfrac{2+3i}{-i}\left(\dfrac{i}{i}\right)$
   $\dfrac{2i+3i^2}{-i^2} = \dfrac{2i-3}{1}$
   $\boxed{-3+2i}$

8. $\dfrac{25}{\sqrt{-25}} \quad \dfrac{25}{5i} \quad \dfrac{5}{1}\left(\dfrac{-i}{-i}\right) \quad \dfrac{-5i}{i^2}$
   $= \boxed{-5i}$

5.2—205

Homework 5.2 Complex Numbers, Division

CHALLENGE PROBLEMS

1. Simplify $\dfrac{(1+i)i}{(1-i)^2}$

2. Solve for the real number, a.

$$(1+ai)\cdot(2-5i) = 17+i$$

5.3. Graphing Vertex Form of Quadratics

Up until this point of the course, we've mainly been dealing with linear functions. Remember linear functions could be expressed in different forms.

	$y = mx + b$
	$y - y_1 = m(x - x_1)$
	$Ax + By = C$

In the linear equations above, the highest power of x is x^1. Quadratic functions will have an x^2 term as the highest power of x. Just as there were several forms for the equation of a line, there are multiple forms of the equation for a quadratic function:
- Vertex Form
- Intercept Form
- Standard Form

In the lessons to come we will look at each form, what information it gives, and how to graph the function.

First let's graph a simple quadratic function $f(x) = x^2$

x	$y = x^2$
-3	
-2	
-1	
0	
1	
2	
3	

Vertex

Axis of Symmetry

Domain	Range

The graph of a quadratic function is a *parabola*. The function $f(x) = x^2$ is the parent function. We can derive all other parabolas from this function.

Vertex Form of a Parabola

$$y = a(x - h)^2 + k$$

The vertex is (h, k).
The axis of symmetry is the vertical line, $x = h$.
If $a > 0$, then the parabola opens upward.
If $a < 0$, then the parabola opens downward.
The larger $|a|$ is, the "thinner" the parabola is.

Graph the parabola, $y = 2(x - 1)^2 + 1$.

Step 1: Make sure the parabola is in Vertex Form. $y = a(x - h)^2 + k$	
Step 2: Plot the vertex (h, k)	
Step 3: Does the parabola open up or down?	
Step 3: The axis of symmetry is the vertical line, $x = h$.	
Step 4: Plot 2 points on one side of the axis of symmetry, then the corresponding symmetrical points on the other side of the AoS.	
Step 5: Domain	Step 6: Range

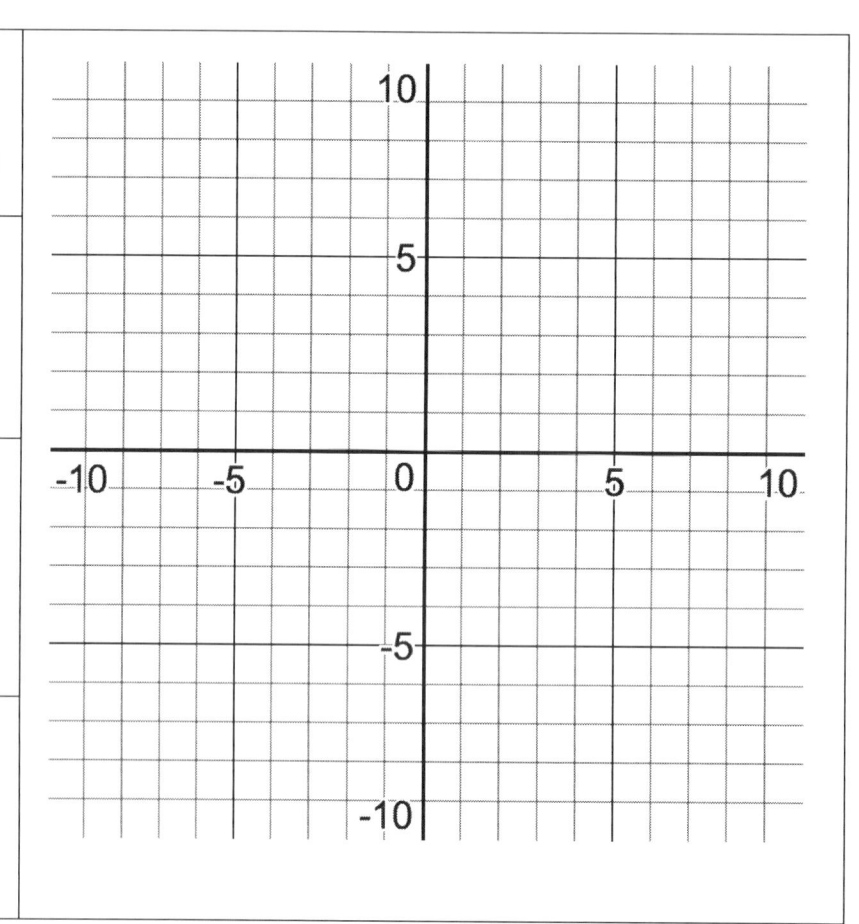

Graph the parabola, $y = -2(x+2)^2 - 1$.

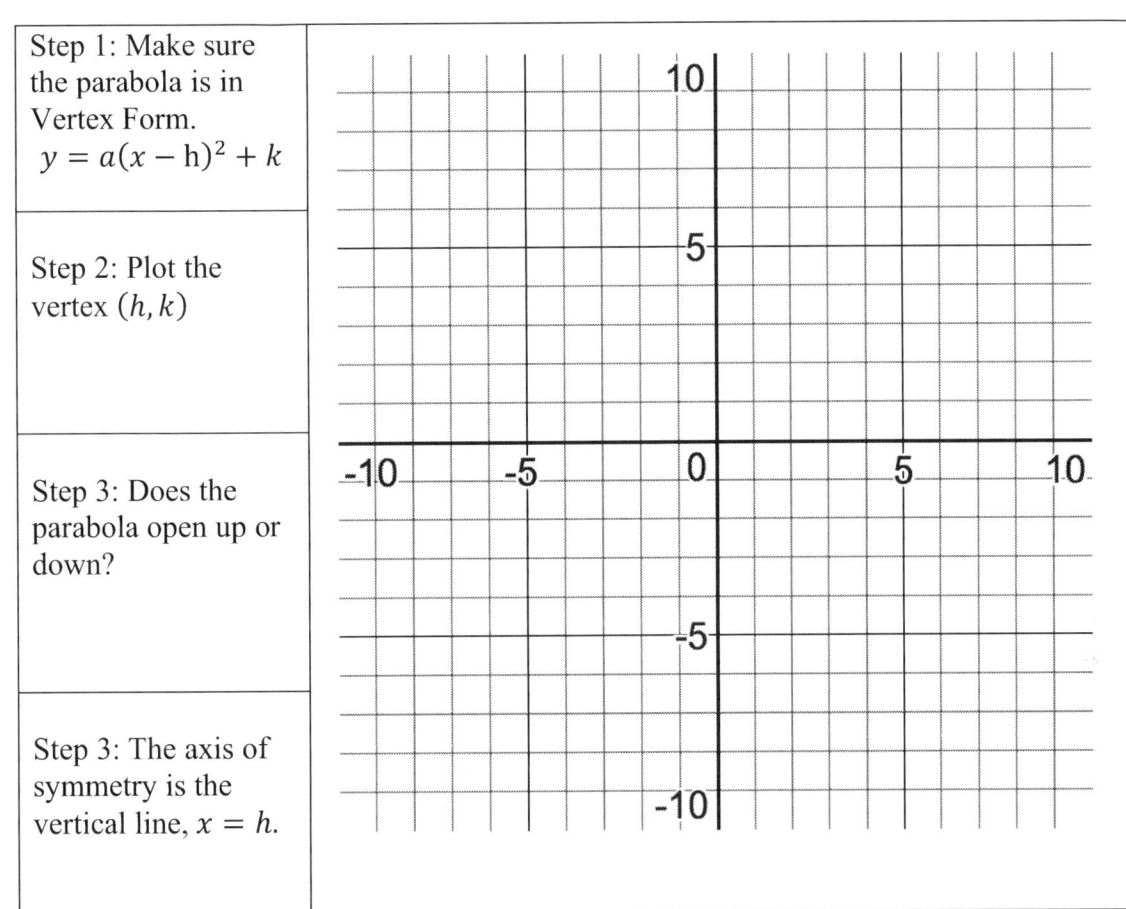

Step 1: Make sure the parabola is in Vertex Form. $y = a(x-h)^2 + k$	
Step 2: Plot the vertex (h, k)	
Step 3: Does the parabola open up or down?	
Step 3: The axis of symmetry is the vertical line, $x = h$.	

Step 4: Plot 2 points on one side of the axis of symmetry, then the corresponding symmetrical points on the other side of the AoS.

Domain

Range

Graph the parabola, $y = -\frac{1}{2}(x+3)^2$

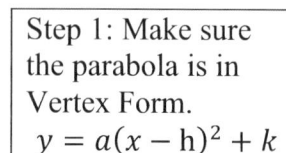

Step 1: Make sure the parabola is in Vertex Form. $y = a(x-h)^2 + k$	
Step 2: Plot the vertex (h, k)	
Step 3: Does the parabola open up or down?	
Step 3: The axis of symmetry is the vertical line, $x = h$.	

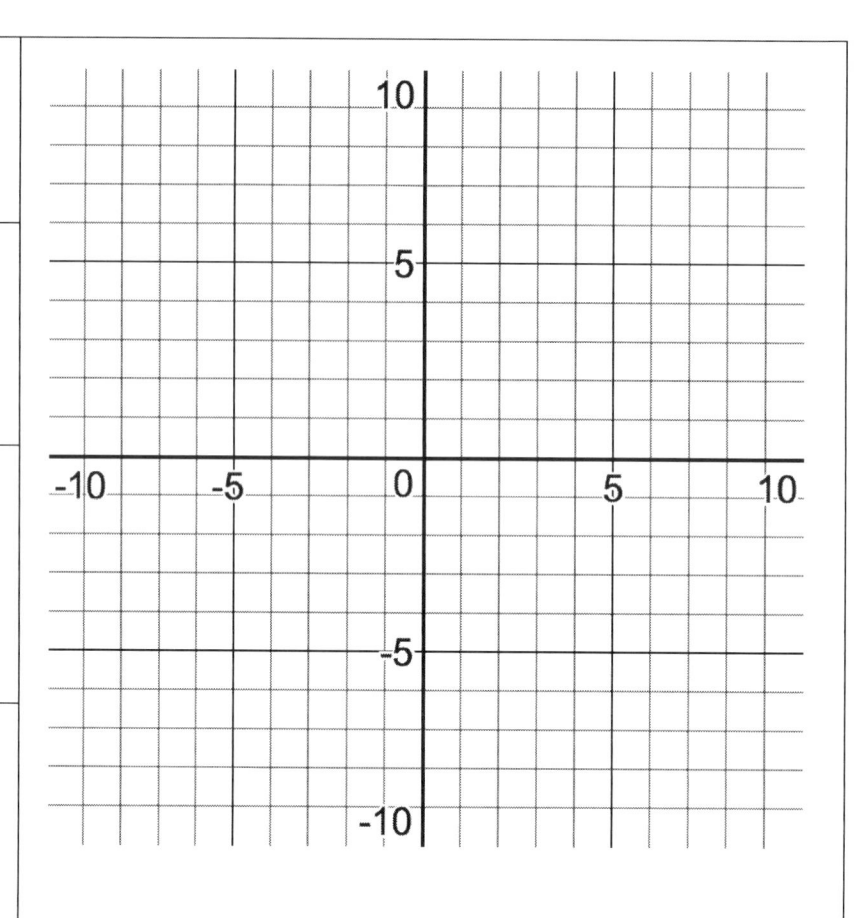

Step 4: Plot 2 points on one side of the axis of symmetry, then the corresponding symmetrical points on the other side of the AoS.

Domain

Range

Graph the parabola, $y = 3(x+4)^2 - 5$

Vertex:

Opens up or down?

Axis of Symmetry

Additional Points

x	y

Domain

Range

Homework 5.3 Graphing Vertex Form of Quadratics

1. Graph the parabola, $y = (x - 3)^2$

Vertex:

Opens up or down?

Axis of Symmetry

Additional Points

x	y

Domain

Range

Homework 5.3 Graphing Vertex Form of Quadratics

2. Graph the parabola, $y = -(x+3)^2 + 5$

Vertex:

Opens up or down?

Axis of Symmetry

Additional Points

x	y

Domain

Range

Homework 5.3 Graphing Vertex Form of Quadratics

3. Graph the parabola, $y = -4(x-2)^2 + 8$

Vertex:	
Opens up or down?	
Axis of Symmetry	

Additional Points

x	y

Domain

Range

Homework 5.3 Graphing Vertex Form of Quadratics

4. Graph the parabola, $y = \frac{1}{2}(x-3)^2 + 2$

Vertex:

Opens up or down?

Axis of Symmetry

Additional Points

x	y

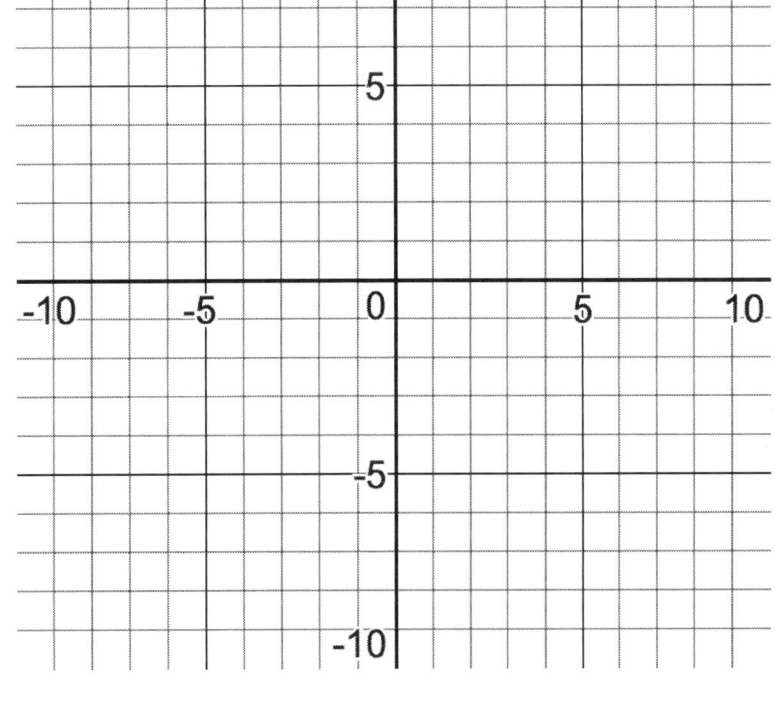

Domain

Range

Homework 5.3 Graphing Vertex Form of Quadratics

CHALLENGE PROBLEMS

$f(x)$ graph	$g(x) = (x-6)^2 - 4$

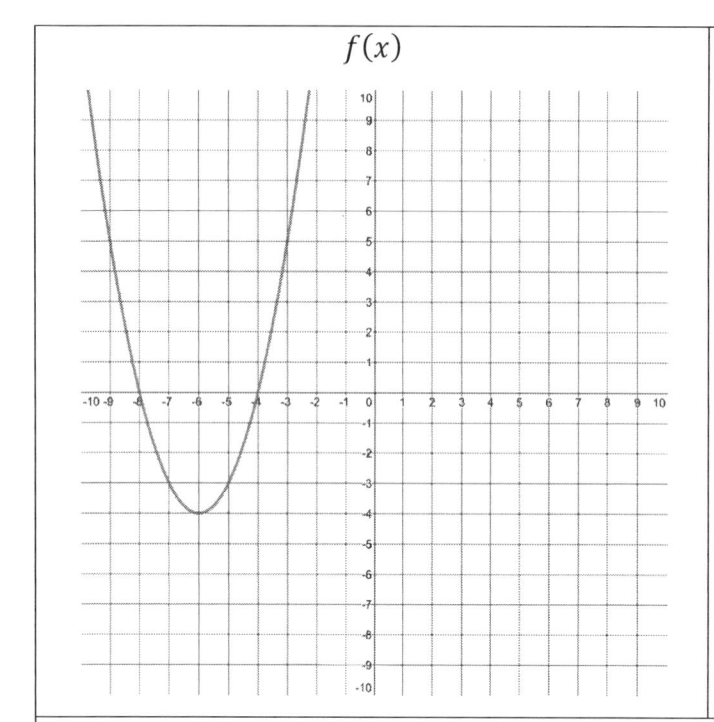

TRUE or FALSE	
f and g have the same vertex	
f and g have the same domain	
f and g have the same range	
$\|x\| = 4$ represents the axes of symmetry for f and g	

Write two inequalities that represent the shaded solution graphed below.

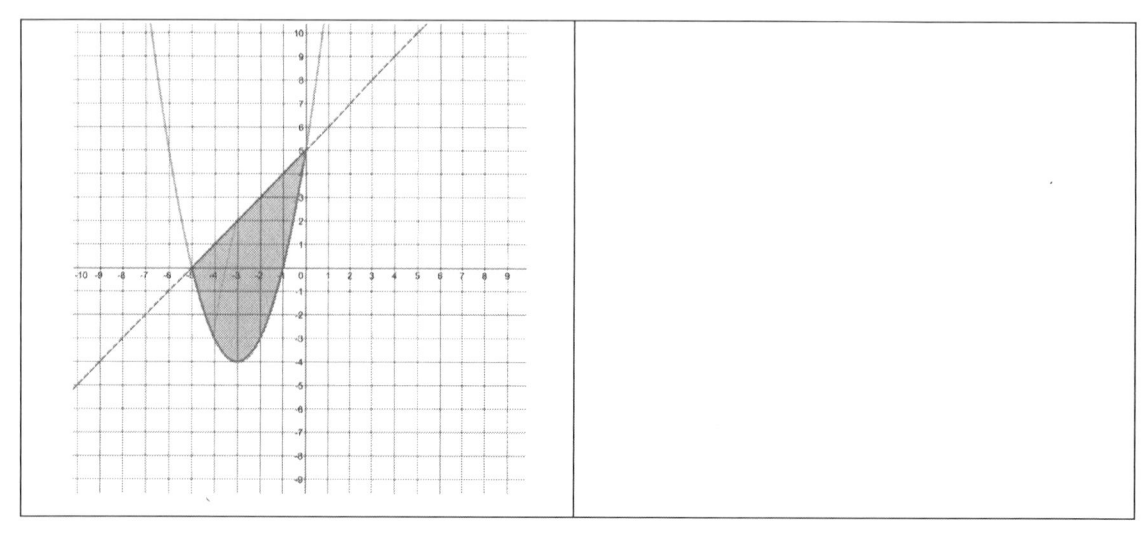

5.4. Graphing Intercept Form of Quadratics

Intercept Form of a Parabola

$$y = a(x - p)(x - q)$$

The x-intercepts are $(p, 0)$ and $(q, 0)$. (Remember the x-intercept is where $y = 0$).
The axis of symmetry is the vertical line, $x = \frac{(p+q)}{2}$
If $a > 0$, then the parabola opens upward.
If $a < 0$, then the parabola opens downward.

Substitute $x = \frac{p+q}{2}$ into the original equation to find the vertex.

Graph $y = -2(x - 3)(x - 7)$

$y = a(x-p)(x-q)$

Make sure the equation is in Intercept Form: $y = a(x-p)(x-q)$	
Opens up or down? **down**	
x-intercepts **(3,0)(7,0)**	
y-intercept $X=0$ $-2(-3)(-7) = y$ **(0,-42)**	
Find/Graph the AoS: $x = \frac{p+q}{2}$ **X = 5**	

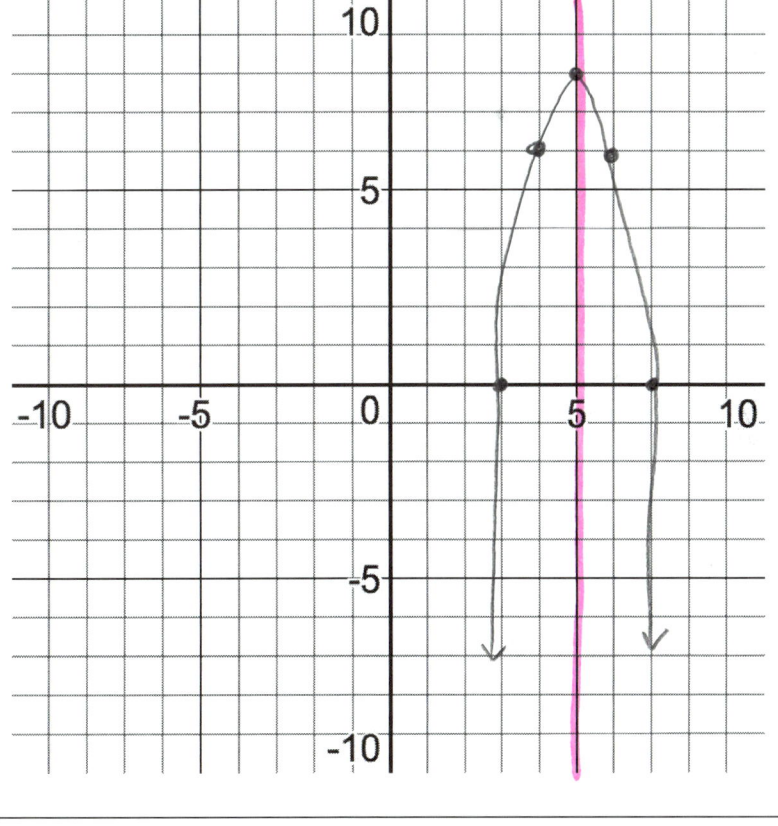

Substitute $x = \frac{p+q}{2}$ into the original equation to find the vertex. Complete the graph.

$y = -2(5-3)(5-7) = -2(2)(-2) = 8 \quad (5, 8)$

Domain	Range
$(-\infty, \infty)$	$(-\infty, 8]$

Graph $y = 2x(x+2)$ $y = 2(x-0)(x+2)$

Make sure the equation is in Intercept Form: $y = a(x-p)(x-q)$
Opens up or down? **UP**
x −intercepts $(0,0)\ (-2,0)$
y −intercept $(0,0)$
Find/Graph the AoS: $x = \dfrac{p+q}{2}$ $x = -1$

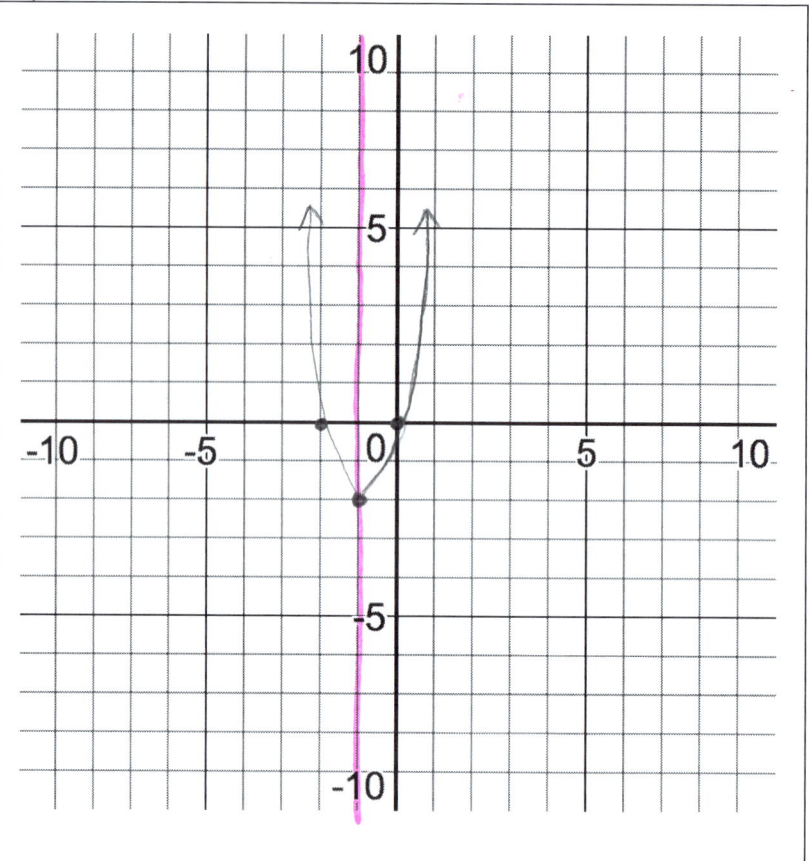

Substitute $x = \dfrac{p+q}{2}$ into the original equation to find the vertex. Complete the graph.

$y = 2(-1)(-1+2)$
$y = -2(1)$
$y = -2$

Vertex $= (-1, -2)$

Domain	Range
$(-\infty, \infty)$	$[-2, \infty)$

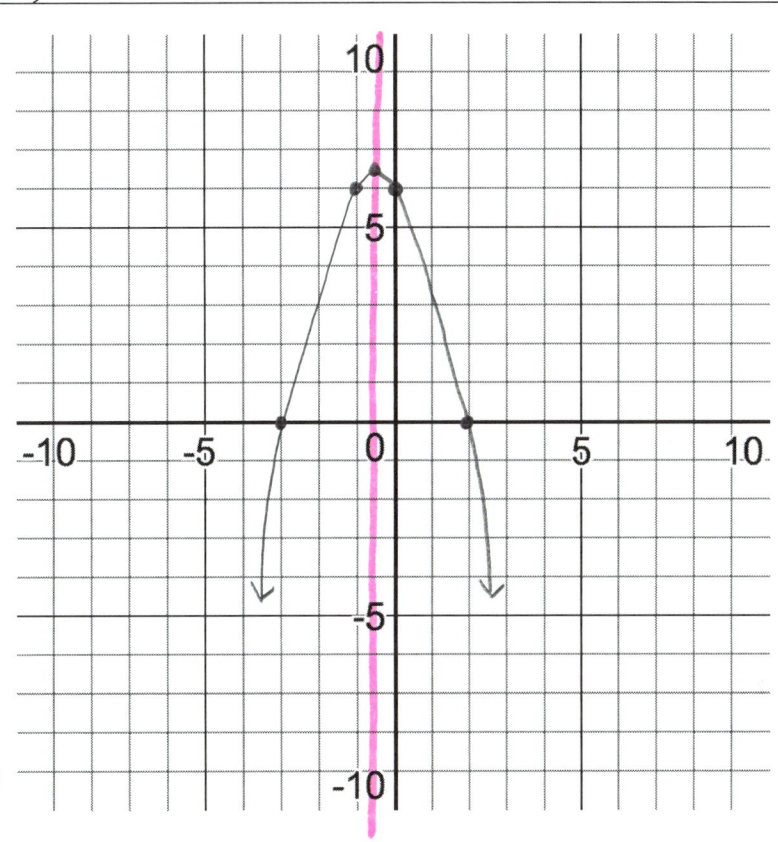

Graph $y = -(x+3)(x-2)$

Opens up or down?
down

x –intercepts
$(-3, 0)(2, 0)$

y –intercept
$x = 0$
$(-3)(-2) = 6$
$(0, 6)$

AoS:
$x = \frac{-3+2}{2}$
$x = -\frac{1}{2}$

Vertex:
$y = -1(-\frac{1}{2}+3)(-\frac{1}{2}-2)$
$y = -1(2.5)(-2.5)$
$x = 6.25$
$(-\frac{1}{2}, 6.25)$

Domain
$(-\infty, \infty)$

Range
$(-\infty, 6.25]$

Homework 5.4 Graphing Intercept Form of Quadratics

1. Graph $y = (x + 3)(x - 3)$

Opens up or down?	UP
x −intercepts	$(-3, 0)$ $(3, 0)$
y −intercept	$x = 0$ $y = (3)(-3)$ $y = -9$ $(0, -9)$
AoS:	$x = 0$
Vertex:	$x = \dfrac{3-3}{2}$ $x = 0$ $(0, -9)$
Domain	$(-\infty, \infty)$
Range	$[-9, \infty)$

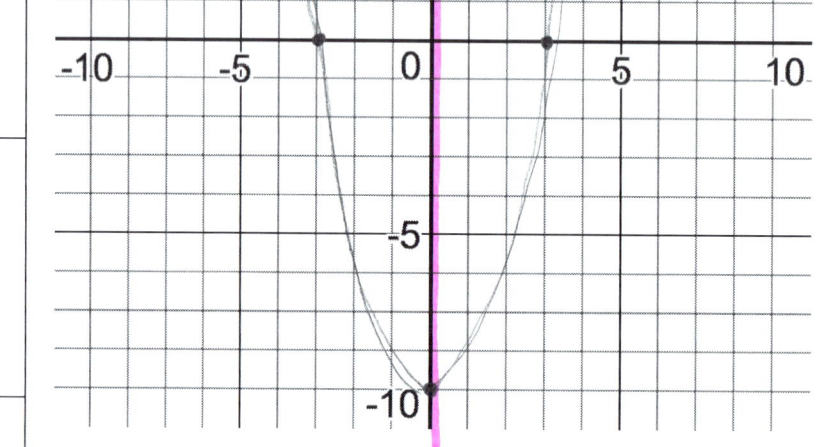

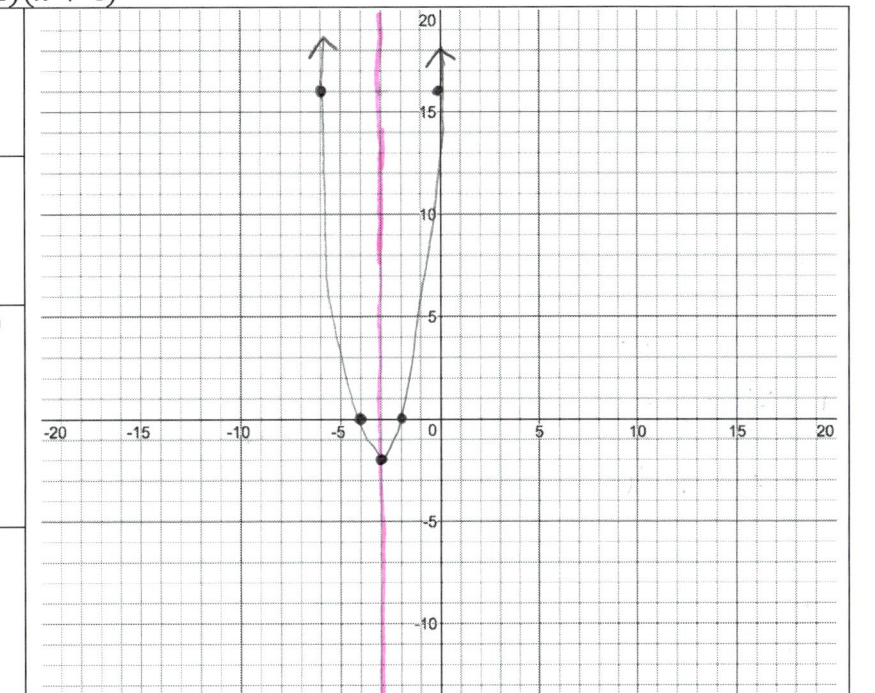

Homework 5.4 Graphing Intercept Form of Quadratics

2. Graph $y = 2(x + 2)(x + 4)$ $y = a(x-p)(x-q)$

Opens up or down?
UP

x −intercepts
(-2, 0) (-4, 0)

y −intercept $x = 0$
$2(2)(4) = 16$
(0, 16)

AoS:
$x = -3$

Vertex:

$x = -3$

$2(-3+2)(-3+4)$ $(-3, -2)$
$2(-1)(1)$
-2

Domain	Range
$(-\infty, \infty)$	$[-2, \infty)$

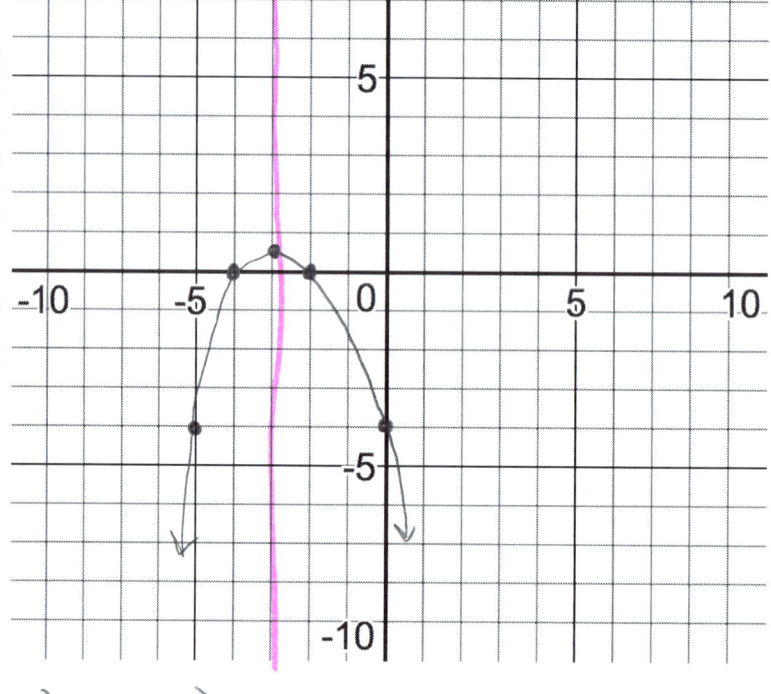

Homework 5.4 Graphing Intercept Form of Quadratics

3. Graph $y = -\frac{1}{2}(x+2)(x+4)$

Opens up or down? down
x –intercepts $(-2,0)(-4,0)$
y –intercept $x=0$ $y = -\frac{1}{2}(2)(4)$ $y = -4$ $(0,-4)$
AoS: $x = -3$
Vertex: $y = -\frac{1}{2}(-3+2)(-3+4)$ $y = -\frac{1}{2}(-1)(1)$ $(-3, \frac{1}{2})$

Domain $(-\infty, \infty)$	Range $(-\infty, \frac{1}{2}]$

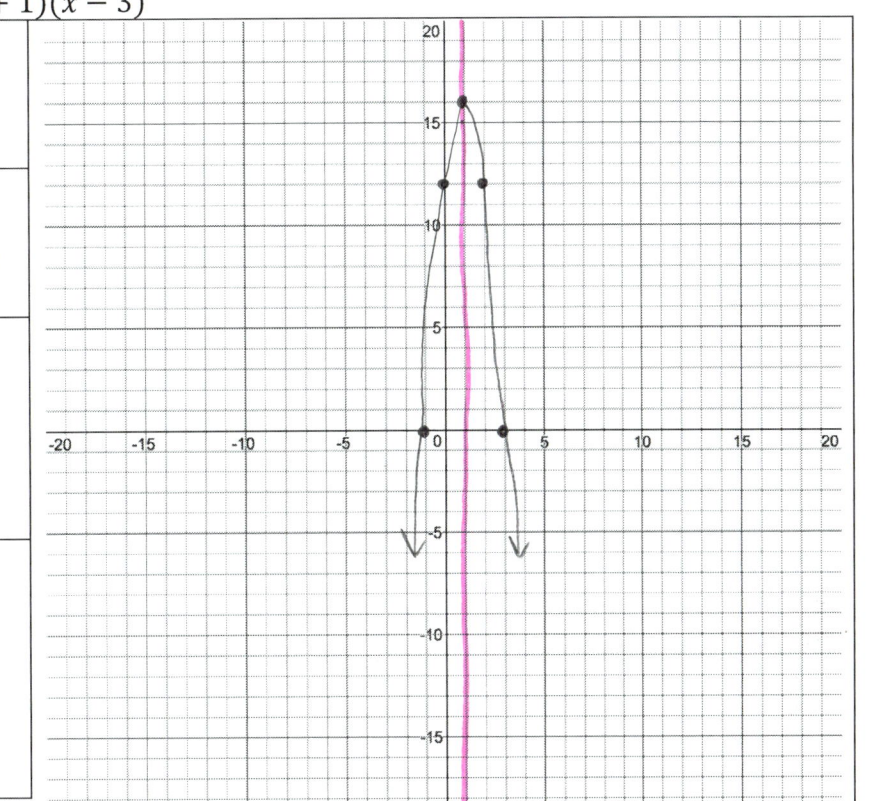

Homework 5.4 Graphing Intercept Form of Quadratics

4. Graph $y = -4(x+1)(x-3)$

Opens up or down?
down

x –intercepts
$(-1,0) (3,0)$

y –intercept $x=0$
$-4(1)(-3)$
$(0,12)$

AoS:
$x=1$

Vertex:
$x=1$
$-4(1+1)(1-3)$
$-4(2)(-2)$
$(1,16)$

Domain	Range
$(-\infty, \infty)$	$(-\infty, 16]$

Homework 5.4 Graphing Intercept Form of Quadratics

CHALLENGE PROBLEMS

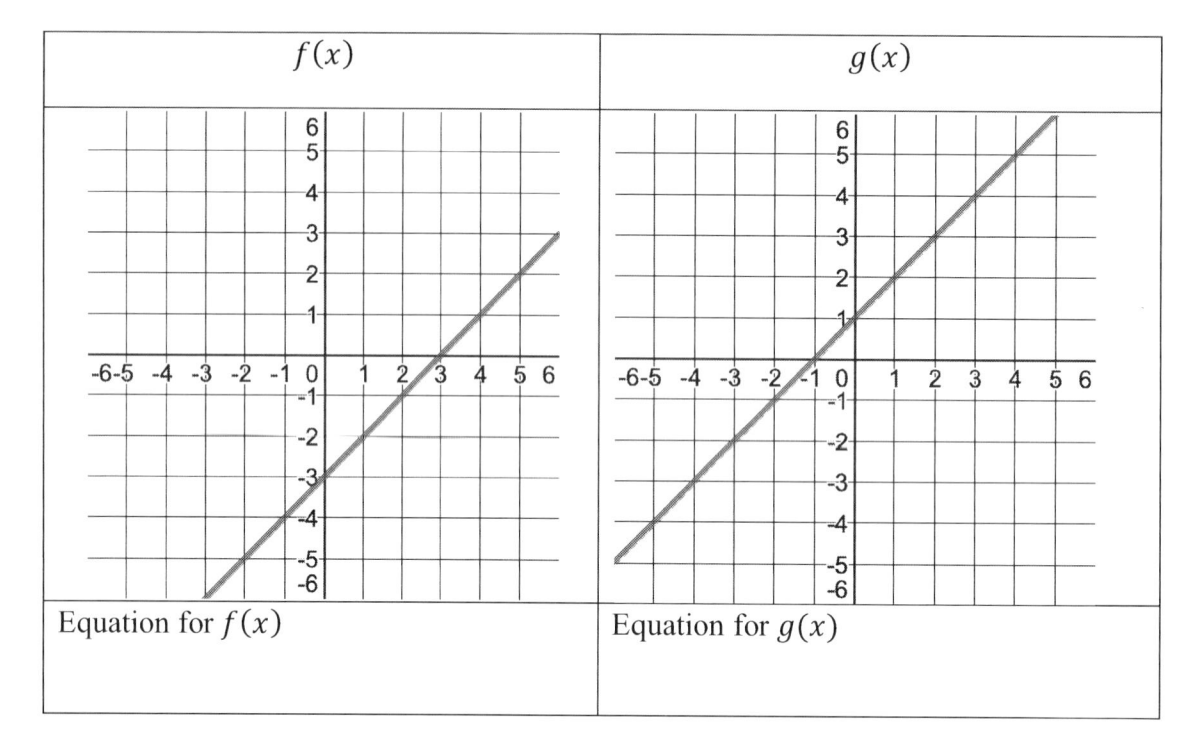

$f(x)$	$g(x)$
[graph of line through (0,-3) with slope 1]	[graph of line through (-2,0) with slope 1]
Equation for $f(x)$	Equation for $g(x)$

Graph the product of $f(x)$ and $g(x)$.

x	$f(x)$	$g(x)$	$f(x) \cdot g(x)$
-2			
-1			
0			
1			
2			
3			

What is the algebraic expression for $f(x) \cdot g(x)$?

What can you conclude about the graph of a $f(x) \cdot g(x)$ from the graph of two lines?

What is the x −intercept of:

$f(x)$	$g(x)$	$f(x) \cdot g(x)$

5.4—224

5.5. Graphing Standard Form of Quadratics, Quadratic Inequalities

We've covered the graphing of quadratic functions in the following two forms:

Vertex Form	Intercept Form
$y = a(x - h)^2 + k$	$y = a(x - p)(x - q)$
Axis of Symmetry $x = h$	Axis of Symmetry $x = \dfrac{p+q}{2}$
Vertex (h, k)	Vertex $\dfrac{p+q}{2}$ is the x-coordinate of the vertex. Substitute into the original equation to find the y-coordinate
	x-intercepts $(p, 0)$ & $(q, 0)$

The third form we will study is **Standard Form**.

The **Standard Form** of a quadratic function is $y = ax^2 + bx + c$.

Standard Form of a Quadratic
$y = ax^2 + bx + c$
Axis of Symmetry
$x = -\dfrac{b}{2a}$
Vertex
$x = -\dfrac{b}{2a}$ is the x-coordinate of the vertex. Substitute $x = -\dfrac{b}{2a}$ into the original equation to find the y-coordinate of the vertex.
y-intercept
Remember the y-intercept is found by setting $x = 0$. The y-intercept is $(0, c)$
If $a > 0$, the parabola opens upward. If $a < 0$, the parabola opens downward.

Graph $y = -x^2 + 4x - 5$.

Step 1: Write the function in Standard Form: $y = ax^2 + bx + c$ $a = -1$ $b = 4$ $c = -5$	
Does this parabola open upward or downward? down	
Step 2: Find the axis of symmetry, $x = -\frac{b}{2a}$, and graph it. $-\frac{4}{2(-1)} = \boxed{2}$	
Step 3: Find the vertex and graph it. $-2^2 + 4(2) - 5$ $-4 + 8 - 5$ $\boxed{-1}$ $(2, -1)$	Find the y-intercept and graph it. If the axis of symmetry is not on the y-axis, use the y-intercept and the parabola's symmetry to match a point on the parabloa that correpsonds to the y-intercept. y-int = $(0, -5)$
Domain $(-\infty, \infty)$	Range $(-\infty, -1]$

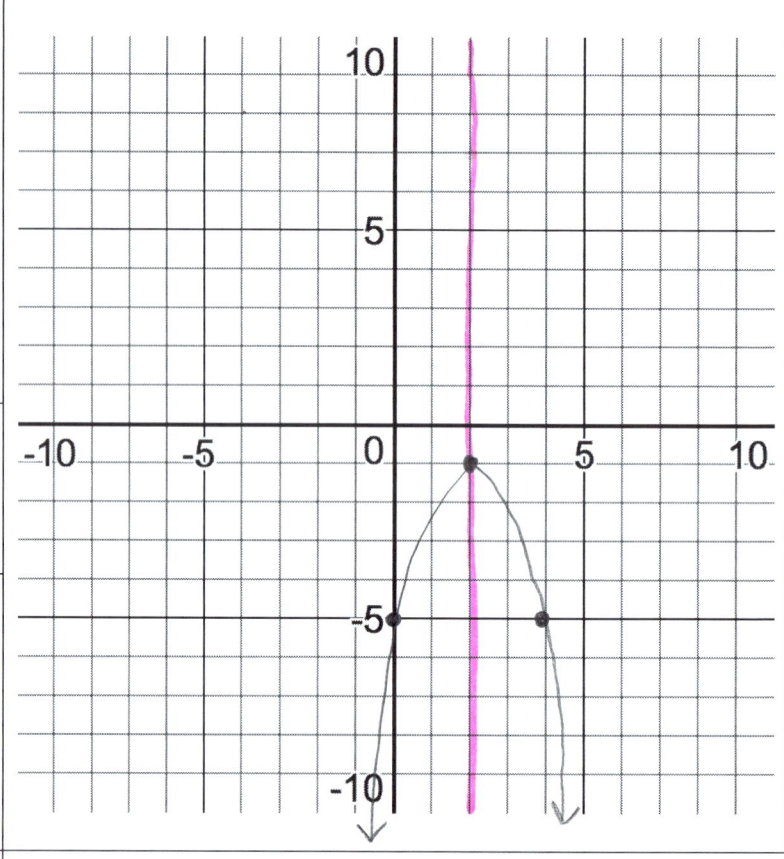

For best results, you want to be sure that other than the vertex, you have at least two more points on the parabola to graph accurately. The y-intercept is one option. Another option is to pick a point near the vertex, and its corresponding symmetrical point.

Graph $y = 2x^2 + 8x - 2$.

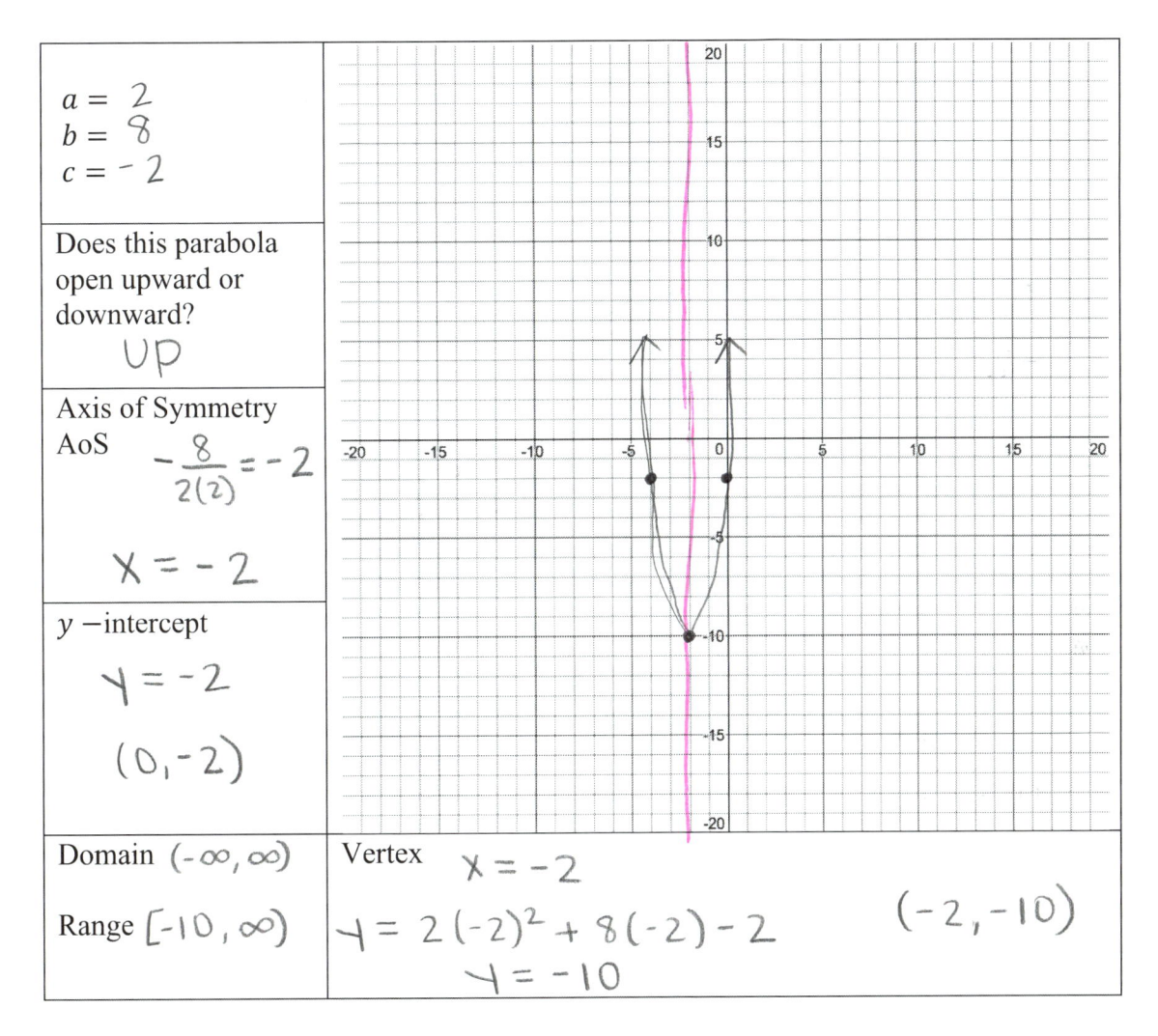

$a = 2$ $b = 8$ $c = -2$	
Does this parabola open upward or downward? **UP**	
Axis of Symmetry AoS $-\frac{8}{2(2)} = -2$ $x = -2$	
y –intercept $y = -2$ $(0, -2)$	
Domain $(-\infty, \infty)$ Range $[-10, \infty)$	Vertex $x = -2$ $y = 2(-2)^2 + 8(-2) - 2$ $y = -10$ $\qquad (-2, -10)$

Once you master graphing parabolas, quadratic equalities, it is a natural progression to graph **quadratic inequalities**.

Quadratic Equality	Quadratic Inequality
$y = 2(x-2)(x+2)$ **intercept** This is a parabola, the set of all y values equal to $2(x-2)(x+2)$	$y \geq 2(x-2)(x+2)$ This is a region, the set of all y values that are greater than or equal to $2(x-2)(x+2)$
$y = -(x+5)^2 - 2$ **vertex** This is a parabola in Vertex Form.	$y < -(x+5)^2 - 2$ This is a region, the set of all y values less than $-(x+5)^2 - 2$

5.5—228

Let's graph the solution to $y > -x^2 - 2x - 2$

(↑ dotted)

Step 1: Graph the parabola (in whatever form it is.) This parabola is in standard from. $y = ax^2 + bx + c$ $a = -1$ $b = -2$ $c = -2$	
Step 2: Does this parabola open upward or downward? **down**	
Step 3: Find the y-intercept and graph it. $(0, -2)$	

Step 4: Find the axis of symmetry, $x = -\frac{b}{2a}$, and graph it. $\frac{-(-2)}{2(-1)}$ $x = -1$	Step 5: Find the vertex and graph it. $x = -1$ $1^2 - 2(-1) - 2$ $1 + 2 - 2$ $3 - 2$ $= 1$ $(-1, 1)$
Step 6: Since the inequality is > and not ≥, we graph the parabola with a (dotted) line. **shade outside the parabola**	Step 7: Pick a point NOT on the parabola to test the inequality. A true statement means the point is in the solution. A false statement means the point is NOT in the solution. $(0,0)$ $x = 0$ $y = 0$ $0 > -0^2 - 2(0) - 2$ $0 > -2$ **YES**

                    solid
                      ↑
Graph $y \leq (x - \boxed{1})^2 + 1$ $y = a(x-h)^2 + k$

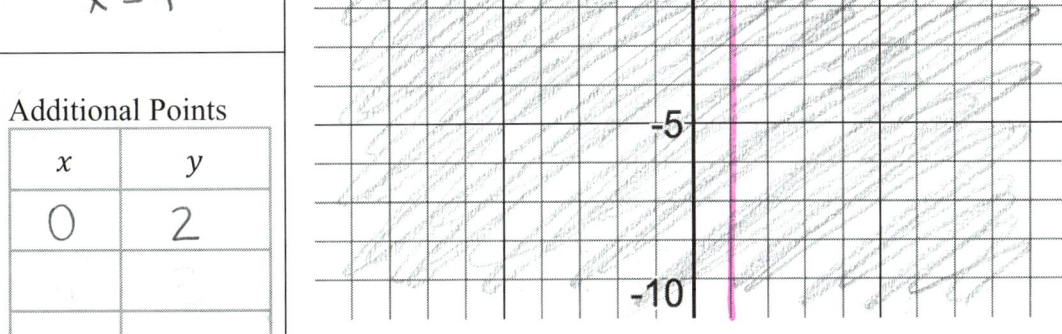

Vertex: (1, 1)
Opens up or down? UP
Graph of curve DOTTED (SOLID)
Axis of Symmetry $x = 1$

Additional Points

x	y
0	2

Test points, then shade the solution set

(0, 0) $0 \leq (0-1)^2 + 1$
          $0 \leq 2$ yes

Homework 5.5 Graphing Standard Form of Quadratics, Quadratic Inequalities

1. Graph the parabola given in Standard Form. $y = -2x^2 + 4x + 3$

$a = -2$
$b = 4$
$c = 3$

Does this parabola open upward or downward? **down**
Axis of Symmetry AoS $-\frac{4}{-2(2)} = 1$
y-intercept $(0, 3)$
Vertex $-2(1)^2 + 4(1) + 3$ $-2 + 4 + 3 = 5$ $(1, 5)$
Domain $(-\infty, \infty)$ Range $(-\infty, 5]$

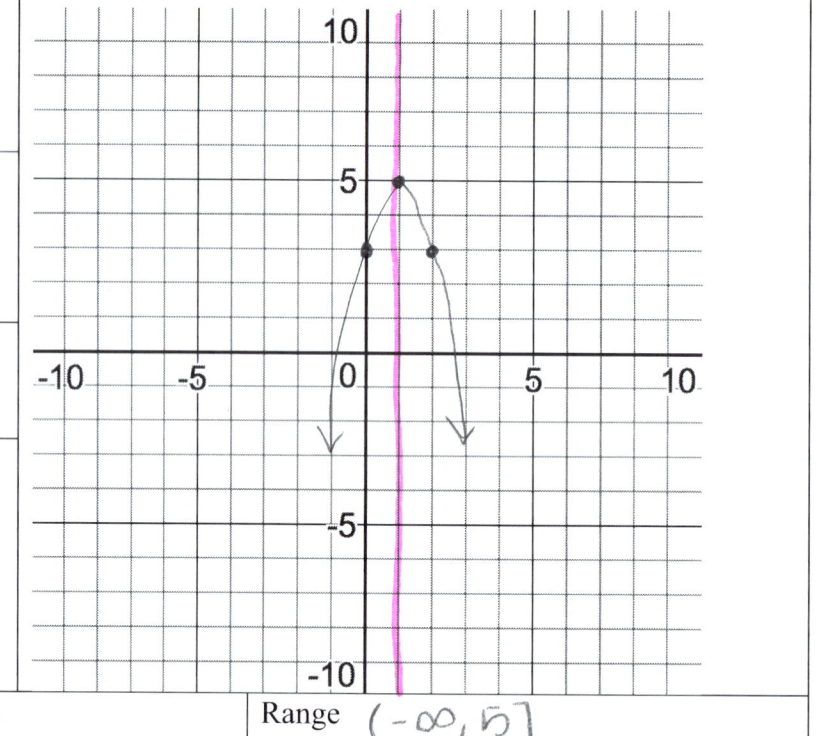

2. Graph the parabola given in Standard Form. $y = 3x^2 - 12x + 5$

$a = 3$
$b = -12$
$c = 5$

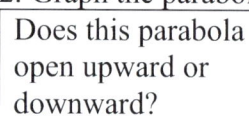

Does this parabola open upward or downward? **UP**
Axis of Symmetry AoS $-\frac{(-12)}{2(3)} = 2$
y-intercept $(0, 5)$
Vertex $3(2^2) - 12(2) + 5$ $12 - 24 + 5 = -7$ $(2, -7)$
Domain $(-\infty, \infty)$ Range $[-7, \infty)$

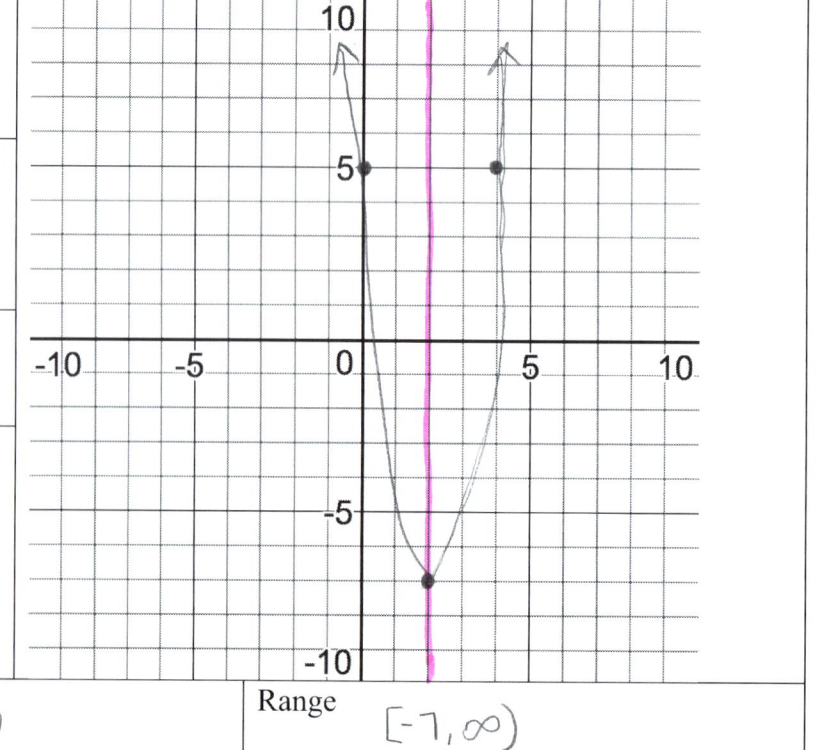

Homework 5.5 Graphing Standard Form of Quadratics, Quadratic Inequalities

3. Graph the parabola given in Standard Form. $y = -4x^2 - 16x - 9$

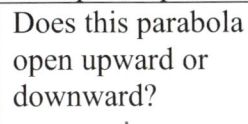

$a = -4$
$b = -16$
$c = -9$

Does this parabola open upward or downward? **down**
Axis of Symmetry AoS $\dfrac{-(-4)}{2(-16)} = -2$
y-intercept $(0, -9)$
Vertex $y = -4(-2^2) - 16(-2) - 9$ $-16 + 32 - 9$ $= 7 \quad (-2, 7)$
Domain $(-\infty, \infty)$ — Range $(-\infty, 7]$

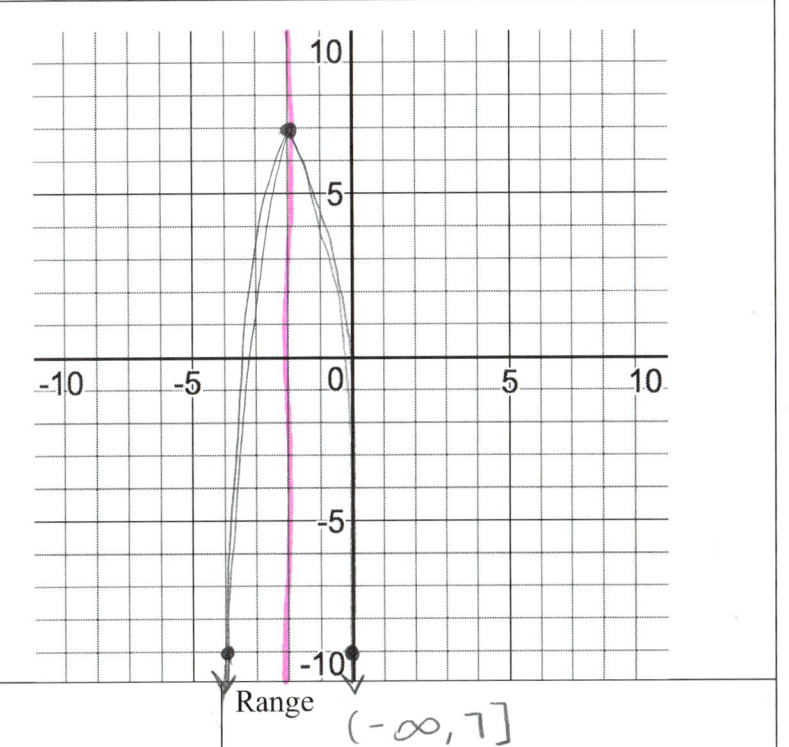

4. Graph the parabola given in Standard Form. $y = \dfrac{1}{2}x^2 - x - 4$

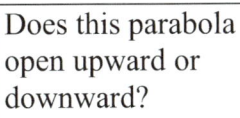

$a = \dfrac{1}{2}$
$b = -1$
$c = -4$

Does this parabola open upward or downward? **up**
Axis of Symmetry AoS $\dfrac{-(-1)}{2(0.5)} = 1$
y-intercept $(0, -4)$
Vertex $y = \dfrac{1}{2}(1)^2 - 1 - 4$ $\dfrac{1}{2} - 1 - 4$ $\left(1, -\dfrac{9}{2}\right)$
Domain $(-\infty, \infty)$ — Range $\left[-\dfrac{9}{2}, \infty\right)$

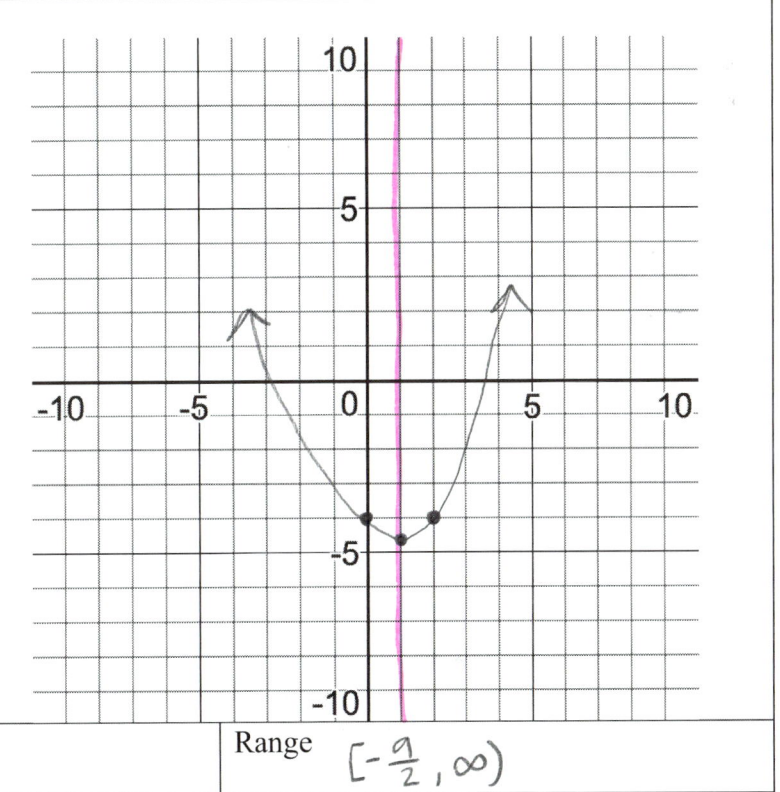

Homework 5.5 Graphing Standard Form of Quadratics, Quadratic Inequalities

5. Graph. $y > \frac{1}{2}(x-1)(x+3)$ intercept, dotted

$y = a(x-p)(x-q)$

UP

x-int →
   (1,0) (-3,0)

AOS = -1

Vertex = (-1,-2)

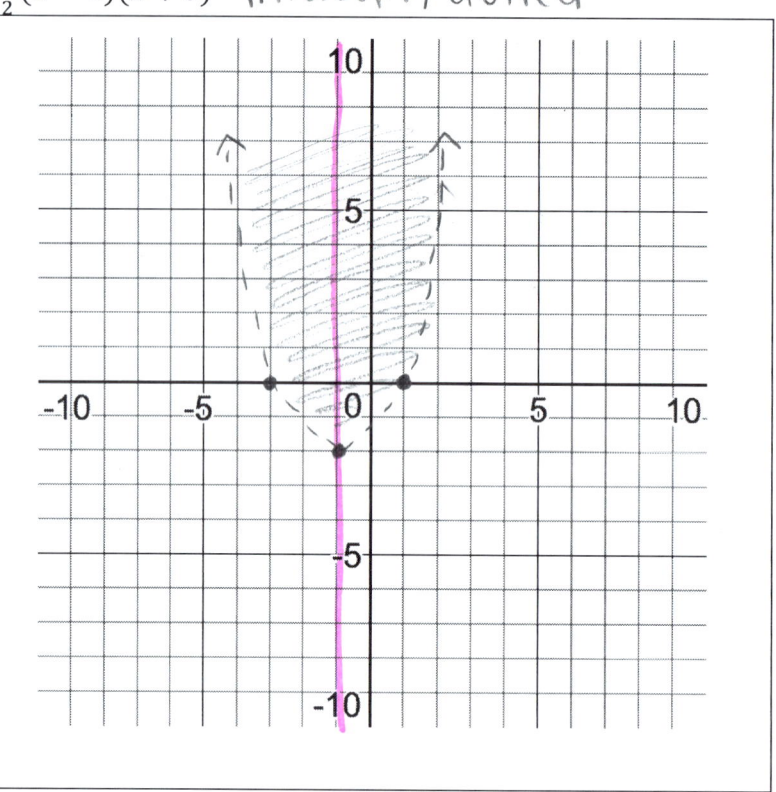

6. $y \leq -(x+2)^2 + 7$ vertex, solid

$y = a(x-h)^2 + k$

down

vertex (-2, 7)

AOS = x = -2

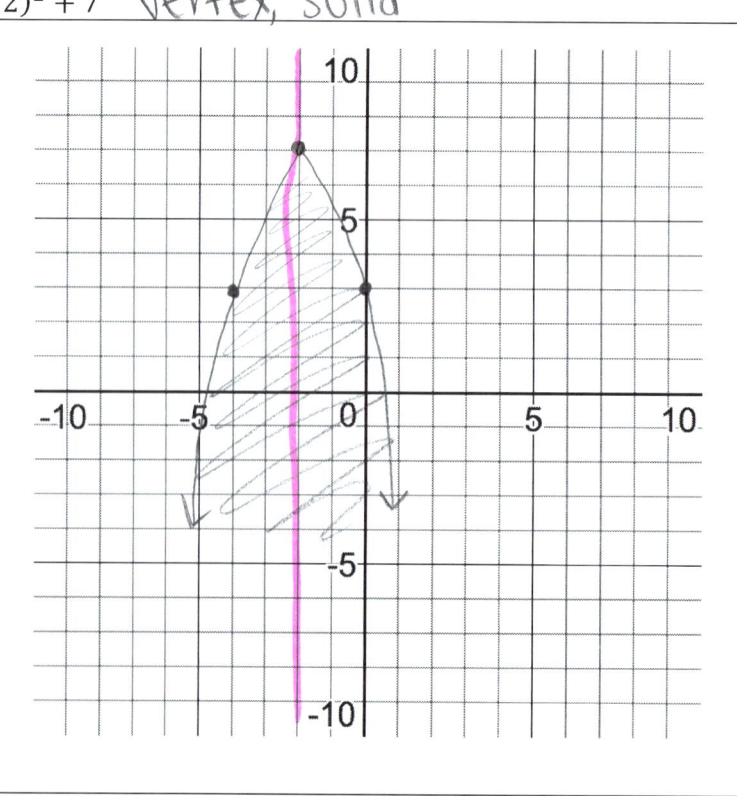

Homework 5.5 Graphing Standard Form of Quadratics, Quadratic Inequalities

7. Graph. $y > -\frac{1}{2}x^2 - x + \frac{3}{2}$ Standard, dotted

$a = -\frac{1}{2}$
$b = -1$
$c = \frac{3}{2}$

down

$AOS = \frac{-(-1)}{2(-\frac{1}{2})}$
$= -1$

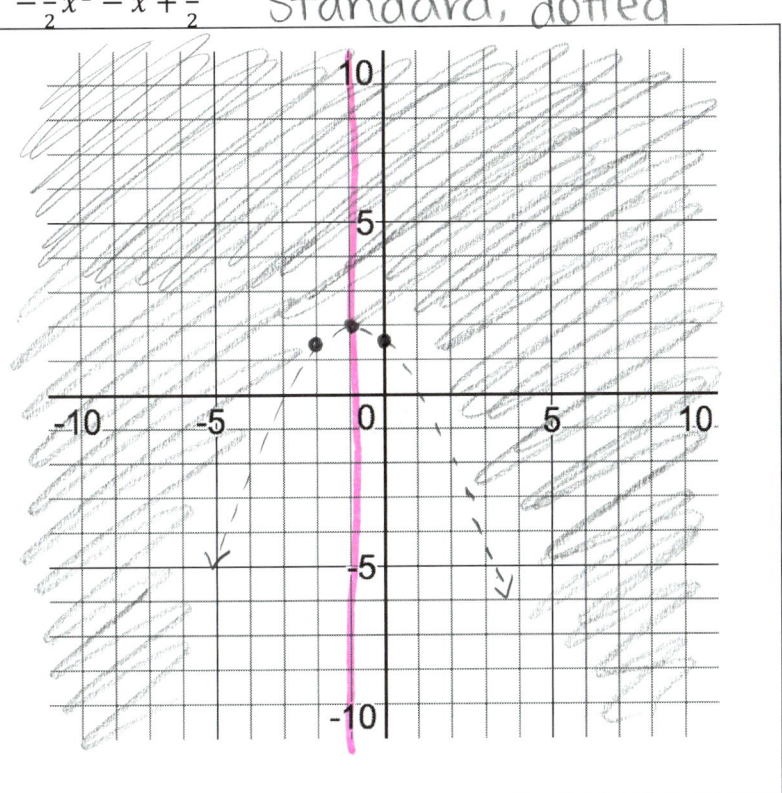

y-int $= (0, \frac{3}{2})$

vertex =

Homework 5.5 Graphing Standard Form of Quadratics, Quadratic Inequalities

CHALLENGE PROBLEMS

1. The function $y = (x - 3)(x + 5)$ is a parabola in the xy-plane. In which of the following equivalent equations do the $x-$ and $y-$coordinates of the vertex of the parabola appear as constants or coefficients?

   a.) $y = x^2 + 2x - 15$
   b.) $y = (x - 3)^2 + 5$
   c.) $y = (x + 1)^2 - 16$
   d.) $y = (x - 1)^2 - 12$
   e.) $y = (x - 1)^2 - 15$

2. The function $y = x^2 + 2x - 8$ is a parabola in the xy-plane. In which of the following equivalent equations do the $x-$ and $y-$ coordinates of the $x-$intercepts appear as constants?

   a.) $y = x^2 - 2x + 8$
   b.) $y = (x + 2)(x - 4)$
   c.) $y = (x - 2)(x + 4)$
   d.) $y = (x - 2)(x - 4)$
   e.) $y = (x - 2)^2 - 9$

3. Graph the requested functions.

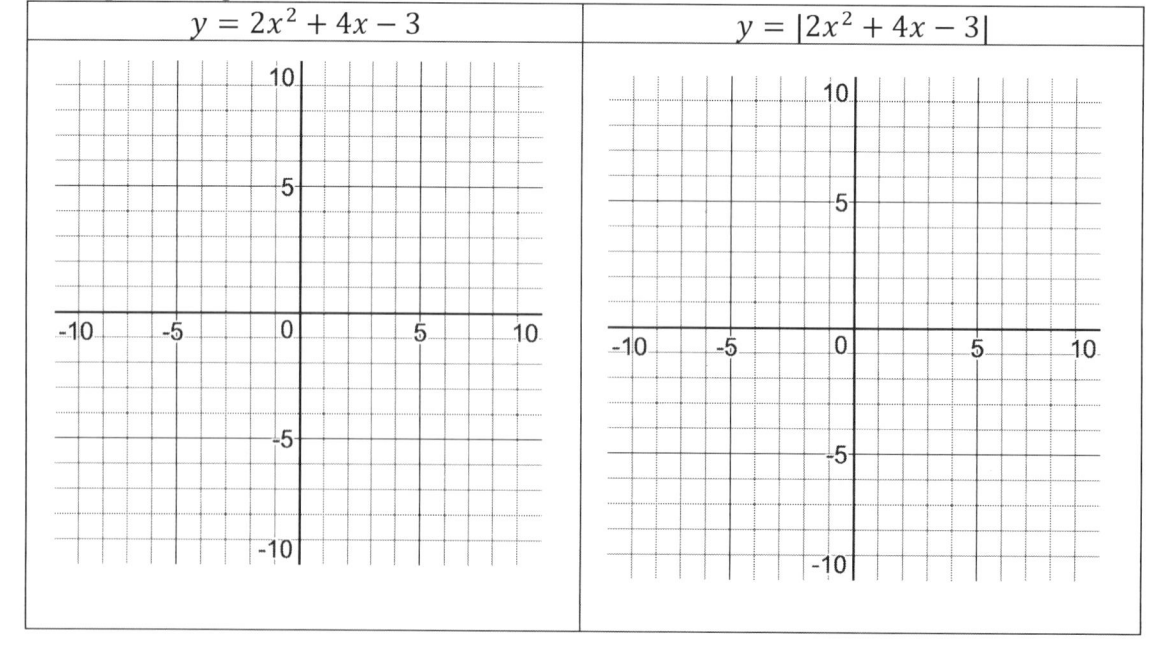

$y = 2x^2 + 4x - 3$	$y = |2x^2 + 4x - 3|$

5.6—235

5.6. Review

Assessment Checklist. Below are the competencies one should master in preparation for an assessment on complex numbers, and quadratic functions.

- [ ] Know the definition of i.
- [ ] Understand the real and imaginary parts of complex numbers.
- [ ] Addition, subtraction and multiplication of complex numbers. **a+bi**
- [ ] Evaluate powers of the complex number, i.
- [ ] Definition of, and how to find, a complex number's complex conjugate
- [ ] Division of complex numbers $= \dfrac{a-bi}{a-bi}$ **complex conjugate**
- [ ] Vertex Form of a quadratic equality and how to graph from the form
- [ ] Intercept Form of a quadratic equality and how to graph from that form
- [ ] Standard Form of a quadratic equality and how to graph from that form
- [ ] From any form of a quadratic, find the vertex, axis of symmetry, y-intercept and direction of opening.
- [ ] Finding the x-intercepts of a parabola from Intercept Form.
- [ ] From any quadratic form, graph quadratic inequalities.

Vertex form = $y = a(x-h)^2 + k$
Intercept form = $y = a(x-q)(x-p)$
Standard form = $y = ax^2 + bx + c$

Inequalities:
$>, <$ = dotted
$\geq, \leq$ = solid
use point 0,0 to find shaded

definition of "i" = $\sqrt{-1}$
$i^2 = -1$ $i^3 = -i$ $i^4 = 1$
$i^6 = -1$ $i^7 = -i$ $i^8 = 1$
$i^{10} = -1$

to find large exponents → ÷4 then × i^{remain}

complex conjugate = $a - bi$

Vertex form =
Vertex = (h, k)
AOS = $x = h$

Intercept form =
x-intercepts = $(p, 0)(q, 0)$
y-intercept = $x = 0$
AOS = $\dfrac{p+q}{2}$
Vertex = $x = $ AOS

Standard form =
AOS = $\dfrac{-b}{2a}$
Vertex = $x = $ AOS
y-int = $y = c$, $x = 0$

Review 5.6

1. Evaluate the following expressions with complex numbers. Your final answer should be a number in the form $a + bi$.

---

a) $(2 - 4i) - \left(1 + \frac{1}{2}i\right)$

$1 - 3\frac{1}{2}i$

---

b) $(4 - 3i) \cdot (i + 7)$

$4i - 21i - 3i^2 + 28$

$\boxed{4i - 21i} \boxed{-3(-1) + 28}$

$-17i + 31$

---

c) $i \cdot (1 - i) \cdot i$

$i^2(1-i) \rightarrow -1(1-i)$

$-1 + i$

---

d) $\sqrt{i^{702}}$

$\begin{array}{r} 175 \\ 4\overline{)702} \\ 4\downarrow \\ \overline{30} \\ 28 \checkmark \\ \overline{22} \\ 20 \\ \overline{(2)} \end{array}$

$\sqrt{(i^4)^{175} \cdot i^2}$

$\sqrt{i^2} \rightarrow \sqrt{-1} = i$

---

e) $(1 + 2i) - (2i -^+ 1)$

$\boxed{2}$

---

f) i^{76}

$\begin{array}{r} 19 \\ 4\overline{)76} \\ 4 \\ \overline{36} \\ 36 \\ \overline{0} \end{array}$

$(i^4)^{19} = 1$

i^4 is always ★ 1 ★

---

g) $\frac{3+i}{2-i} \left(\frac{2+i}{2+i}\right) = \frac{6 + 3i + 2i + i^2}{4 - 2i + 2i + i^2}$

$\frac{6 + 5i + \boxed{i^2}^{-1}}{4\boxed{-i^2}^{-1}} \quad \frac{5 + 5i}{5}$

$\boxed{1 + i}$

---

h) $\frac{2}{1+i}\left(\frac{1-i}{1-i}\right) = \frac{2 - 2i}{1\boxed{-i^2}\boxed{1}} = \frac{2 - 2i}{2}$

$\boxed{1 - i}$

Review 5.6 vertex form = $y = a(x-h)^2 + k$

2. Give the requested information and graph the quadratic. $y = -\frac{1}{2}(x-3)^2 + 4$

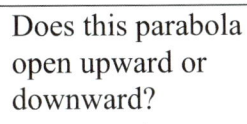

Does this parabola open upward or downward? **down**
Axis of Symmetry AoS $x = 3$
Vertex $(3, 4)$
y –intercept $x = 0$ $-\frac{1}{2}(0-3)^2 + 4$ $-\frac{1}{2}(9) + 4$ $-4.5 + 4 = -\frac{1}{2}$

Domain	Range
$(-\infty, \infty)$	$(-\infty, 4]$

vertex form = $y = a(x-h)^2 + k$

3. Using the information above, graph $y < -\frac{1}{2}(x-3)^2 + 4$.

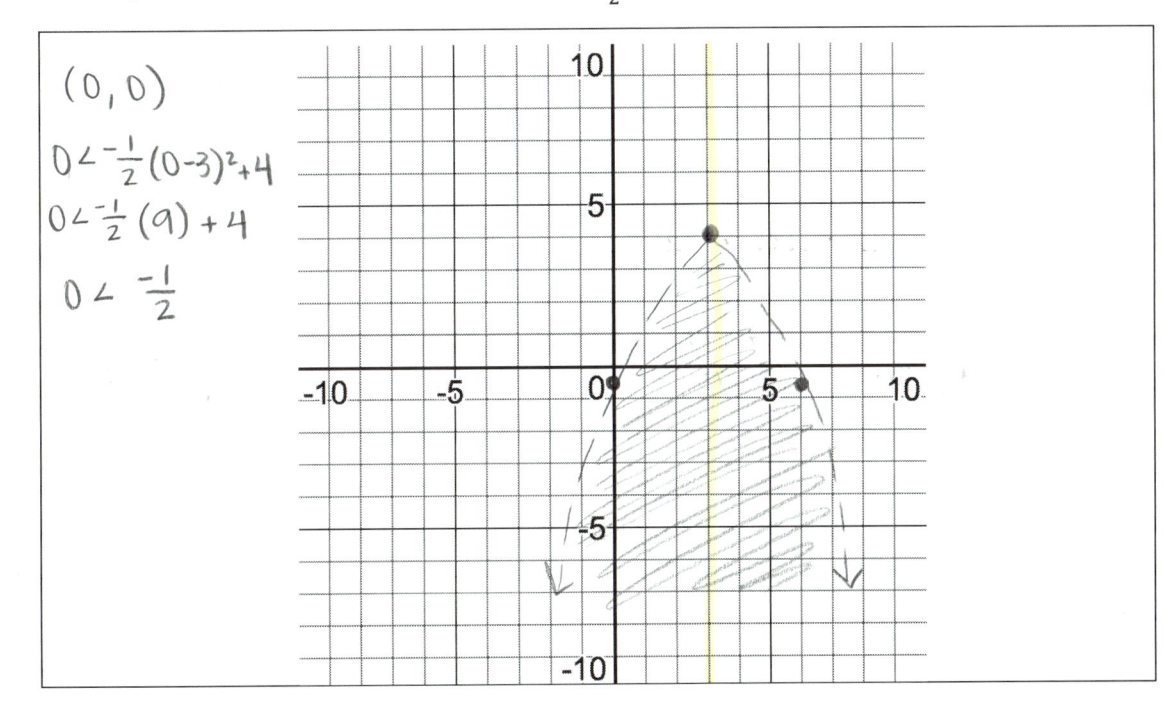

$(0, 0)$
$0 < -\frac{1}{2}(0-3)^2 + 4$
$0 < -\frac{1}{2}(9) + 4$
$0 < -\frac{1}{2}$

Review 5.6   Intercept form = $y = a(x-p)(x-q)$

4. Give the requested information and graph the quadratic. $y = \frac{1}{2}(x-2)(x+6)$

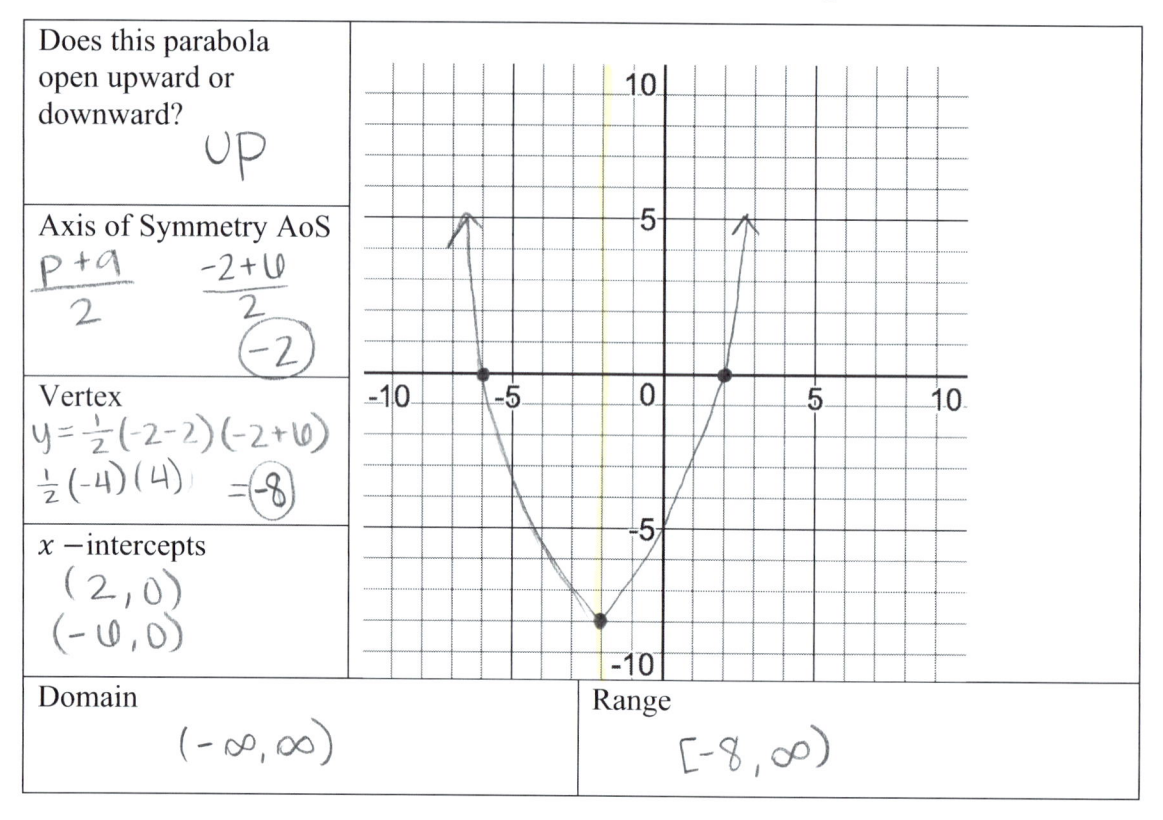

Does this parabola open upward or downward? **UP**	
Axis of Symmetry AoS $\frac{p+q}{2} \quad \frac{-2+6}{2}$ $\boxed{-2}$	
Vertex $y = \frac{1}{2}(-2-2)(-2+6)$ $\frac{1}{2}(-4)(4) = \boxed{-8}$	
x–intercepts $(2, 0)$ $(-6, 0)$	
Domain $(-\infty, \infty)$	Range $[-8, \infty)$

5. Using the information above, graph $y > \frac{1}{2}(x-2)(x+6)$

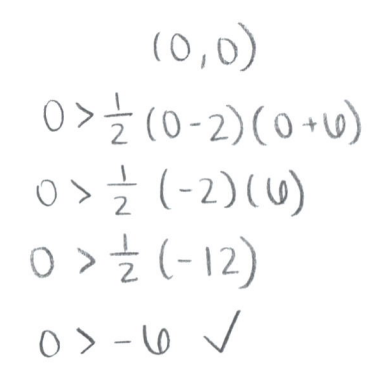

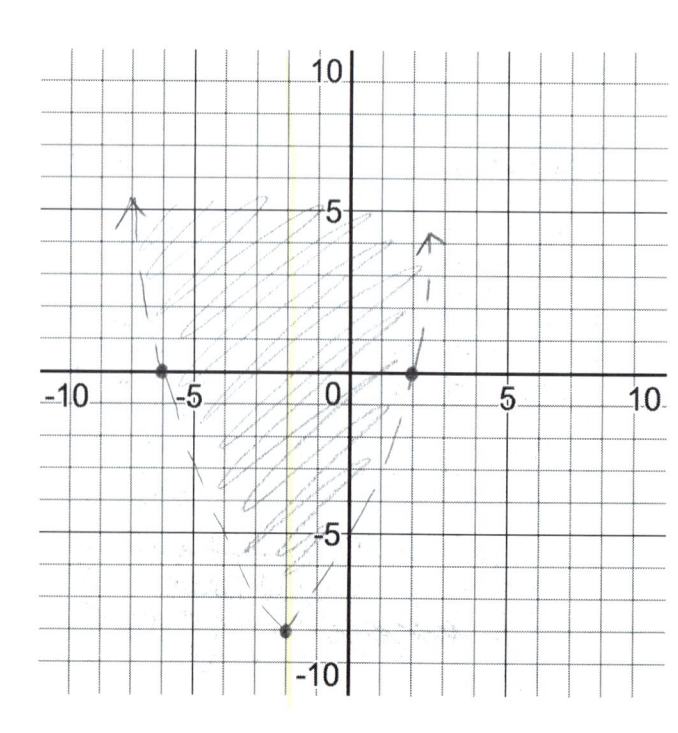

$(0, 0)$
$0 > \frac{1}{2}(0-2)(0+6)$
$0 > \frac{1}{2}(-2)(6)$
$0 > \frac{1}{2}(-12)$
$0 > -6$ ✓

5.6—239

$a = 1$
$b = -4$
$c = -1$

Review 5.6 Standard form = $y = ax^2 + bx + c$

6. Give the requested information below and graph $y = x^2 - 4x - 1$.

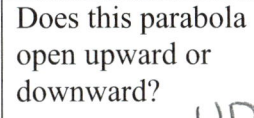

Does this parabola open upward or downward? **UP**	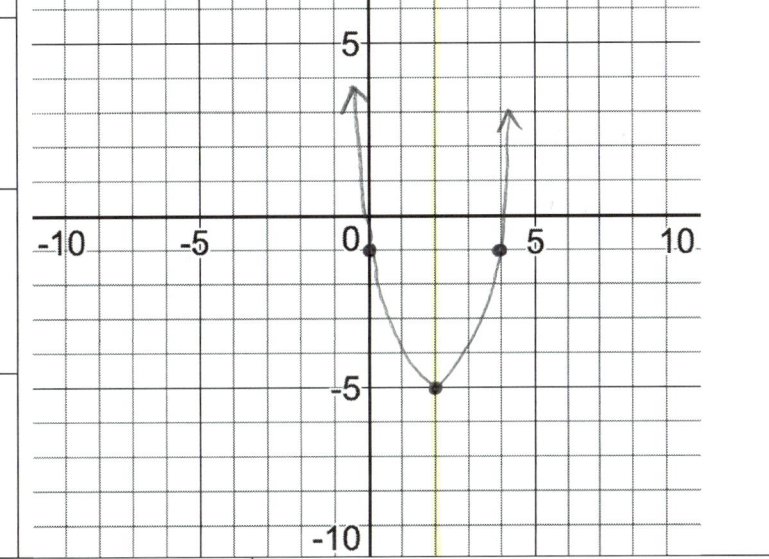
Axis of Symmetry AoS $\dfrac{-b}{2a}$ $\dfrac{4}{2(1)} = \boxed{2}$	
Vertex $2^2 - 4(2) - 1$ $4 - 8 - 1$ $\boxed{-5}$	
y-intercept $x = 0$ $y = -1$	

Domain	Range
$(-\infty, \infty)$	$[-5, \infty)$

7. Graph $y \leq x^2 - 4x - 1$.

$(0, 0)$
$0 \leq 0^2 - 4(0) - 1$
$0 \leq -1$ NO

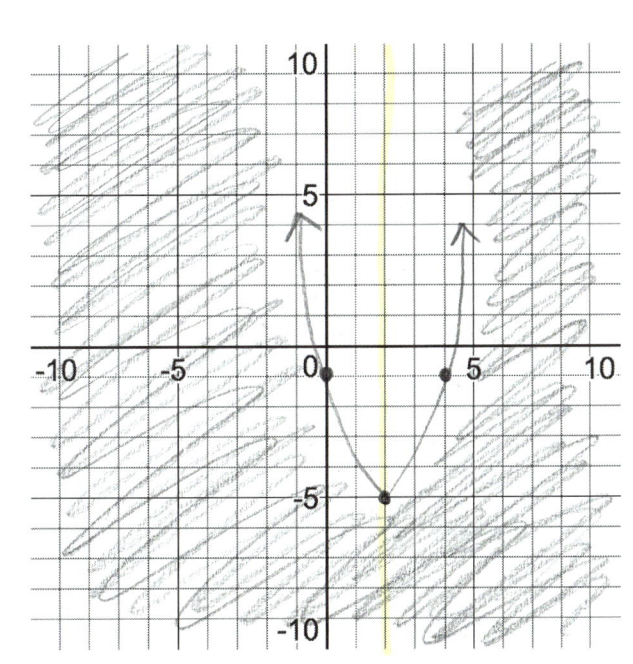

6. QUADRATIC EQUATIONS

We've spent time graphing quadratic functions. By this point you should be able to confidently graph quadratic equalities and quadratic inequalities in various forms

Vertex	(h, k) $x = h$	$y = a(x - h)^2 + k$
Intercept	$(p, 0)$ $\frac{p+q}{2}$ $(q, 0)$	$y = a(x - p)(x - q)$
Standard	$(0, C)$ $\frac{-b}{2a}$	$y = ax^2 + bx + c$

In the equations above, you have two variables, x and y. In each case there is an x^2 term, an x term, and a constant.

The solution in each case has been to draw a graphical representation of the relationship between x and y.

In this unit we will look at methods for solving quadratics for specific values of y.

Consider each form of the quadratic equation and how we might solve for x if $y = 0$.

Vertex	Intercept	Standard
$0 = a(x - h)^2 + k$	$0 = a(x - p)(x - q)$	$0 = ax^2 + bx + c$

The easiest of these to solve is Intercept Form.

$$0 = a(x - p)(x - q)$$

Then

$$x - p = 0 \text{ or } x - q = 0$$
$$x = p \text{ or } x = q$$

The solution is the set of points: $\{(p, 0), (q, 0)\}$

By setting $y = 0$, and solving for x, we find the x-intercepts. In this unit we will look at how to solve for x in quadratic equations.

6.1. Solving Quadratic Equations by Square Roots

Consider how we might solve the quadratic equation from Vertex Form.
$$y = a(x - h)^2 + k$$

To solve for $y = 0$
$$0 = a(x - h)^2 + k$$

We could isolate the $(x - h)^2$ term
$$(x - h)^2 = -\frac{k}{a}$$

==We could then solve this by taking the square root of both sides.== Before we do that, we need to review a bit about the properties of square roots, and variables.

Product Property of Radicals
$$\sqrt{a} \cdot \sqrt{b} = \sqrt{ab}$$

$\sqrt{2} \cdot \sqrt{5} = \sqrt{10}$

$\sqrt{32} = \sqrt{2 \cdot 16}$

Quotient Property of Radicals
$$\frac{\sqrt{a}}{\sqrt{b}} = \sqrt{\frac{a}{b}}$$

We can use these properties to simplify expressions with radicals.

Simplify $\sqrt{50} = \sqrt{2 \cdot 25} = 5\sqrt{2}$

$\sqrt{2} \cdot \sqrt{25}$

Simplify $\sqrt{27} = \sqrt{3 \cdot 9} = 3\sqrt{3}$

$\sqrt{3} \cdot \sqrt{9}$

6.1—242

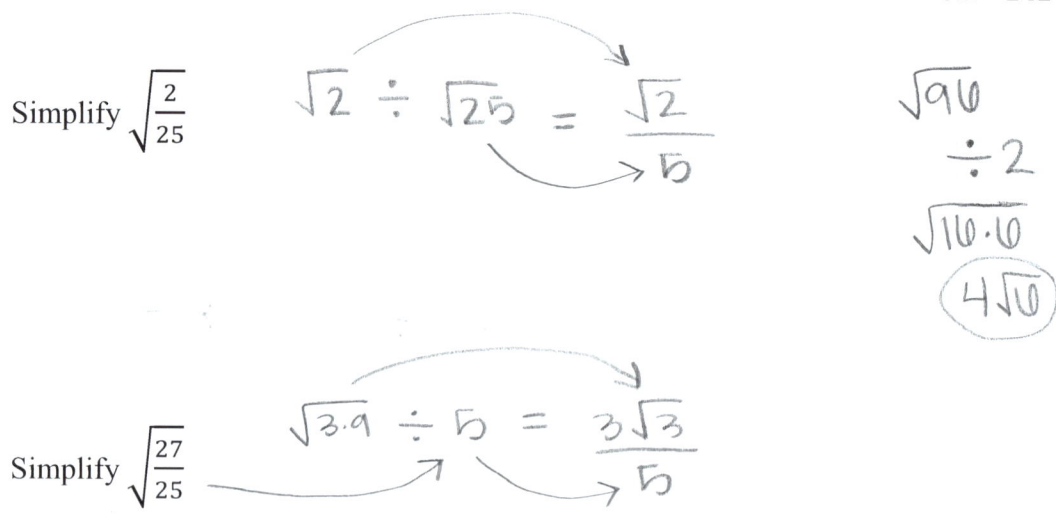

Simplify $\sqrt{\dfrac{2}{25}}$ $\sqrt{2} \div \sqrt{25} = \dfrac{\sqrt{2}}{5}$

$\sqrt{96}$
$\div 2$
$\sqrt{16 \cdot 6}$
$\boxed{4\sqrt{6}}$

Simplify $\sqrt{\dfrac{27}{25}}$ $\sqrt{3 \cdot 9} \div 5 = \dfrac{3\sqrt{3}}{5}$

Another concept in simplifying expressions with radicals is rationalizing the denominator.

To **rationalize the denominator** is to write an expression in a form such that there are no radicals in the denominator. simplify

Simplify the following expressions, leaving no radicals in the denominator.

$\dfrac{2}{\sqrt{2}} \quad \dfrac{2}{\sqrt{2}}\left(\dfrac{\sqrt{2}}{\sqrt{2}}\right) = \dfrac{2\sqrt{2}}{(\sqrt{2})^2} = \dfrac{2\sqrt{2}}{2} = \boxed{\sqrt{2}}$

$\dfrac{1}{\sqrt{7}}\left(\dfrac{\sqrt{7}}{\sqrt{7}}\right) = \dfrac{1\sqrt{7}}{(\sqrt{7})^2} = \boxed{\dfrac{\sqrt{7}}{7}}$

$\dfrac{1}{1+\sqrt{2}}\left(\dfrac{1-\sqrt{2}}{1-\sqrt{2}}\right) = \dfrac{1-\sqrt{2}}{1-2} = \boxed{-1+\sqrt{2}}$

$\dfrac{3}{3-\sqrt{3}}\left(\dfrac{3+\sqrt{3}}{3+\sqrt{3}}\right) = \dfrac{9+3\sqrt{3}}{9-3} = \dfrac{9+3\sqrt{3}}{6} \quad \boxed{\dfrac{3}{2}+\dfrac{1}{2}\sqrt{3}}$

Now, consider the equation below
$$x^2 = 4$$

What are all the numbers that satisfy that equation?

What about $x^2 = 9$? What are the solutions? 3, -3 $X^2 = 25$ $X = 5, -5$

In both cases we see that there are positive and negative numbers that make the equation a true statement.

For that reason, you must be cautious whenever you evaluate expressions such as $\sqrt{x^2}$ to account for both positive and negative solutions.

The correct evaluation of $\sqrt{x^2}$ is below: ☆ absolute value ☆
$$\sqrt{x^2} = |x|$$

So, to solve
$$x^2 = 4$$

Take the square root of both sides
$$\sqrt{x^2} = \sqrt{4}$$
$$|x| = 2$$
$$x = \pm 2$$

It is common to simply skip the step with the absolute value, and just know that taking the square root of an exponent squared will require a $\pm$ in your answer.

$$x^2 = 9$$
$$\sqrt{x^2} = \pm\sqrt{9}$$
$$x = \pm 3$$

Solve $2x^2 - 15 = 65$
$ +15 \; +15$

$$\frac{2x^2 = 80}{2}$$

$\sqrt{x^2} = \pm\sqrt{40} \rightarrow x = \pm\sqrt{40}$
$\boxed{x = \pm 2\sqrt{10}}$

Solve $\frac{1}{4}(x-3)^2 = 16 \; \left(\frac{4}{1}\right)$

$(x-3)^2 = 64$
$x - 3 = \pm\sqrt{64}$
$x - 3 = \pm 8$
$x = 3 \pm 8 \quad \boxed{x = 11, -5}$

Solve $2x^2 + 72 = -18$
$ -72 \; -72$

$$\frac{2x^2 = -90}{2}$$

$\sqrt{x^2} = \pm\sqrt{-45}$
$x = \pm\sqrt{-45} = \boxed{\pm 3i\sqrt{5}}$

Solve $3x^2 - 7 = -31$
$ +7 \; +7$

$$\frac{3x^2 = -24}{3}$$

$\sqrt{x^2} = \sqrt{-8} \quad \boxed{x = \pm i \, 2\sqrt{2}}$

Consider the quadratic equation $y = 2(x-1)^2 - 6$.

Does this parabola open upward or downward?	UP
Axis of Symmetry AoS	$x = 1$
Vertex	$(1, -6)$
y-intercept	$x = 0$ $2(0-1)^2 - 6$ $y = -4$ $(0, -4)$

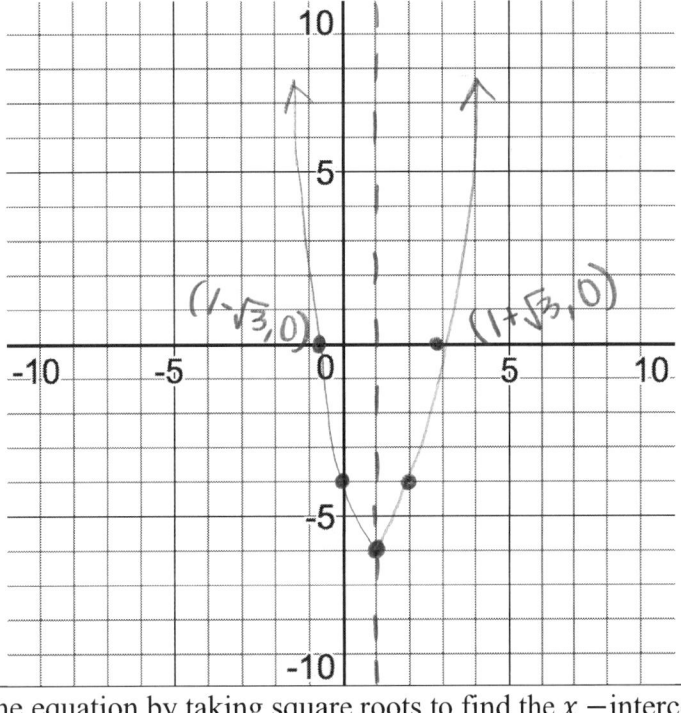

Set $y = 0$ and solve the equation by taking square roots to find the x-intercepts.

$$0 = 2(x-1)^2 - 6$$
$$+6 \qquad\qquad +6$$
$$\frac{2(x-1)^2}{2} = \frac{6}{2}$$
$$\sqrt{x-1^2} = \sqrt{3}$$
$$x - 1 = \pm\sqrt{3}$$
$$\boxed{x = 1 \pm \sqrt{3}}$$

$x = 1 + \sqrt{3} \approx 2.7$

$x = 1 - \sqrt{3} \approx -0.7$

Homework 6.1 Solving Quadratic Equations by Square Roots

Simplify the following expressions with radicals. Leave no radicals in the denominator.

1. $\sqrt{-45}$

   $x = \pm i\sqrt{45} \to \sqrt{9} \cdot \sqrt{5}$

   $\boxed{x = i3\sqrt{5}}$

2. $\sqrt{24} \to \sqrt{6} \cdot \sqrt{4}$

   $\boxed{x = 2\sqrt{6}}$

3. $\sqrt{12} \to \sqrt{4} \cdot \sqrt{3}$

   $\boxed{x = 2\sqrt{3}}$

4. $\sqrt{\dfrac{4}{9}}$ $\sqrt{4} \div \sqrt{9}$

   $2 \div 3 = \boxed{\dfrac{2}{3}}$

5. $\sqrt{\dfrac{2}{49}}$ $\sqrt{2} \div \sqrt{49}$

   $\boxed{\dfrac{\sqrt{2}}{7}}$

6. $\dfrac{1}{\sqrt{7}}\left(\dfrac{\sqrt{7}}{\sqrt{7}}\right) = \dfrac{\sqrt{7}}{(\sqrt{7})^2} = \boxed{\dfrac{\sqrt{7}}{7}}$

7. $\dfrac{1}{1-\sqrt{5}}\left(\dfrac{1+\sqrt{5}}{1+\sqrt{5}}\right) = \dfrac{1+\sqrt{5}}{1-(\sqrt{5})^2}$

   $\dfrac{1+\sqrt{5}}{1-5}$ $\boxed{\dfrac{1+\sqrt{5}}{-4}}$

8. $\dfrac{2}{2+\sqrt{11}}\left(\dfrac{2-\sqrt{11}}{2-\sqrt{11}}\right) = \dfrac{4-2\sqrt{11}}{4-11}$

   $\boxed{\dfrac{-4}{7} + \dfrac{2}{7}\sqrt{11}}$

Homework 6.1 Solving Quadratic Equations by Square Roots

Solve each equation by taking square roots. Simplify all radicals.

9. $x^2 = -68$	10. $4x^2 + 1 = 65$
$\sqrt{x^2} = \sqrt{-68} \to \sqrt{4} \cdot \sqrt{17}$	$\quad -1 \quad -1$
$\boxed{x = \pm i\, 2\sqrt{17}}$	$\dfrac{4x^2}{4} = \dfrac{64}{4}$
	$\sqrt{x^2} = \sqrt{16}$
	$\boxed{x = \pm 4}$
11. $2x^2 + 10 = 154$	12. $6x^2 + 9 = -69$
$\quad -10 \quad -10$	$\quad -9 \quad -9$
$2x^2 = 144$	$6x^2 = -78$
$x^2 = 72 \to \sqrt{9} \cdot \sqrt{8}$	$\sqrt{x^2} = \sqrt{-13} \to$ cannot be simplified
$\boxed{x = \pm\, 6\sqrt{2}}$	$x = \pm i\sqrt{13}$
13. $(x-4)^2 + 1 = 0$	14. $9v^2 + 10 = -22$
$\quad -1 \; -1$	$\quad -10 \quad -10$
$\sqrt{(x-4)^2} = \sqrt{-1}$	$\dfrac{9v^2 = -32}{9}$ $\boxed{v = \pm \dfrac{4}{3} i\sqrt{2}}$
$x - 4 = \sqrt{-1} = i$	$\sqrt{v^2} = \sqrt{\dfrac{-32}{9}}$
$\boxed{x = 4 \pm i}$	
15. $\frac{1}{2}(x+1)^2 - 3 = 0$	16. $-10x^2 = 90$
$\quad +3 \; +3$	$\quad -10$
$\frac{1}{2}(x+1)^2 = 3\,(2)$	$\sqrt{x^2} = \sqrt{-9}$
$\sqrt{(x+1)^2} = \sqrt{6}$	$\boxed{x = \pm i\, 3}$
$x + 1 = \sqrt{6}$	
$\boxed{x = \pm 1\, i\sqrt{6}}$	

Homework 6.1 Solving Quadratic Equations by Square Roots

CHALLENGE PROBLEMS
1. Rationalize the expression
$$\frac{1}{(1+\sqrt{2})(1-\sqrt{3})}$$

2. Solve by using square roots, $\frac{1}{4}(2x-1)^2 + \frac{9}{16} = 0$

6.2. Solve by factoring

In the last lesson we discussed solving quadratic equations by taking the square root of both sides of an equation. This method works well <mark>when one side of the equation can be expressed as a perfect square:</mark>

$$(x - 1)^2 = 8$$
$$x^2 = -25$$

$\sqrt{x^2} = \sqrt{-25}$
$x = \pm \sqrt{51}$

You will often encounter quadratic equations where we do not have perfect squares. Suppose you were asked to solve

$$x^2 + 3x = 0$$

How do we proceed?

$\sqrt{x^2} = \sqrt{-3x}$ ✗ DID NOT WORK!
$x = \sqrt{-3x}$

$x(x+3) = 0$
$x = 0$ OR $x+3 = 0$
$\boxed{x = 0, -3}$ use factoring!

In this section we will cover a few examples of using factoring to solve quadratic equations. There are four factoring techniques we need to make sure we can perform:

- Greatest Common Factor
- Grouping
- Factoring Trinomials, $x^2 + bx + c$
- Factoring Trinomials, $ax^2 + bx + c$

Factoring the Greatest Common Factor

This is a factoring technique that can often simplify an equation quickly. Find the greatest common factor, common across all terms of the equation to write the equation in factored form.

$3x + 300$ $3(x+100)$

$-7v - 77v^2$ $-7v(1+11v)$ factoring $\boxed{\text{solve} = \text{set} = 0}$

$4x^2 - 2x$ $2x(2x-1)$

$14y^2 - 21y$ $7y(2y-3)$

Factor by grouping
In cases where there is not a common factor across all terms, but some terms share common factors, factoring by grouping is an approach to try.

Factor the following expression by grouping.

$xz + yz + 5x + 5y$

$z(x+y) + 5(x+y)$

$= (x+y)(z+5)$

$5x^3 + 4x^2 + 10x + 8$

$x^2(5x+4) + 2(5x+4)$

$= (5x+4)(x^2+2)$

$4x^3 + x^2 + 20x + 5$

$x^2(4x+1) + 5(4x+1)$

$= (4x+1)(x^2+5)$

Factoring trinomials, $x^2 + bx + c$. $= (x+p)(x+q)$
For trinomials where the coefficient of the x^2 term is one, we can factor using the following steps.
Find two numbers that multiply to give c, and combine through addition/subtraction to give the value of b. If p and q are those numbers, then the factorization is

$(x+p)(x+q) = x^2 + px + qx + pq$

$c = pq \quad b = p+q$

Use factoring to solve: $x^2 - x - 42 = 0$.

$(x+6)(x-7) = 0$

$x = -6, 7$

2 #s
mult: $-42 \leftarrow c$
add: $-1 \leftarrow b$

$6, -7$

Use factoring to solve $2x^2 + 6x - 56 = 0 \ = (x+p)(x+q)$

$2(x^2 + 3x - 28) = 0$

$2 \#$
mult: -28
add: 3

$-4, 7$

$2(x-4)(x+7) = 0$ 2 IS NEVER $= 0$

$\boxed{x = 4, -7}$

Factor and solve, $x^2 + 5x - 24 = 0$

$2 \#$
$\times : -24$
$+ : 5$

$-3, 8$

$(x-3)(x+8) = 0$

$\boxed{x = 3, -8}$

Factoring trinomials, $ax^2 + bx + c$.

In cases where the leading coefficient of the x^2 term is not 1, find two numbers that multiply to give ac and add to give b. Use these numbers to rewrite the original equation so that it can be factored by grouping.

multiply → ac
add → b

Solve by factoring: $6x^2 + 11x + 4 = 0$

$a = 6$
$b = 11$
$c = 4$

$2 \# S$
$\times : 24$
$+ : 11$

$8, 3$

$\underline{6x^2 + 8x} + \underline{3x + 4} = 0$

$2x(3x+4) + (3x+4) = 0$

$(2x+1)(3x+4) = 0$

$2x+1 = 0 \qquad 3x+4 = 0$

$\boxed{x = -\frac{1}{2}} \qquad \boxed{x = -\frac{4}{3}}$

Solve by factoring: $4x^2 - 17x - 15 = 0$

2#
$\times : -60$
$+ : -17$
$\boxed{3, -20}$

$4x^2 + 3x - 20x - 15 = 0$
$x(4x+3) - 5(4x+3) = 0$
$(x-5)(4x+3) = 0$
$x - 5 = 0 \qquad 4x + 3 = 0$
$\boxed{x = 5} \qquad \boxed{x = -\dfrac{3}{4}}$

Solve by factoring: $6x^2 + 26x = -8 \rightarrow 6x^2 + 26x + 8 = 0$

2#
$\times : 12$
$+ : 13$
$\boxed{1, 12}$

$2(3x^2 + 13x + 4) = 0$
$2(3x^2 + x + 12x + 4) = 0$
$2(x(3x+1) + 4(3x+1)) = 0$
$2(x+4)(3x+1) = 0$
$x + 4 = 0 \qquad 3x + 1 = 0$
$\boxed{x = -4} \qquad \boxed{x = \dfrac{-1}{3}}$

Solve by factoring: $3x^2 - 6x - 9 = 0$

2#
$\times : -3$
$+ : -2$
$\boxed{-3, 1}$

$3(x^2 - 2x - 3) = 0$
$3(x-3)(x+1) = 0$
$\boxed{x = 3, -1}$

Homework 6.2 Solve by factoring

Solve the following equations by factoring.

1. $n^2 - 8n = 0$	2. $x^2 - 10 = -3x$
$n(n-8) = 0$ $n = 0 \quad n = 8$	$x^2 + 3x - 10 = 0$ 2# $x: -10$ $+: 3$ $\boxed{-2, 5}$ $(x-2)(x+5) = 0$ $\boxed{x = 2, -5}$
3. $3x^2 = -24 - 27x$ $-3x^2 - 27x - 24 = 0$ $-3(x^2 + 9x + 8) = 0$ 2#S $x: 8$ $+: 9$ $\boxed{1, 8}$ $-3(x+1)(x+8) = 0$ $\boxed{x = -1, -8}$	4. $k^2 = -14k - 48$ $k^2 + 14k + 48 = 0$ $(k+6)(k+8) = 0$ $\boxed{k = -6, -8}$ 2# $x: 48$ $+: 14$ $\boxed{6, 8}$
5. $3n^2 + 11n = 4$ $3n^2 + 11n - 4 = 0$ $3n^2 - n + 12n - 4 = 0$ $n(3n-1) + 4(3n-1) = 0$ $(3n-1)(n+4) = 0$ $\boxed{\frac{1}{3}, -4}$ 2# $-12: x$ $11: x$ $\boxed{-1, 12}$	6. $4r^2 - 17r - 49 = -7$ $4r^2 - 17r - 42 = 0$ $4r^2 - 24r + 7r - 42 = 0$ $4r(r-6) + 7(r-6) = 0$ $(4r+7)(r-6) = 0$ $\boxed{r = 6, -\frac{7}{4}}$ 2# $m: 168$ $a: 10$ $\boxed{-24, 7}$

Homework 6.2 Solve by factoring

Solve the following equations by factoring.

---

7. $4x^2 - 2x - 14 = -8$

$4x^2 - 2x - 6 = 0$

$4x^2 - 6x + 4x - 6 = 0$

$2x(2x-3) + 2(2x-3) = 0$

$(2x+2)(2x-3) = 0$

$2x+2=0 \quad 2x-3=0$
$2x=-2 \quad 2x=3$
$\boxed{x=-1} \quad \boxed{x=\frac{3}{2}}$

2#
m: -24
a: -2
$\boxed{-6, 4}$

---

8. $6x^2 + 7x = -2$

$6x^2 + 7x + 2 = 0$

$6x^2 + 3x + 4x + 2 = 0$

$3x(2x+1) + 2(2x+1)$

$(3x+2) + (2x+1) = 0$

$3x+2=0 \quad 2x+1=0$
$\boxed{x=\frac{-2}{3}} \quad \boxed{-\frac{1}{2}}$

2#
m: 12
a: 7
$\boxed{3, 4}$

---

9. $-16n = -5n^2 + 2n - 9$

$5n^2 - 18n - 9 = 0$

$5n^2 - 3n - 15n - 9 = 0$

$n(5n-3) - 3(5n-3) = 0$

$(n-3)(5n-3) = 0$

$\boxed{n = 3, \frac{3}{5}}$

#
m: 45
a: -18
$\boxed{-3, -15}$

---

10. $16n^2 + 16 = n^2 + 46n$

$16n^2 - n^2 - 46n + 16 = 0$

$15n^2 - 46n + 16 = 0$

$15n^2 - 6n - 40n + 16 = 0$

$3n(5n-2) - 8(5n-2) = 0$

$(3n-8)(5n-2) = 0$

$\boxed{x = \frac{8}{3}, \frac{2}{5}}$

2#
m: 240
a: -46
$\boxed{-6, -40}$

Homework 6.2 Solve by factoring

CHALLENGE PROBLEMS

1. If a and b are constants, what is the value of a?
$$x^2 + ax + 6 = (x+2)(x+b)$$

2. For constants a, b, and c, all non-zero real numbers, find all possible value of a, b, and c such that
$$(ax+b)(x-c) = 2x^2 - 7x - 15$$

6.3. Solve by Completing the Square

We discovered in section 6.1 that it is straightforward to solve by taking the square root of both sides of a quadratic equation when the term with the variable is a perfect square.

$x^2 = 64$	$(x-1)^2 = 64$	$\dfrac{2(x+4)^2}{2} = \dfrac{128}{2}$
$\sqrt{x^2} = \pm\sqrt{64}$	$x - 1 = \pm\sqrt{64}$	$(x+4)^2 = 64$
$x = \pm 8$	$x = 1 \pm 8$	$x + 4 = \pm\sqrt{64}$
	$x = 9, -7$	$x = 4 \pm 8$
		$x = 12, -4$

Consider the equation

$$x^2 + 4x + 2 = 0$$

Is there a way that we can represent the left side of the equation as a perfect square?

We know that geometrically, x^2, is a perfect square. Draw a square with the side length of x.

Now draw the geometric representation of $x^2 + 4x$

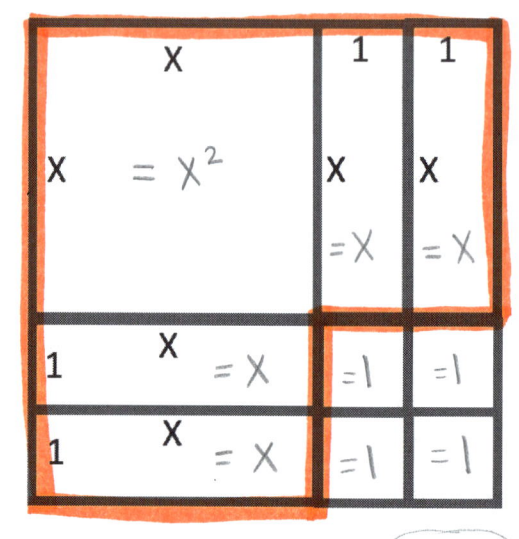

How many 1x1 boxes are needed above to make $x^2 + 4x$ a perfect square?

We can see that $x^2 + 4x + 4$ is a perfect square by factoring it:

$$x^2 + 4x + 4 = (x+2)^2$$

Now let's return to solving

$$x^2 + 4x + 2 = 0$$

Subtract 2 from both sides	$x^2 + 4x + 4 = -2 + 4$ (with $-2\ -2$ shown above)
Add 4 to both sides so that the left side is a perfect square	$(x+2)^2 = 2$
Factor the left side	$\sqrt{(x+2)^2} = \sqrt{2}$
Solve by taking the square root of both sides	$x + 2 = \pm\sqrt{2}$ $x = -2 \pm \sqrt{2}$

What does it mean for an expression like $(x+k)^2$ to be a perfect square?

Expand

$$(x+k)^2 = (x+k)(x+k) \quad \text{FOIL}$$
$$x^2 + kx + kx + k^2 = \boxed{x^2 + 2kx + k^2}$$

Compare

$$x^2 + 2kx + k^2$$

$x^2 + bx + c$	$x^2 + 2kx + k^2$
To factor this, you need two numbers that Multiply to : c Add to: b	To factor this as a perfect square you need one number that: Squared gives: k^2 Add to itself: $2k$

So, take an expression, $x^2 + bx$, and complete the square, the constant must be $\left(\dfrac{b}{2}\right)^2$

$$x^2 + 2kx + k^2 \quad \div 2 + x^2$$

$\sqrt{} \cdot 2$

Factor $x^2 + bx + \left(\frac{b}{2}\right)^2$

What number squared gives you $\left(\frac{b}{2}\right)^2$ and added to itself gives you b?

$$x^2 + bx + \left(\frac{b}{2}\right)^2 = \left(x + \frac{b}{2}\right)^2$$

Solve $x^2 + 2x - 3 = 0$

Move the constant to the right hand side.	$x^2 + 2x = 3$
Complete the square by adding 1/2 of the coefficient of the x term (2) squared to BOTH sides.	$x^2 + 2x + 1^2 = 3 + 1^2$ $(x+1)^2 = 4$
Factor the left side as a perfect square.	$x + 1 = \pm\sqrt{4}$ $x = -1 \pm 2$ $x = 1, -3$
Solve by taking square roots.	$x = 1, -3$

Solve $x^2 - 2x - 7 = 0$

Step	Work
Move the constant to the right hand side.	$x^2 - 2x = 7$
Complete the square by adding 1/2 of the coefficient of the x term (-2) squared to BOTH sides.	$x^2 - 2x + \boxed{1^2} = 7 + \boxed{1^2}$ (x-1)(x-1) FOIL $(x-1)^2 = 8$
Factor the left side as a perfect square.	$x - 1 = \pm\sqrt{8}$
Solve by taking square roots.	$x = 1 \pm \sqrt{8} = \sqrt{2 \cdot 4}$ $\boxed{x = 1 \pm 2\sqrt{2}}$

$\left[\begin{array}{c}\text{When we need to complete the square for an equation such as}\\ ax^2 + bx + c = 0 \\ \text{Where } a \neq 1\text{, we must first factor the value of } a \text{ BEFORE completing the square.}\end{array}\right]$

$x^2 - 6x - 1 = 0$
$x^2 - 6x = 1$
$x^2 - 6x + 3^2 = 1 + 3^2$
$(x-3)^2 = 10$
$x - 3 = \pm\sqrt{10}$
$\boxed{x = 3 \pm \sqrt{10}}$

6.3—260

Solve $2x^2 + 4x - 3 = 0$ (not a one! — annotation pointing at the 2)

Move the constant to the right hand side.	$2x^2 + 4x = 3$ $2(x^2 + 2x) = 3$
Factor a 2 on the left hand side of the equation	$2(x^2 + 2x + \boxed{1^2}) = 3$ $=2$ $2(x^2 + 2x + 1^2) = 3 + 2$
Complete the square in x paying close attention to the value that must be added on both sides.	$2(x+1)^2 = 5$ $(x+1)^2 = \dfrac{5}{2}$
Factor the left side as a perfect square.	$x + 1 = \pm\sqrt{\dfrac{5}{2}}$ $x = -1 \pm \sqrt{\dfrac{5}{2}}$
Solve by taking square roots.	$\boxed{x = -1 \pm \sqrt{\dfrac{5}{2}}}$

$3x^2 + 6x - 1 = 0$

$3x^2 + 6x = 1$

$3(x^2 + 2x + 1^2) = 1 + 3$

$3(x+1)^2 = 4$

$(x+1)^2 = \dfrac{4}{3}$

$x + 1 = \pm\sqrt{\dfrac{4}{3}}$

$\boxed{x = -1 \pm \dfrac{2}{\sqrt{3}}}$

Solve by completing the square.

$$9x^2 - 6x + 5 = 0$$

$9x^2 - 6x = -5$

$9\left(x^2 - \frac{6}{9}x \right) = -5$

$9\left(x^2 - \frac{2}{3}x + \left(\frac{1}{3}\right)^2\right) = -5 + 9\left(\frac{1}{3}\right)^2$

$9\left(x - \frac{1}{3}\right)^2 = -4$

$\left(x - \frac{1}{3}\right)^2 = \frac{-4}{9}$

$x - \frac{1}{3} = \pm\sqrt{\frac{-4}{9}} \rightarrow \quad \sqrt{-4} = 2i$

$\sqrt{9} = 3$

$x - \frac{1}{3} = \pm \frac{2}{3}i$

$\boxed{x = \frac{1}{3} \pm \frac{2}{3}i}$

Homework 6.3 Solve by Completing the Square

Solve by completing the square.

---

1. $x^2 + 4x - 10 = 0$

$$x^2 + 4x = 10$$
$$x^2 + 4x + 2^2 = 10 + 2^2$$
$$(x+2)^2 = 14$$
$$x + 2 = \pm\sqrt{14}$$
$$x = -2 \pm \sqrt{14}$$

---

2. $x^2 + 12x + 18 = 0$

$$x^2 + 12x = -18$$
$$x^2 + 12x + 6^2 = -18 + 6^2$$
$$(x+6)^2 = 18$$
$$x + 6 = \pm\sqrt{18}$$
$$x = -6 \pm 3\sqrt{2}$$

---

3. $x^2 - 2x + 25 = 0$

$$x^2 - 2x = -25$$
$$x^2 - 2x + 1^2 = -25 + 1^2$$
$$(x-1)^2 = -24$$
$$x - 1 = \pm\sqrt{-24}$$
$$x - 1 = \pm 2i\sqrt{6}$$
$$x = 1 \pm 2i\sqrt{6}$$

---

4. $4x^2 - 40x - 11 = 0$

$$4x^2 - 40x = 11$$
$$4(x^2 - 10x + 5^2) = 11 + (4 \cdot 5^2)$$
$$4(x-5)^2 = 111$$
$$(x-5)^2 = \frac{111}{4}$$
$$x - 5 = \pm\sqrt{\frac{111}{4}}$$
$$x = 5 \pm \frac{\sqrt{111}}{2}$$

Homework 6.3 Solve by Completing the Square

Solve by completing the square.

9. $x^2 + 6x = -7$

$x^2 + 6x + 3^2 = -7 + 3^2$

$(x+3)^2 = 2$

$x + 3 = \pm\sqrt{2}$

$x = 3 \pm \sqrt{2}$

10. $0 = 3x^2 + 9x + 1$

$3x^2 + 9x = -1$

$3(x^2 + 3x + \frac{3}{2}^2) = -1 + (\frac{3}{2}^2) 3$

$3(x + \frac{3}{2})^2 = \frac{23}{4}$

$(x + \frac{3}{2})^2 = \frac{23}{12}$

$x + \frac{3}{2} = \pm\sqrt{\frac{23}{12}}$

$x = -\frac{3}{2} \pm \frac{\sqrt{23}}{2\sqrt{3}}$

11. $-2x^2 + 4x - 7 = 0$

$-2x^2 + 4x = 7$

$-2(x^2 - 2x + 1^2) = 7 - 2$

$-2(x-1)^2 = 5$

$(x-1)^2 = \frac{-5}{2}$

$x - 1 = \pm\sqrt{\frac{-5}{2}}$

$x = 1 \pm \sqrt{\frac{-5}{2}}$

12. $0 = 5x^2 + 9x + 1$

$5x^2 + 9x = -1$

$5(x^2 + \frac{9}{5}x + \frac{9^2}{10}) = -1 + 5 \cdot (\frac{9}{10})^2$

$5(x + \frac{9}{10})^2 = 1 + \frac{405}{100}$

$5(x + \frac{9}{10})^2 = \frac{305}{100} \cdot 5$

$(x + \frac{9}{10})^2 = \frac{61}{100}$

$x + \frac{9}{10} = \pm\sqrt{\frac{61}{100}}$

$x + \frac{9}{10} = \pm\frac{\sqrt{61}}{10}$

$x = \frac{-9}{10} \pm \frac{\sqrt{61}}{10}$

Homework 6.3 Solve by Completing the Square

CHALLENGE PROBLEM

Use completing the square to solve
$$ax^2 + bx + c = 0$$

First, divide both side by a to get
$$x^2 + \frac{b}{a}x + \frac{c}{a} = 0$$
Then use the method of completing the square from this section.

6.4. Solve by quadratic formula

So, in solving quadratic equations, so far, we've looked at two methods:

Solving by taking square roots

Example: $(x+1)^2 = 7$	
$\sqrt{(x+1)^2} = \sqrt{7}$ $x+1 = \pm\sqrt{7}$ $\boxed{x = -1 \pm \sqrt{7}}$	
When will this method work? one side squared (variables isolated)	When will this method not work? do not have perfect squares $(x+1)^2 = 7x$

Solving by factoring

Example: $10x^2 - x - 3 = 0$		
$10x^2 - 6x + 5x - 3 = 0$ $2x(5x-3) + (5x-3) = 0$ $(5x-3) + (2x+1) = 0$	$5x - 3 = 0$ $5x = 3$ $x = \frac{3}{5}$ $2x + 1 = 0$ $x = \frac{-1}{2}$ $\boxed{x = \frac{3}{5}, \frac{-1}{2}}$	2 # m: -30 a: -1 $\bigcirc$ -6, 5
When will this method work? quadratic is able to be factored	When will this method not work? $10x^2 - 14x - 3 = 0$ doesn't factor	

We need a method that will work every time for each quadratic equation.

Quadratic Formula: For a quadratic equation of the form

$$ax^2 + bx + c = 0$$

The quadratic formula gives the solutions as:

$$x = \frac{-b \pm \sqrt{b^2 - 4ac}}{2a}$$

The quadratic formula can be used for every quadratic equation. You will need to know this formula and how to work with it. If you did the challenge homework from the prior homework section, this formula is simply derived by completing the square.

6.4—266

The quadratic formula: $x = \dfrac{-b \pm \sqrt{b^2-4ac}}{2a}$

Left side of the equation	Right side of the equation
x	$\dfrac{-b \pm \sqrt{b^2 - 4ac}}{2a}$
This is the algebra. Pretty simple.	This is the arithmetic. Care and caution must be applied to eliminate careless arithmetic mistakes.

Solve the following quadratic equation using the quadratic formula.

$x^2 - 5x = 7$ $x^2 - 5x - 7 = 0$ $a = 1$ $b = -5$ $c = -7$	Write the quadratic formula: $x = \dfrac{-b \pm \sqrt{b^2 - 4ac}}{2a}$
Substitute the values of $a, b, \& c$ into the formula.	$x = \dfrac{5 \pm \sqrt{-5^2 - 4(1)(-7)}}{2(1)}$ ← $b^2 - 4ac \rightarrow -5^2 - 4(1)(-7) = 53$
Simplify.	$x = \dfrac{5 \pm \sqrt{53}}{2(1)}$ $x = \dfrac{5 \pm \sqrt{53}}{2}$
Solution $x = \dfrac{5 \pm \sqrt{53}}{2}$	Number of solutions 2 $x = \dfrac{5 + \sqrt{53}}{2}$ $x = \dfrac{5 - \sqrt{53}}{2}$	Type of solution(s) (Real) Complex

Solve the following quadratic equation using the quadratic formula.

$4x^2 - 10x = 2x - 9$	Write the quadratic formula:
$4x^2 - 10x - 2x + 9 = 0$ $4x^2 - 12x + 9 = 0$ $a = 4$ $b = -12$ $c = 9$	$x = \dfrac{-b \pm \sqrt{b^2 - 4ac}}{2a}$ $b^2 - 4ac \rightarrow -12^2 - 4(4)(9)$ $= 0$
Substitute the values of $a, b,$ & c into the formula.	$x = \dfrac{12 \pm \sqrt{0}}{2(4)}$
Simplify.	$x = \dfrac{12 \pm 0}{8} = \dfrac{12}{8} \div 4$ $x = \dfrac{3}{2}$
Solution $x = \dfrac{3}{2}$	Number of solutions one	Type of solution(s) (Real) Complex

Solve the following quadratic equation using the quadratic formula.

$x^2 - 6x + 10 = 0$	Write the quadratic formula:
$a = 1$ $b = -6$ $c = 10$	$x = \dfrac{-b \pm \sqrt{b^2 - 4ac}}{2a}$
Substitute the values of $a, b,$ & c into the formula.	$x = \dfrac{6 \pm \sqrt{-6^2 - 4(1)(10)}}{2(1)}$
Simplify.	$x = \dfrac{6 \pm \sqrt{36 - 40}}{2} = \dfrac{6 \pm \sqrt{-4}}{2}$ $x = \dfrac{6 \pm 2i}{2} \quad x = 3 \pm i$

Solution	Number of solutions	Type of solution
$x = 3 \pm i$	two	Real (Complex)

Homework 6.3 Solve by Quadratic Formula

Solve each of the quadratic equations below. Your final answer should be written in the form $x = \{\quad\}$, where the brackets contain the solution(s) to the equation. Also, specify how many solutions you obtain, and if they are real or complex.

For example, $10x^2 - x - 3 = 0$. Solution $x = \{-\frac{1}{2}, \frac{3}{5}\}$, 2 real solutions.

1. $2x^2 + 6x - 5 = 0$	5. $6x^2 - x = 15$	6. $8x^2 + 4x + 16 = 4 + 3x^2$
	$6x^2 - x - 15 = 0$	$8x^2 + 4x + 16 - 4 - 3x^2$
		$5x^2 + 4x + 12 = 0$
$a=2\ b=6\ c=-5$	$a=6\ b=-1\ c=-15$	$a=5\ b=4\ c=12$
$x = \frac{-b \pm \sqrt{b^2 - 4ac}}{2a}$	$x = \frac{1 \pm \sqrt{-1^2 - 4(6)(-15)}}{2(6)}$	$x = \frac{-4 \pm \sqrt{4^2 - 4(5)(12)}}{2(5)}$
$x = \frac{-6 \pm \sqrt{6^2 - 4(2)(-5)}}{2(2)}$	$-1^2 - 4(6)(-15)$	$4^2 - 4(5)(12)$
	$-1 - 360 = -361$	$16 - 240 = -224$
$6^2 - 4(2)(-5)$		$\frac{-4 \pm \sqrt{-224}}{10}$ ★
$36 - 40 = -4$	$x = \frac{1 \pm \sqrt{361}}{12}$	
$x = \frac{-6 \pm \sqrt{-4}}{4}$	$\frac{1 \pm 19}{12} =$	$\frac{-4 \pm \sqrt{(-1)(16)(14)}}{10}$
$\frac{-6 \pm 2i}{4}\ \{-\frac{3}{2} \pm \frac{1}{2}i\}$	$\frac{1+19}{12}\quad \frac{1-19}{12}$	$\frac{-4 \pm 4i\sqrt{14}}{10}$
	$\frac{12}{12}\qquad \frac{12}{12}$	$= -\frac{2}{5} \pm \frac{2}{5}i\sqrt{14}$
	$= \frac{5}{3}\qquad \frac{-3}{2}$	
Solution $x = \{-\frac{3}{2} \pm \frac{1}{2}i\}$	Solution $x = \frac{5}{3}, \frac{-3}{2}$	Solution $x = \{-\frac{2}{5} \pm \frac{2}{5}i\sqrt{14}\}$
Number of Solutions two	Number of Solutions two	Number of Solutions two
Type of Solution complex	Type of Solution real	Type of Solution complex

Homework 6.3 Solve by Quadratic Formula

Solve each of the quadratic equations below.

7. $35x^2 + x = 1$ $35x^2 + x - 1 = 0$

$a = 35$
$b = 1$
$c = -1$

$b^2 - 4ac$
$1^2 - 4(35)(-1)$
$= 141$

$$x = \frac{-1 \pm \sqrt{1^2 - 4(35)(-1)}}{2(35)}$$

$$x = \frac{-1 \pm \sqrt{141}}{70}$$

8. $2x^2 + 32 = 16x$ $2x^2 + 32 - 16x = 0$

$a = 2$
$b = -16$
$c = 32$

$$x = \frac{16 \pm \sqrt{-16^2 - 4(2)(32)}}{2(2)}$$

$$x = \frac{16 \pm \sqrt{256 - 256}}{4}$$

$$x = \frac{16}{4} = 4$$

9. $-2x^2 - 5x + 6 = 0$

$a = -2$
$b = -5$
$c = 6$

$$x = \frac{5 \pm \sqrt{-5^2 - 4(-2)(6)}}{2(-2)}$$

$$\frac{5 \pm \sqrt{25 + 48}}{-4} = \frac{5 \pm \sqrt{73}}{-4}$$

$$= \frac{-5}{4} \pm \frac{\sqrt{73}}{-4}$$

10. $8x^2 + 6x = 9$

$a = 8$
$b = 6$
$c = -9$

$$x = \frac{-6 \pm \sqrt{6^2 - 4(8)(-9)}}{2(8)}$$

$$\frac{-6 \pm \sqrt{36 + 288}}{16}$$

$$\frac{-6 \pm 18}{16}$$

$\frac{-6 + 18}{16} = \frac{3}{4}$ $\frac{-6 - 18}{16} = \frac{-3}{2}$

11. $3x^2 = 4x - 5$ $-3x^2 + 4x - 5 = 0$

$a = -3$
$b = 4$
$c = -5$

$$\frac{-4 \pm \sqrt{4^2 - 4(-3)(-5)}}{2(-3)}$$

$$\frac{-4 \pm \sqrt{16 - 60}}{-6} = \frac{-4 \pm \sqrt{-44}}{-6}$$

$$\frac{-4 \pm i\sqrt{44}}{}$$

$$x = \frac{-4 \pm 2i\sqrt{11}}{-6}$$

12. $x^2 - 6x + 10 = 0$

$a = 1$
$b = -6$
$c = 10$

$$\frac{6 \pm \sqrt{-6^2 - 4(1)(10)}}{2(1)}$$

$$\frac{6 \pm \sqrt{-4}}{2} = 2i$$

$$\frac{6 \pm 2i}{2}$$

$$x = 3 \pm i$$

Homework 6.3 Solve by Quadratic Formula

CHALLENGE PROBLEM

1. Take the quadratic formula and derive the solution for all values of x when the quadratic equation is of the form
$$ax^2 + c = 0$$

2. Give the formula of the solution to quadratic equations of the form:

$$x^2 + 2bx + b = 0$$

Simplify the equation completely.

$$x = \frac{-b \pm \sqrt{b^2-4ac}}{2a}$$

6.5. The Discriminant

Today we will continue with the quadratic formula.

When graphing a parabola, $y = ax^2 + bx + c$, we know how to find the vertex, the axis of symmetry, determine the direction of opening, and plot the graph.

When solving $ax^2 + bx + c = 0$, we are setting $y = 0$, and solving for x, or finding the roots or $x-$intercepts.

What are all the possibilities when solving $ax^2 + bx + c = 0$ standard form

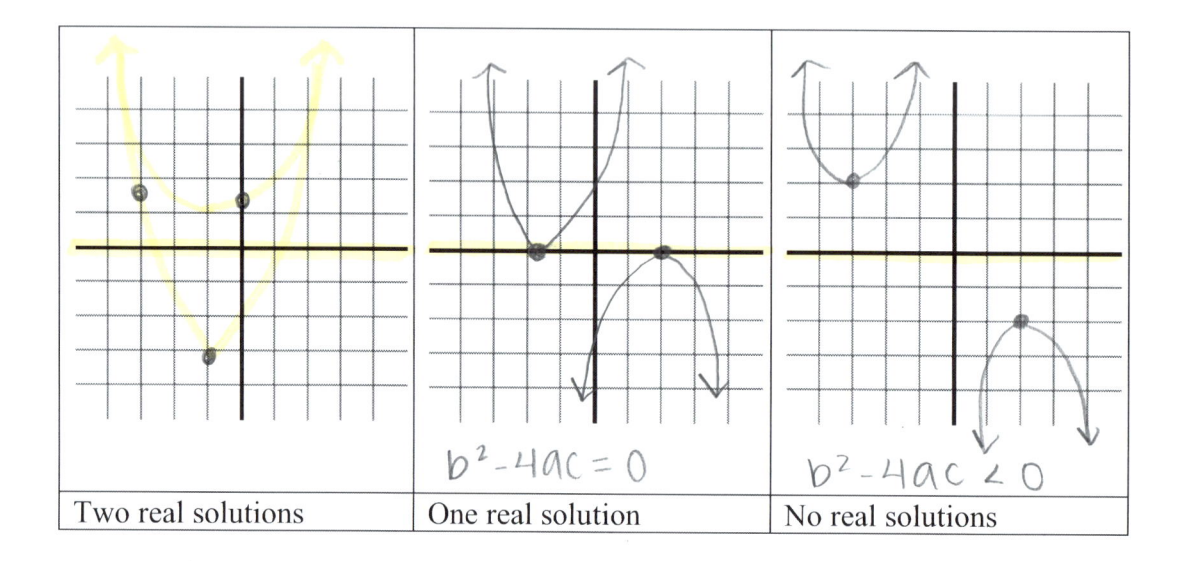

Two real solutions	One real solution $b^2 - 4ac = 0$	No real solutions $b^2 - 4ac < 0$

In the quadratic formula, $x = \frac{-b \pm \sqrt{b^2-4ac}}{2a}$, the ***discriminant*** is the term $b^2 - 4ac$. By evaluating the discriminant, we can determine the type and number of solutions yielded by the quadratic formula.

Discriminant		Solutions
$b^2 - 4ac > 0$, and a perfect square	$\frac{-b \pm \sqrt{4}}{2a}$	1 2 real rational roots
$b^2 - 4ac > 0$, not a perfect square	$\frac{-b \pm \sqrt{3}}{2a}$	2 2 real irrational roots
$b^2 - 4ac = 0$	$x = \frac{-b}{2a}$	3 1 real rational root
$b^2 - 4ac < 0$	$\sqrt{-\ } = i$	4 2 complex roots no real roots

6.5—273

$$\frac{-b \pm \sqrt{b^2-4ac}}{2a}$$

Consider $x^2 - x - 6 = 0$. $ax^2 + bx + c = 0$

$a = 1$
$b = -1$
$c = -6$

Find the value of the discriminant: $b^2 - 4ac = (-1)^2 - 4(1)(-6)$ $= 25$	Number of solutions 2	Type of solutions real rational roots **1**
Find the solution(s): $x = \frac{-b \pm \sqrt{b^2-4ac}}{2a}$ $x = \frac{-(-1) \pm \sqrt{25}}{2(1)} = \frac{1 \pm 5}{2} = \frac{1}{2} \pm \frac{5}{2}$ $\frac{6}{2}, \frac{-4}{2}$ $\boxed{x = 3, -2}$ real, rational		

Consider $x^2 - x - \frac{1}{4} = 0$.

$a = 1$
$b = -1$
$c = -\frac{1}{4}$

Find the value of the discriminant: $b^2 - 4ac$ $(-1)^2 - 4(1)\left(-\frac{1}{4}\right) = 2$	Number of solutions 2	Type of solutions real, irrational roots **2**
Find the solution(s): $x = \frac{-b \pm \sqrt{b^2-4ac}}{2a}$ $x = \frac{-(-1) \pm \sqrt{2}}{2(1)} = \frac{1 \pm \sqrt{2}}{2} \quad \boxed{\frac{1}{2} \pm \frac{\sqrt{2}}{2}}$		

Consider $x^2 + 14x + 49 = 0$.

Find the value of the discriminant: $b^2 - 4ac$ $(14)^2 - 4(1)(49) = 0$	Number of solutions 1	Type of solutions real, rational 3
Find the solution(s): $x = \dfrac{-b \pm \sqrt{b^2-4ac}}{2a}$ $x = \dfrac{-(14) \pm \sqrt{0}}{2(1)} = \boxed{-7}$		

Consider $x^2 - x + 2 = 0$.

Find the value of the discriminant: $b^2 - 4ac$ $(-1)^2 - 4(1)(2) = -7$	Number of solutions 2	Type of solutions complex 4
Find the solution(s): $x = \dfrac{-(-1) \pm \sqrt{-7}}{2(1)} = \dfrac{1 \pm i\sqrt{7}}{2}$		

discriminant = $b^2 - 4ac$

Homework 6.5 The Discriminant

For each problem below, evaluate the discriminant, then give the information as to the number and type of roots for the quadratic equation. Then find the roots for each equation using the quadratic formula.

Equation	1. $x^2 + x - 2 = 0$	Equation	2. $3x^2 - 22x + 7 = 0$
Value of the discriminant: $a = 1$ $b = 1$ $c = -2$	$b^2 - 4ac$ $1^2 - 4(1)(-2)$ $1 + 8 = \boxed{9}$	Value of the discriminant: $a = 3$ $b = -22$ $c = 7$	$b^2 - 4ac$ $(-22)^2 - 4(3)(7)$ $= 400$
Number of solutions	☐ 0 ☐ 1 ☑ 2 Check one	Number of solutions	☐ 0 ☐ 1 ☑ 2 Check one
Type of solutions	☑ Real ☐ Complex ☑ Rational ☐ Irrational Check all that apply	Type of solutions	☑ Real ☐ Complex ☑ Rational ☐ Irrational Check all that apply
Solution	$x = \dfrac{-b \pm \sqrt{b^2-4ac}}{2a}$ $x = \dfrac{-1 \pm \sqrt{9}}{2(1)} = \dfrac{-1 \pm 3}{2}$ $\dfrac{-1}{2} + \dfrac{3}{2}$ $\boxed{1, -2}$	Solution	$x = \dfrac{22 \pm \sqrt{400}}{2(3)}$ $\dfrac{22 \pm 20}{6}$ $\dfrac{11}{3} \pm \dfrac{10}{3}$ $\boxed{7, \dfrac{1}{3}}$

Homework 6.5 The Discriminant

Equation	3. $2x^2 - 9x + 13 = 0$	Equation	4. $4x^2 - 12x + 9 = 0$
Value of the discriminant: $a = 2$ $b = -9$ $c = 13$	$b^2 - 4ac$ $(-9)^2 - 4(2)(13)$ $= -23$	Value of the discriminant: $a = 4$ $b = -12$ $c = 9$	$b^2 - 4ac$ $(-12)^2 - 4(4)(9)$ $= 0$
Number of solutions	☐ 0 ☐ 1 ☒ 2 Check one	Number of solutions	☐ 0 ☒ 1 ☐ 2 Check one
Type of solutions	☐ Real ☒ Complex ☐ Rational ☐ Irrational Check all that apply	Type of solutions	☒ Real ☐ Complex ☒ Rational ☐ Irrational Check all that apply
Solution	$x = \dfrac{-(-9) \pm \sqrt{-23}}{2(2)}$ $\dfrac{9 \pm i\sqrt{23}}{4}$	Solution	$x = \dfrac{-(-12) \pm \sqrt{0}}{2(4)}$ $\dfrac{12}{8} = \dfrac{3}{2}$

Homework 6.5 The Discriminant

For each problem below, evaluate the discriminant, then give the information as to the number and type of roots for the quadratic equation. Then find the roots for each equation using the quadratic formula.

Equation	5. $4x^2 + 20x + 25 = 0$	Equation	6. $2x^2 + 6x - 9 = 0$
Value of the discriminant:	$b^2 - 4ac$ $(20)^2 - 4(4)(25)$ $= 0$	Value of the discriminant:	$b^2 - 4ac$ $(6)^2 - 4(2)(-9)$ $= 108$
Number of solutions	☐0 ☑1 ☐2 Check one	Number of solutions	☐0 ☐1 ☑2 Check one
Type of solutions	☑ Real ☐ Complex ☑ Rational ☐ Irrational Check all that apply	Type of solutions	☑ Real ☐ Complex ☐ Rational ☑ Irrational Check all that apply
Solution	$a = 4$ $b = 20$ $c = 25$ $\dfrac{-20 \pm \sqrt{0}}{2(4)}$ $\dfrac{-20}{8} = \dfrac{-5}{2}$	Solution	$a = 2$ $b = 6$ $c = -9$ $\dfrac{-6 \pm \sqrt{108}}{2(2)}$ $\dfrac{-6 \pm 6\sqrt{3}}{4}$ $-\dfrac{3}{2} \pm \dfrac{3}{2}\sqrt{3}$

Homework 6.5 The Discriminant

For each problem below, evaluate the discriminant, then give the information as to the number and type of roots for the quadratic equation. Then find the roots for each equation using the quadratic formula.

Equation	7. $16x^2 - 8x + 14 = 0$	Equation	8. $6x^2 - 11x - 21 = 0$
Value of the discriminant:	$b^2 - 4ac$ $(-8)^2 - 4(16)(14)$ -832	Value of the discriminant:	$b^2 - 4ac$ $(-11)^2 - 4(6)(-21)$ $= 625$
Number of solutions	☐ 0 ☐ 1 ☑ 2 Check one	Number of solutions	☐ 0 ☐ 1 ☑ 2 Check one
Type of solutions	☐ Real ☑ Complex ☐ Rational ☐ Irrational Check all that apply	Type of solutions	☑ Real ☐ Complex ☑ Rational ☐ Irrational Check all that apply
Solution $a = 16$ $b = -8$ $c = 14$	$\dfrac{-(-8) \pm \sqrt{-832}}{2(16)}$ $\dfrac{8 \pm i\sqrt{832}}{32}$ $\dfrac{8 \pm i 8\sqrt{13}}{32}$	Solution $a = 6$ $b = -11$ $c = -21$	$\dfrac{11 \pm \sqrt{625}}{2(6)}$ $\dfrac{11 \pm 25}{12}$ $\dfrac{11}{12} \pm \dfrac{25}{12}$ $3, \dfrac{-7}{6}$

Homework 6.5 The Discriminant

CHALLENGE PROBLEMS

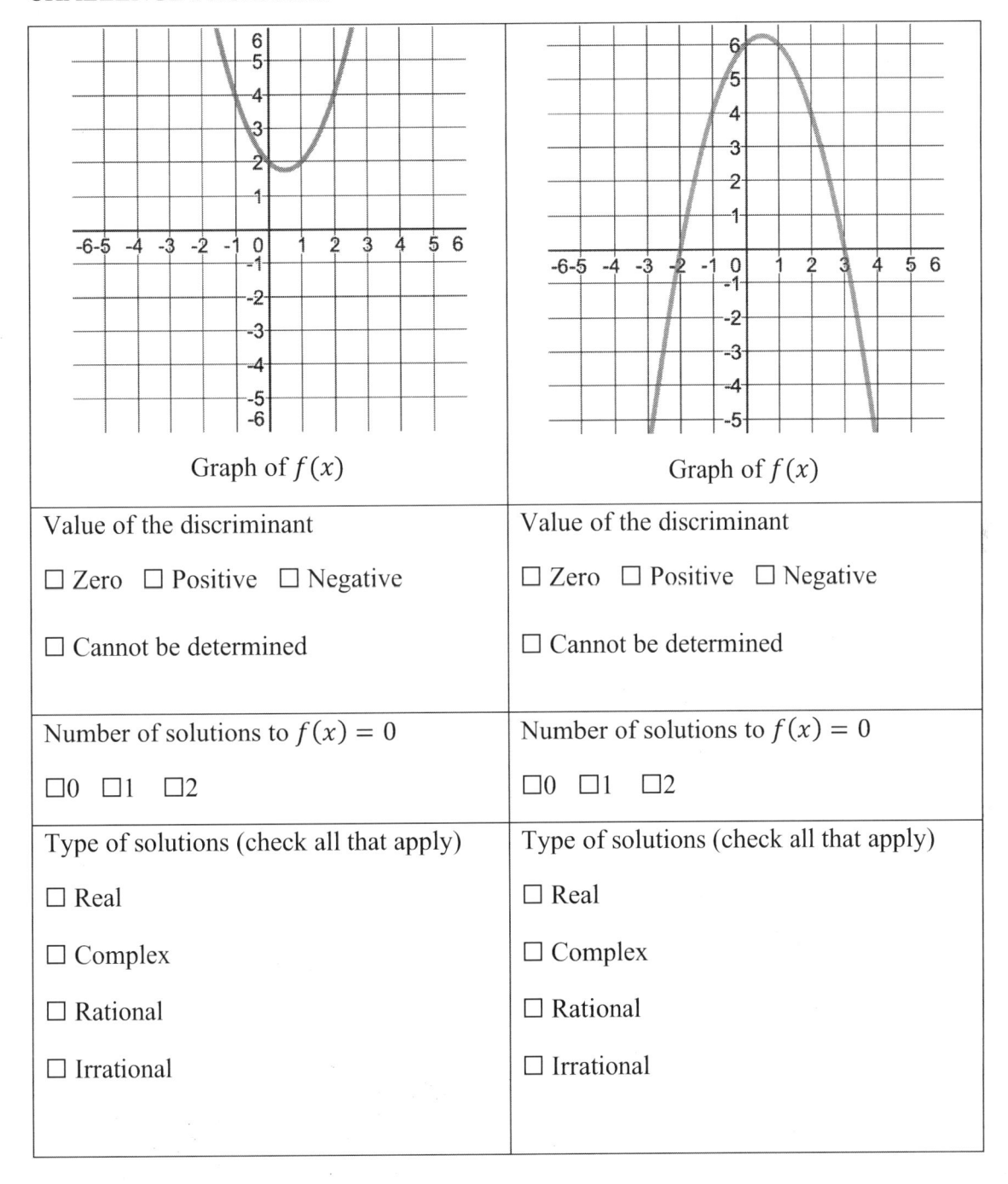

Graph of $f(x)$	Graph of $f(x)$
Value of the discriminant ☐ Zero ☐ Positive ☐ Negative ☐ Cannot be determined	Value of the discriminant ☐ Zero ☐ Positive ☐ Negative ☐ Cannot be determined
Number of solutions to $f(x) = 0$ ☐ 0 ☐ 1 ☐ 2	Number of solutions to $f(x) = 0$ ☐ 0 ☐ 1 ☐ 2
Type of solutions (check all that apply) ☐ Real ☐ Complex ☐ Rational ☐ Irrational	Type of solutions (check all that apply) ☐ Real ☐ Complex ☐ Rational ☐ Irrational

Homework 6.5 The Discriminant

CHALLENGE PROBLEMS

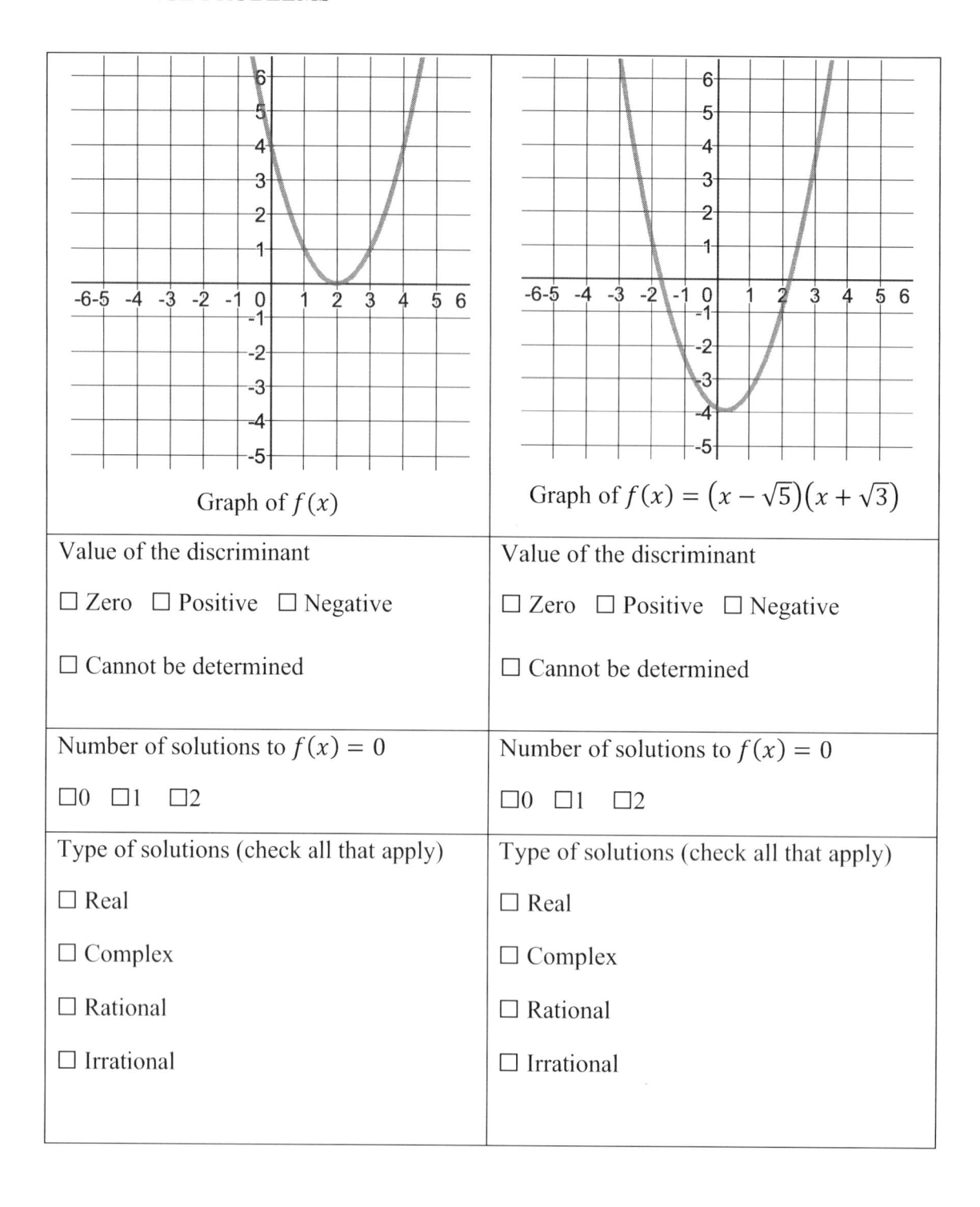

Graph of $f(x)$	Graph of $f(x) = (x - \sqrt{5})(x + \sqrt{3})$
Value of the discriminant ☐ Zero ☐ Positive ☐ Negative ☐ Cannot be determined	Value of the discriminant ☐ Zero ☐ Positive ☐ Negative ☐ Cannot be determined
Number of solutions to $f(x) = 0$ ☐ 0 ☐ 1 ☐ 2	Number of solutions to $f(x) = 0$ ☐ 0 ☐ 1 ☐ 2
Type of solutions (check all that apply) ☐ Real ☐ Complex ☐ Rational ☐ Irrational	Type of solutions (check all that apply) ☐ Real ☐ Complex ☐ Rational ☐ Irrational

6.6. Solve by any method

So, having covered 4 methods for solving quadratic equations, how do we know which one to use?

Method	When to use
Solving by square roots	One side of your equation can be expressed as a perfect square. $$2(x+3)^2 + 5 = 0$$ $(x+3)^2 = -\frac{5}{2}$ $x + 3 = \pm\sqrt{-\frac{5}{2}}$ $x = -3 \pm i\frac{\sqrt{5}}{\sqrt{2}}$
Solving by factoring.	When the quadratic is factorable. It has rational solutions, so $b^2 - 4ac$ is a **positive** perfect square. $$x^2 - 3x - 10 = 0$$ $(x-5)(x+2) = 0$ $x = 5, -2$ 2 # m: -10 a: -3 -5, 2
Solve by quadratic formula	Always works for any quadratic equation. $$x = \frac{-b \pm \sqrt{b^2 - 4ac}}{2a}$$
Solve by completing the square.	Always works for any quadratic equation. Can be overkill for something that easily factors. Once you complete the square this becomes the same method as taking square roots of both sides.

Can you think of pros and cons for each method?

Method	Pro	Con
Solving by square roots $$2x^2 = 50$$ $$x^2 = 25$$ $$x = \pm 5$$	– quick if something is squared	– not always in the form of something squared
Solving by factoring	– fairly quick if factorable	– not all are factorable
Solving by quadratic formula	– always works!	– arithmetic
Solve by completing the square	– always works!	– arithmetic & algebra

If we can solve via square roots or factoring, generally speaking, those methods are quicker, and involve a bit less arithmetic. So, if we can quickly determine if those methods will or will not work, we can save some time in finding our solutions.

For the rest of this lesson, let's practice solving equations by any method.

Solve using any method: $(x+5)^2 = 25$ **square roots**

$$x+5 = \pm\sqrt{25}$$
$$x+5 = \pm 5$$
$$x = -5 \pm 5$$
$$\boxed{x = 0, -10}$$

Solve using any method: $x^2 - 6x + 7 = 0$ **complete square**

2#
m: 7
a: -6
1, 7
-1, 7
cannot be factored

$$\frac{-(-6) \pm \sqrt{8}}{2(1)}$$

$$x = \frac{6 \pm 2\sqrt{2}}{2}$$

$$\boxed{x = 3 \pm \sqrt{2}}$$

quadratic formula

$$x^2 - 6x = -7$$
$$x^2 - 6x + 3^2 = -7 + 3^2$$
$$(x-3)^2 = 2$$
$$x - 3 = \pm\sqrt{2}$$
$$\boxed{x = 3 \pm \sqrt{2}}$$

Solve using any method: $5p^2 - 10p + 24 = 0$ **quadratic formula**

$$b^2 - 4ac$$
$$(-10)^2 - 4(5)(24)$$
$$100 - 480$$
$$-380$$

$$x = \frac{-(-10)^2 \pm \sqrt{-380}}{2(5)} \rightarrow 4 \cdot 5 \cdot 19$$

$$x = \frac{10 \pm i\sqrt{4 \cdot 5 \cdot 19}}{10}$$

$$x = \frac{10}{10} \pm \frac{2}{10}i\sqrt{95}$$

$$\boxed{x = 1 \pm \frac{1}{5}i\sqrt{95}}$$

1st step
(does it factor)
↓

2 #
m: 24
a: -11
-8, -3

Solve using any method: $6x^2 - 11x + 4 = 0$ factoring

$6x^2 - 8x - 3x + 4 = 0$

$2x(3x-4) - (3x-4) = 0$

$(2x-1)(3x-4) = 0$

$x = \frac{1}{2}, \frac{4}{3}$

Solve using any method: $x^2 + 3x - 7 = 0$ quadratic form.

$x = \frac{-3 \pm \sqrt{3^2 - 4(1)(-7)}}{2}$

$\frac{-3 \pm \sqrt{37}}{2} = \frac{-3}{2} \pm \frac{\sqrt{37}}{2}$

Solve using any method: $x^2 + 5x - 50 = 0$

2 #
m: -50
a: 5
-5, 10

$(x-5)(x+10) = 0$

$x = 5, -10$

Solve using any method: $x^2 + 7x = 0$ factor

$$x(x+7) = 0$$
$$\boxed{x = 0, -7}$$

Solve using any method: $x^2 - x + 1 = 0$ quadratic formula

$$x = \frac{-(-1) \pm \sqrt{(-1)^2 - 4(1)(1)}}{2(1)}$$

$$x = \frac{1 \pm \sqrt{1-4}}{2} \quad \boxed{x = \frac{1 \pm i\sqrt{3}}{2}}$$

Solve using any method: $(x + 7)^2 = -16$ square roots

$$x + 7 = \pm\sqrt{-16}$$
$$x + 7 = \pm 4i$$
$$\boxed{x = -7 \pm 4i}$$

Homework 6.6 Solve by any method

1. $\frac{1}{2}(x+7)^2 = 12 \cdot 2$

$\sqrt{(x+7)^2} = \sqrt{24}$

$x+7 = \pm\sqrt{24} = 2\sqrt{6}$

$\boxed{x = -7 \pm 2\sqrt{6}}$

2. $6x^2 - 5x - 4 = 0$

$x = \dfrac{-(-5) \pm \sqrt{-5^2 - 4(6)(-4)}}{2(6)}$

$x = \dfrac{5 \pm \sqrt{121}}{12}$

$x = \dfrac{5}{12} \pm \dfrac{11}{12} = \boxed{\dfrac{4}{3}, \dfrac{-1}{2}}$

3. $x^2 + 2x + 3 = 0$

2#
m: 3
a: 2
N/a

$x = \dfrac{-2 \pm \sqrt{2^2 - 4(1)(3)}}{2(1)}$

$x = \dfrac{-2 \pm \sqrt{-8}}{2}$

$x = \dfrac{-1 \pm i \, 2\sqrt{2}}{2}$

$\boxed{-1 \pm i\sqrt{2}}$

4. $4x^2 - 4x + 1 = 0$

2#
m: 4
a: -4
N/a

$x = \dfrac{4 \pm \sqrt{(-4)^2 - 4(4)(1)}}{2(4)}$

$x = \dfrac{4 \pm \sqrt{0}}{8}$

$\boxed{x = \dfrac{1}{2}}$

5. $\dfrac{3(x-4)^2}{3} = \dfrac{15}{3}$

$\sqrt{(x-4)^2} = \sqrt{5}$

$x - 4 = \pm\sqrt{5}$

$\boxed{x = 4 \pm \sqrt{5}}$

Homework 6.6 Solve by any method

6. $x^2 - 6x = 7$ → $x^2 - 6x - 7 = 0$

2#
m: -7
a: -6
$\boxed{1, -7}$

$x^2 + x - 7x - 7 = 0$
$x(x+1) - 7(x+1) = 0$
$(x-7)(x+1) = 0$
$x = -1, 7$

7. $2x^2 - 12 = 0$ → $2x^2 - 12 - 0 = 0$

$$\frac{-(-12) \pm \sqrt{12^2 - 4(2)}}{2(2)}$$

$$\frac{12 \pm \sqrt{136}}{4}$$

$$\frac{3 \pm 2\sqrt{34}}{4}$$

8. $3x^2 + 2x = -4$ → $3x^2 + 2x + 4 = 0$

2#
m: 12
a: 2
N/a

$$\frac{-2 \pm \sqrt{2^2 - 4(3)(4)}}{2(3)}$$

$$\frac{-2 \pm \sqrt{-44}}{6}$$

$$\frac{-2 \pm i\sqrt{44}}{6}$$

$$\frac{-2 \pm i2\sqrt{11}}{6}$$

$-\frac{1}{3} \pm i\frac{1}{3}\sqrt{11}$

9. $6x^2 + 23x + 21 = 0$

$$\frac{-23 \pm \sqrt{23^2 - 4(6)(21)}}{2(6)}$$

$$\frac{-23 \pm \sqrt{25}}{12} = \frac{-23 \pm 5}{12}$$

$\frac{-23}{12} \pm \frac{5}{12}$

10. $2x^2 + 2x + 5 = 0$

$$\frac{-2 \pm \sqrt{2^2 - 4(2)(5)}}{2(2)} = \frac{-2 \pm \sqrt{-36}}{4} = \frac{-2 \pm i\sqrt{36}}{4}$$

$\frac{-2 \pm 6i}{4}$ $\left(-\frac{1}{2} \pm \frac{3}{2}i\right)$

6.7. Review

Assessment Checklist. Below are the competencies one should master in preparation for an assessment on solving quadratic equations.

☐ Solving equations by taking the square root of both sides.

☐ Understanding the $\pm$ solutions that arise from taking the square roots of both sides of a quadratic equation.

☐ Ability to factor and solve quadratic equations by factoring the greatest common factor.

☐ Ability to factor and solve quadratic equations by factoring trinomials of the form $x^2 + bx + c = 0$.

☐ Ability to factor and solve quadratic equations by factoring trinomials of the form $ax^2 + bx + c = 0$.

☐ Ability to perform factor by grouping in solving quadratic equations.

☐ Ability to solve $x^2 + bx + c = 0$ and $ax^2 + bx + c = 0$ by completing the square.

☐ Know and use the quadratic formula, $x = \frac{-b \pm \sqrt{b^2 - 4ac}}{2a}$, fearlessly and accurately.

☐ Know how to calculate the discriminant and interpret what the discriminant means to the solution of a quadratic equation.

☐ Ability to solve quadratic equations using any method: square roots, factoring, quadratic formula, completing the square

☐ rationalizing denominator (∪ !)

$$\left(\frac{\sqrt{11}}{\sqrt{11}}\right)\frac{2}{\sqrt{11}}$$

$$= \frac{2\sqrt{11}}{11}$$

Review 6.7

1. Solve the following by factoring by grouping $\underbrace{6x^2 + 15x}_{} + \underbrace{4x + 10}_{} = 0$

   $3x(2x+5) + 2(2x+5)$
   $(3x+2)(2x+5) = 0$
   $$\boxed{x = -\frac{2}{3}, -\frac{5}{2}}$$

2. Solve the following by factoring.

a. $x^2 - 4x - 45 = 0$	b. $15x^2 + 11x + 2 = 0$
$(x-9)(x+5) = 0$	$15x^2 + 6x + 5x + 2 = 0$
$\boxed{x = -5, 9}$	$3x(5x+2) + (5x+2) = 0$
	$(3x+1)(5x+2) = 0$
	$\boxed{x = -\frac{1}{3}, -\frac{2}{5}}$

   2 #
   m: -45
   a: -4
   -9, 5

   2 #
   m: 30
   a: 11
   6, 5

3. Solve using the quadratic formula.

a. $5x^2 + 9x = -7$	b. $2x^2 - 5x - 7 = 0$
$5x^2 + 9x + 7 = 0$	$x = \dfrac{5 \pm \sqrt{(-5)^2 - 4(2)(-7)}}{2(2)}$
$x = \dfrac{-9 \pm \sqrt{9^2 - 4(5)(7)}}{2(5)}$	$x = \dfrac{5 \pm \sqrt{81}}{4}$
$x = \dfrac{-9 \pm \sqrt{-59}}{10}$	$x = \dfrac{5 \pm 9}{4} = \dfrac{5}{4} \pm \dfrac{9}{4}$
$\boxed{x = \dfrac{-9 \pm i\sqrt{59}}{10}}$	$\boxed{x = \dfrac{7}{2}, -1}$

Review 6.7

4. Solve by square roots.

a. $3x^2 - 7 = -28$	b. $(x-2)^2 = 16$
$3x^2 = -21$	$x - 2 = \pm\sqrt{16}$
$\sqrt{x^2} = \sqrt{-7}$	$x - 2 = \pm 4$
$\boxed{x = \pm i\sqrt{7}}$	$x = 2 \pm 4$
	$\boxed{x = -2, 6}$

5. Complete the information below.

Equation	$4x^2 + 20x + 25 = 0$	Equation	$x^2 - 2x - 2 = 0$
Value of the discriminant:	$b^2 - 4ac = \boxed{0}$ $(20^2) - 4(4)(25)$	Value of the discriminant:	$b^2 - 4ac = \boxed{12}$ $(-2)^2 - 4(1)(-2)$
Number of solutions	☐0 ☑1 ☐2 Check one	Number of solutions	☐0 ☐1 ☑2 Check one
Type of solutions	☑ Real ☐ Complex ☑ Rational ☐ Irrational Check all that apply	Type of solutions	☑ Real ☐ Complex ☐ Rational ☑ Irrational Check all that apply
Equation	$x^2 + 2x - 3 = 0$	Equation	$x^2 + 2x + 5$
Value of the discriminant:	$(2)^2 - 4(1)(-3) = \boxed{16}$	Value of the discriminant:	$(2)^2 - 4(5) = -16$
Number of solutions	☐0 ☐1 ☑2 Check one	Number of solutions	☐0 ☐1 ☑2 Check one
Type of solutions	☑ Real ☐ Complex ☑ Rational ☐ Irrational Check all that apply	Type of solutions	☐ Real ☑ Complex ☐ Rational ☐ Irrational Check all that apply

$$\frac{-2 \pm \sqrt{16}}{2} = \frac{-2 \pm 4}{2}$$

Review 6.7
6. Solve by any method:

a. $\sqrt{(x-4)^2} = \sqrt{17}$

$x - 4 = \pm\sqrt{17}$

$\boxed{x = 4 \pm \sqrt{17}}$

b. $x^2 + x - 2 = 0$

$\dfrac{-1 \pm \sqrt{1^2 - 4(-2)}}{2(1)}$

$\dfrac{-1 \pm \sqrt{9}}{2}$

$\dfrac{-1 \pm 3}{2}$

$-\dfrac{1}{2} \pm \dfrac{3}{2}$

$\boxed{x = 1, -2}$

c. $x^2 + x + 1 = 0$

$\dfrac{-1 \pm \sqrt{1^2 - 4(1)}}{2(1)}$

$\dfrac{-1 \pm \sqrt{-3}}{2}$

$\boxed{\dfrac{-1 \pm i\sqrt{3}}{2}}$

2#
m: 1
a: 1

d. $6x^2 + 11x + 3 = 0$

m: 18
a: 11
2, 9

$\underline{6x^2 + 2x} + \underline{9x + 3} = 0$

$2x(3x+1) + 3(3x+1) = 0$

$(2x+3)(3x+1) = 0$

$\boxed{x = -\dfrac{3}{2}, -\dfrac{1}{3}}$

Review 6.7
7. Solve by completing the square.

$x^2 + 3x - 1 = 0$	$2x^2 + 4x - 1 = 0$
$x^2 + 3x = 1$	$2x^2 + 4x = 1$
$x^2 + 3x + \left(\frac{3}{2}\right)^2 = 1 + \left(\frac{3}{2}\right)^2$	$2x^2 + 4x + 1^2 = 1 + 2$
$\left(x + \frac{3}{2}\right)^2 = 1 + \frac{9}{4}$	$2(x+1)^2 = 3$
$\sqrt{\left(x + \frac{3}{2}\right)^2} = \sqrt{\frac{13}{4}}$	$\sqrt{(x+1)^2} = \sqrt{\frac{3}{2}}$
$x + \frac{3}{2} = \pm\sqrt{\frac{13}{4}}$	$x + 1 = \pm\sqrt{\frac{3}{2}}$
$\boxed{x = -\frac{3}{2} \pm \sqrt{\frac{13}{4}}}$	$\boxed{x = -1 \pm \sqrt{\frac{3}{2}}}$

6.8. Writing Quadratic equations

In preparation for our next unit where we will be looking at applications of quadratic modeling, we will need to spend some time writing the equations of quadratic function given data points. This is similar to what we did in data analysis, when we were modeling datasets with lines, but in this case, we will be using quadratic functions.

Today we will look at 3 different scenarios, and how you will write the quadratic equation in each.

Scenario 1	Scenario 1	Scenario 3
Given the vertex, and one other point on the parabola.	Given the x −intercepts and one other point on the parabola.	Given any 3 points on the parabola.

Scenario 1: Given the vertex and one other point on the parabola.
Use the Vertex Form of the quadratic function. $$y = a(x - h)^2 + k$$
You know the vertex. Use the other point you are given to solve for a.

Find the equation of the quadratic function with vertex (1,5) and passing through (−1,1).

$h = 1$
$k = 5$

$y = a(x-h)^2 + k$
$y = a(x-1)^2 + 5$
$1 = a(-1-1)^2 + 5$ sub. $x = -1$, $y = 1$
$1 = a(-2)^2 + 5$
$1 = 4a + 5$
$-4 = 4a$
$\boxed{-1 = a}$ ← solve for value of "a"

parabola equation: $y = -1(x-1)^2 + 5$

6.8—294

> **Scenario 2:** Given the x-intercepts and one other point on the parabola.

Use the Intercept Form of the quadratic function.
$$y = a(x-p)(x-q)$$

You know the values of p and q. Use the other point you are given to solve for a.

Find the equation of the parabola passing through (−4,0), (3,0), [x-int] and (1,4).

$y = a(x-p)(x-q)$] sub x-int (-4) & (3)

$y = a(x+4)(x-3)$

$4 = a(1+4)(1-3)$

$4 = a(5)(-2)$ $4 = -10a$
                  $\overline{-10}$

$a = \dfrac{-4}{10} \left(\dfrac{-2}{5}\right)$

equation: $y = -\dfrac{2}{5}(x+4)(x-3)$

Find the equation of the parabola with a vertex (1, −4) passing through (0, 10).

vertex   x = 0
              y = 10

$y = a(x-h)^2 + k$

$h = 1$ $y = a(x-1)^2 - 4$
$k = -4$
              $10 = a(0-1)^2 - 4$] subst. (0, 10)

              $10 = a(-1)^2 - 4$

              $a = 14$

equation: $y = 14(x-1)^2 - 4$

$p = 2$ $q = -4$

6.8—295

Find the parabola passing through (2,0), (−4,0), and (3,14).   [int.]

$x = 3$
$y = 14$

$y = a(x-p)(x-q)$
$y = a(x-2)(x+4)$
$14 = a(3-2)(3+4)$
$14 = a(1)(7)$
$14 = 7a$
$a = 2$

equation:
$y = 2(x-2)(x+4)$

So, in both of the scenarios covered, you have to have very specific points, either the vertex, or x-intercepts and one other point in order to derive the equation.

The final scenario is when you are given ANY three points on the parabola, not necessarily the vertex or x-intercepts.

Given any 2 points, you can find the equation of the __line__ passing through those two points.

Given any 3 points that are not co-linear, you can determine the equation of the __parabola__ passing through those 3 points.

Plot any 3 points below to see how only one parabola can fit through the points.

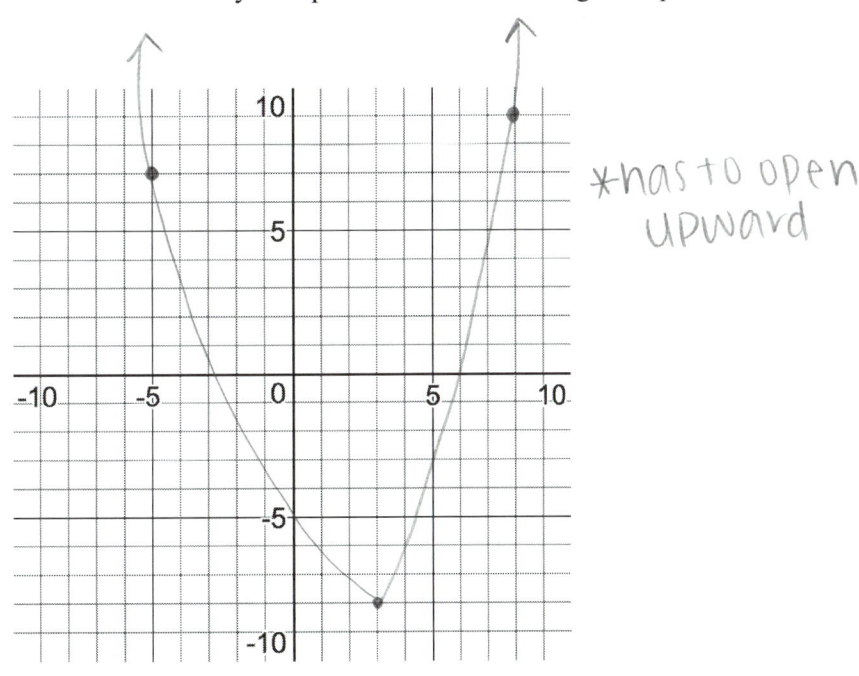

*has to open upward

Scenario 3: Given any 3 points on a parabola. *NOT ON TEST*
Use the TI-Nspire to input the data points, and then quadratic regression to determine the quadratic function.

Find the quadratic function passing through $(2,2), (3,3)$, and $(0,-2)$.

TI-Nspire	
New Document 4: Add Lists and Spreadsheets Enter the values, and label the columns	Spreadsheet with columns xval (2, 3, 0) and yval (2, 3, -2)
Create a Data and Statistics Page CTRL-DOC 5: Add Data and Statistics Assign the values you entered from your spreadsheet to the $x-$ and $y-$ axes.	Scatter plot of the three points

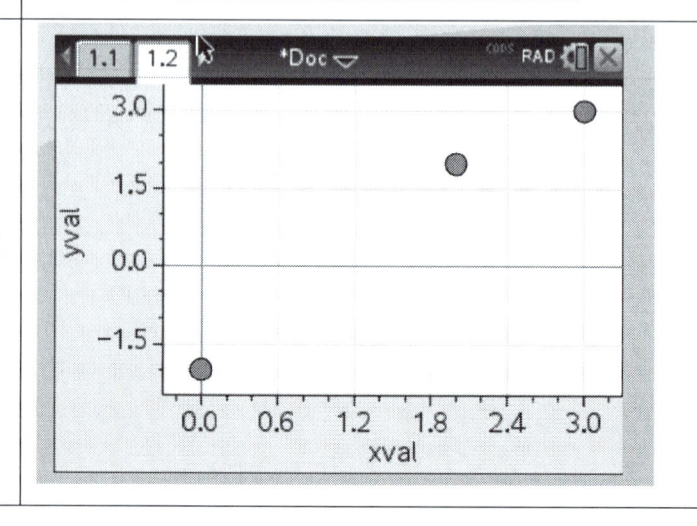

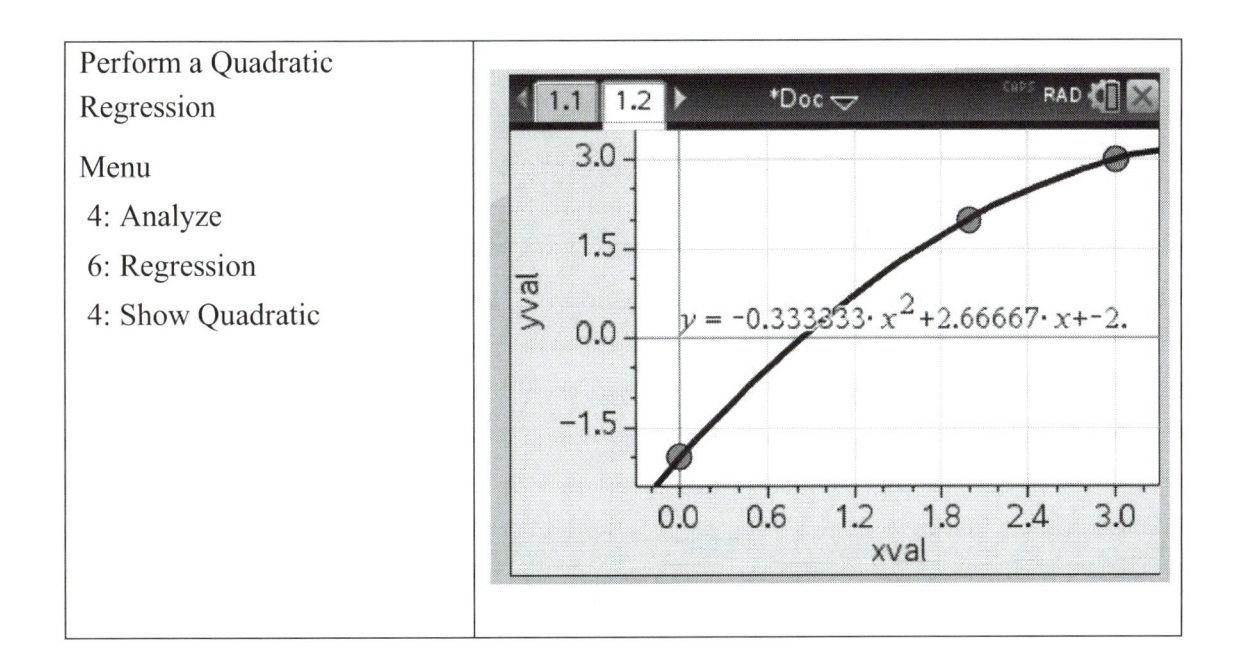

Perform a Quadratic Regression Menu 4: Analyze 6: Regression 4: Show Quadratic	*(calculator screenshot showing scatter plot with quadratic curve and equation $y = -0.333333 \cdot x^2 + 2.66667 \cdot x + -2$)*

Remember when we performed linear lines of best fit, and the correlation, or r values? If we had more than 3 data points, our quadratic would be a regression, a best fit, not passing through all points. Consequently, there would be a correlation value. By only giving 3 points, our quadratic will be an exact fit and have a correlation $r = 1$.
↳ perfect fit

Homework 6.8 Writing Quadratic Equations

1. Write the equation of the quadratic function passing through the described points.

Vertex: (0,0) Point: (2,4)	Vertex: (2,1) Point (4,5)	Vertex: (4,−5) Point (1,13)
$y = a(x-h)^2 + k$ $y = a(x-0)^2 + 0$ $4 = a(2-0)^2 + 0$ $4 = a(2)^2$ $4 = 4a$ $\boxed{a = 1}$ $y = 1(x-0)^2 + 0$	$y = a(x-2)^2 + 1$ $5 = a(4-2)^2 + 1$ $5 = a(2)^2 + 1$ $5 = 4a + 1$ $4 = 4a$ $\boxed{a = 1}$ $y = 1(x-2)^2 + 1$	$y = a(x-4)^2 - 5$ $13 = a(1-4)^2 - 5$ $13 = a(-3)^2 - 5$ $13 = 9a - 5$ $18 = 9a$ $\boxed{a = 2}$ $y = 2(x-4)^2 - 5$

Points: (2,0), (3,0), (4,2)	Points: (−4,0), (1,0), (−3,−4)	Points: (−5,0), (5,0), (6,15)
$y = a(x-p)(x-q)$ $y = a(x-2)(x-3)$ $2 = a(4-2)(4-3)$ $2 = a(2)(1)$ $2 = 2a$ $\boxed{a = 1}$ $y = 1(x-2)(x-3)$	$y = a(x+4)(x-1)$ $-4 = a(-3+4)(-3-1)$ $-4 = a(1)(-4)$ $-4 = -4a$ $\boxed{a = 1}$ $y = 1(x+4)(x-1)$	$y = a(x+5)(x-5)$ $15 = a(6+5)(6-5)$ $15 = a(11)(1)$ $15 = 11a$ $\boxed{a = \frac{15}{11}}$ $y = \frac{15}{11}(x+5)(x-5)$

Homework 6.8 Writing Quadratic Equations

2. Write the equation of the quadratic function passing through the described points.

Points: $(-4,-3), (0,-2), (1,7)$	Points: $(-2,-4), (0,-10), (3,-7)$
$y = 1.75x^2 + 7.25x - 2$	$y = 0.8x^2 + 1.4x - 10$

3. Write the equation of the graphs below:

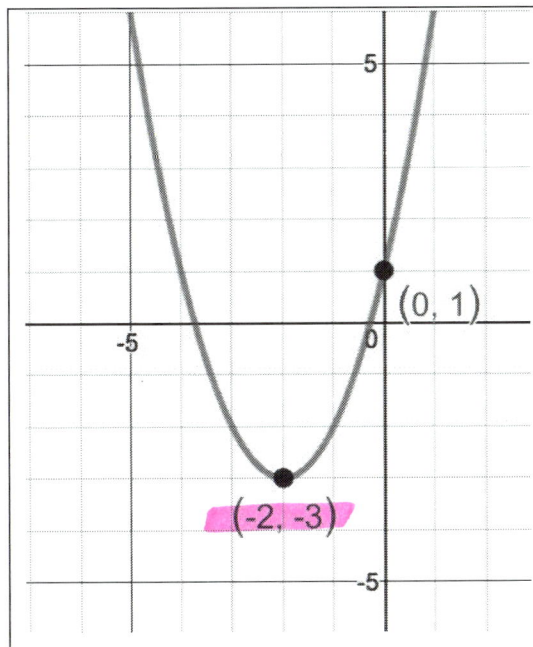

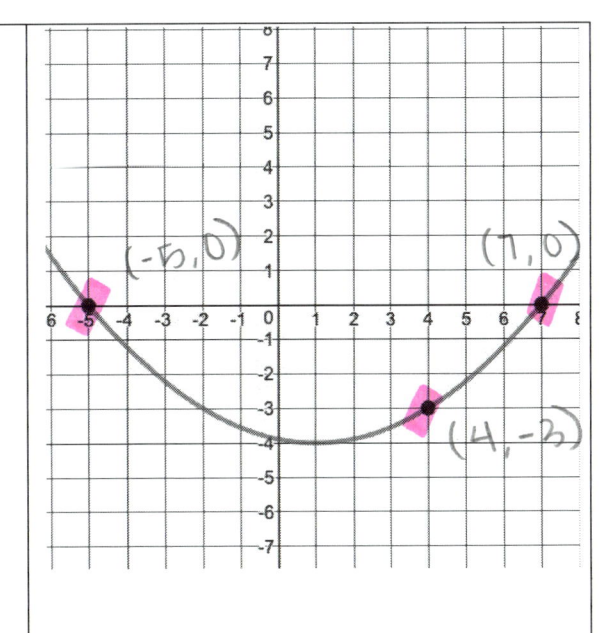

Graph 1: vertex $(-2, -3)$, point $(0, 1)$

Graph 2: x-intercepts $(-5, 0)$ and $(7, 0)$, point $(4, -3)$

vertex → $y = a(x-h)^2 + k$	intercept → $y = a(x-a)(x-p)$
$y = a(x+2)^2 - 3$	$y = a(x+5)(x-7)$
$1 = a(0+2)^2 - 3$	$-3 = a(4+5)(4-7)$
$1 = 4a - 3$	$-3 = a(9)(-3)$
$4 = 4a$ $\boxed{a=1}$	$-3 = -27a$
$y = 1(x+2)^2 - 3$	$a = \frac{1}{9}$
	$y = \frac{1}{9}(x+5)(x-7)$

Homework 6.8 Writing Quadratic Equations

CHALLENGE PROBLEMS

Use our prior work with matrices to find the exact parabola passing through
$$(1,6) \ (-1,10) \ (2,13)$$
You are looking for a function of the form
$$f(x) = ax^2 + bx + c$$
You know the values of $f(1), f(-1),$ and $f(2)$. Use those values to create a 3x3 system of matrices, and solve that equation for $a, b,$ and c.

Equation 1	$a(1)^2 + b(1) + c = 6$ $a + b + c = 6$
Equation 2	
Equation 3	

Use your TI once you get your 3x3 matrix equation.

Final answer, $f(x) = $ _____

6.9. Converting between forms for Quadratic equations

In our work with the three forms of quadratic equations, each form gives us specific information about the graph without any calculations being made. What does each form of the equation tell us?

Vertex Form	Intercept Form	Standard Form
$y = a(x-h)^2 + k$	$y = a(x-p)(x-q)$	$y = ax^2 + bx + c$
DoO (Direction of Opening) $a > 0 \to$ up $a < 0 \to$ down	DoO $a > 0 \to$ up $a < 0 \to$ down	DoO $a > 0 \to$ up $a < 0 \to$ down
AoS $x = h$	AoS $x = \frac{p+q}{2}$	AoS $x = \frac{-b}{2a}$
Vertex (h, k)	$x-$intercepts $(p, 0) \ (q, 0)$ vertex: $x = $ AoS	$y-$intercept $(0, c)$ $x = $ AoS

In today's lesson, we will cover methods for converting between the various forms.

By the end of the lesson, you should be comfortable in converting:

From	To
Vertex Form and Intercept Form	Standard Form
Standard Form	Vertex Form and Intercept Form

Converting from Vertex Form to Standard Form

How do we convert from $y = a(x-h)^2 + k$ to $y = ax^2 + bx + c$?

You simply perform the indicated operation in the Vertex Form and gather like terms.

Example. Convert $y = -2(x + 3)^2 - 5$ to Standard Form. $y = ax^2 + bx + c$

$y = -2(x+3)(x+3) - 5$

$y = -2(x^2 + 3x + 3x + 9) - 5$

$y = -2(x^2 + 6x + 9) - 5$

$y = -2x^2 - 12x \boxed{-18 - 5}$

~~$y = -2x^2 - 12x - 23$~~

x-int
solve →
$-2x^2 - 12x - 23 = 0$
$\boxed{\text{NO X-INT}}$

Converting from Intercept Form to Standard Form

How do we convert from $y = a(x - p)(x - q)$ to $y = ax^2 + bx + c$?

Very similar to what we did with Vertex to standard, perform the indicated binomial multiplication and gather like terms.

Example. Convert $y = 3\overbrace{(x - 4)(x + 6)}^{FOIL}$ to Standard Form.

$y = 3(x^2 + 6x - 4x - 24)$

$y = 3(x^2 + 2x - 24)$

~~$y = 3x^2 + 6x - 72$~~

x-int →
$(4, 0)(-6, 0)$

y-int →
$(0, -72)$

AOS →
$x = \frac{4 - 6}{2}$
$x = -1$

open → $\boxed{UP}$

vertex → plug in -1
$(-1, -75)$

Converting from Standard Form to Intercept Form

How do we convert from $y = ax^2 + bx + c$ to $y = a(x-p)(x-q)$? What method do you think we will use?

To convert from standard to Intercept Form, the equation must be factored. *must be factorable*

Example. Convert $y = x^2 - 3x - 10$ to Intercept Form.

2#
m: -10
a: -3
-5, 2

$y = (x-5)(x+2)$

x-int → (-2, 0) (5, 0)
y-int → (0, -10)

Example. Convert $y = -3x^2 - 21x - 36$ to Intercept Form.

2#
m: 12
a: 7
3, 4

$y = -3(x^2 + 7x + 12)$
$y = -3(x+3)(x+4)$

x-int → (-3, 0) (-4, 0)
y-int → (0, -36)

Example. Convert $y = 2x^2 + 15x + 7$ to Intercept Form.

2#
m: 14
a: 15
1, 14

$y = 2x^2 + x + 14x + 7$
$x(2x+1) + 7(2x+1)$
$y = (x+7)(2x+1)$

x-int → (-7, 0) ($-\frac{1}{2}$, 0)
y-int → (0, 7)

Converting Standard Form to Vertex Form

How do we convert from $y = ax^2 + bx + c$ to $y = a(x-h)^2 + k$? What do you think we need first in order to write something in Vertex Form?

Example. Convert $y = x^2 - 4x + 7$ to Vertex Form.

Step 1: Find the AoS and vertex. You know the axis of symmetry. Use this to find the vertex	$x = \frac{-b}{2a}$ $x = \frac{4}{2(1)}$ $x = \boxed{2}$ AoS $y = (2)^2 - 4(2) + 7$ $y = 4 - 8 + 7$ $\boxed{y = 3}$ (2, 3)
Step 2: You already know the value of a from the original equation. Substitute for a, h, k to write your Vertex Form equation final answer. $y = a(x-h)^2 + k$	$y = 1(x-2)^2 + 3$ $h = $ AoS $(h, k) = $ vertex $a = 1$

Example. Convert $y = 2x^2 + 8x + 7$ to Vertex Form.

Step 1: Find the AoS and vertex. You know the axis of symmetry. Use this to find the vertex $a = 2$ $b = 8$ $c = 7$	AoS → $x = \frac{-b}{2a}$ $x = \frac{-8}{2(2)}$ $\boxed{x = -2}$ vertex → $y = 2(-2)^2 + 8(-2) + 7$ $y = 2(4) - 16 + 7$ $\boxed{y = -1}$ (-2, -1)
Step 2: You already know the value of a from the original equation. Substitute for a, h, k to write your Vertex Form equation final answer. $h = -2$ $(h, k) = (-2, -1)$ $a = 2$	$y = 2(x+2)^2 - 1$

Converting Standard Form to Vertex Form By Completing the Square

Another way to convert from standard to vertex is to simply use completing the square. By completing the square, the resulting equation will end up in vertex form.

Convert $f(x) = 2x^2 + 8x + 1$ to vertex form.

Factor the 2 from the first two terms	$f(x) = 2(x^2 + 4x) + 1$
Complete the square	$f(x) = 2(x^2 + 4x + 2^2) + 1 - (2 \cdot 2^2)$ $f(x) = 2(x+2)^2 + 1 - 8$
Final answer is in vertex form	$f(x) = 2(x+2)^2 - 7$
What is the vertex of the parabola?	$(h, k) = (-2, -7)$ AOS $= x = -2$

Convert $f(x) = -2x^2 + 6x - 3$ to vertex form by completing the square.

$f(x) = -2(x^2 - 3x) - 3$

$f(x) = -2\left(x^2 - 3x + \left(\frac{3}{2}\right)^2\right) - 3 + \left(2 \cdot \frac{3}{2}^2\right)$

$= -2\left(x - \frac{3}{2}\right)^2 + \frac{3}{2}$

vertex $= \left(\frac{3}{2}, \frac{3}{2}\right)$

AOS $= x = \frac{3}{2}$

Homework 6.9 Converting between forms for quadratic equations

Convert the following to Standard Form.

1. $y = -2(x-5)(x+3)$

$y = -2(x^2 + 3x - 5x - 15)$
$\boxed{y = -2x^2 - 2x - 15}$

2. $y = -(x+5)^2 + 2$

$y = -1(x+5)(x+5) + 2$
$y = -1(x^2 + 5x + 5x + 25) + 2$
$y = -1(x^2 + 10x + 25) + 2$
$\boxed{y = -x^2 - 10x - 23}$

---

1. $y - 3 = 2(x-4)^2$

$y = 2(x-4)^2 + 3$
$y = 2(x-4)(x-4) + 3$
$y = 2(x^2 - 4x - 4x + 16) + 3$
$y = 2x^2 - 32x + 32 + 3$
$\boxed{y = 2x^2 - 32x + 35}$

2. $y = 2\left(x + \frac{1}{2}\right)(x+1)$

$y = 2\left(x^2 + x + \frac{1}{2}x + \frac{1}{2}\right)$
$\boxed{y = 2x^2 + 3x + 1}$

---

5. $y = -3(x+1)(x-1)$

$y = -3(x^2 - x + x - 1)$
$y = -3(x^2 - 1)$
$\boxed{y = -3x^2 + 3}$

Homework 6.9 Converting between forms for quadratic equations

Rewrite the following equations in the specified form.

6. $y = x^2 + 6x + 8$ (intercept)	7. $y = 3x^2 - 15x - 18$ (intercept)
2# m: 8 a: 6 2, 4 $y = (x+2)(x+4)$	2# $y = 3(x^2 - 5x - 6)$ m: -6 a: -5 $y = 3(x-6)(x+1)$ -6, 1
8. $y = 2x^2 - 128$ (intercept) $y = 2(x^2 - 64)$ $y = 2(x-8)(x+8)$	9. $y = x^2 - 8x + 19$ (vertex) AOS = $\frac{-(-8)}{2(1)}$ $y = 4^2 - 8(4) + 19$ AOS(x) = 4 $y = 16 - 32 + 19$ vertex → $y = 3$ x = 4 $y = (x-4)^2 + 3$ (4, 3)
13. $y = 2x^2 + 24x + 25$ (vertex) AOS $\frac{-24}{2(2)}$ $y = 2(-6)^2 + 24(-6) + 25$ $y = 2(36) - 144 + 25$ x = -6 $y = -47$ (-6, -47) $y = 2(x+6)^2 - 47$	14. $y = 5x^2 + 10x + 7$ (vertex) AOS $y = 5(-1)^2 + 10(-1) + 7$ $\frac{-10}{2(5)}$ $y = 5 - 10 + 7$ x = -1 $y = 2$ (-1, 2) $y = 5(x+1)^2 + 2$

Homework 6.9 Converting between forms for quadratic equations

CHALLENGE PROBLEMS

Give the equation of the parabola passing through $\left(-\frac{1}{2}, 0\right)$, $(3, 0)$, $(0, -3)$, and $(1, -6)$ in standard, vertex, and intercept form.

Standard	Vertex	Intercept

Homework 6.9 Converting between forms for quadratic equations

CHALLENGE PROBLEMS (Continued)

Graph the parabola $f(x)$

Vertex
x −intercept(s)
y −intercept
Value of the discriminant

What is the solution to $f(x) > 0$?

6.10. Review

Assessment Checklist. Below are the competencies one should master in preparation for an assessment on writing quadratic equations and converting between forms of quadratic equations.

- ☐ Writing the equation of a quadratic given the vertex and one other point, either as ordered pairs, or on a graph.
- ☐ Writing the equation of a quadratic given the x-intercepts and one other point, either as ordered pairs, or on a graph.
- ☐ Writing the equation of a quadratic given ANY 3 points. If you are not given the vertex or x-intercept this is done using the TI-Nspire and quadratic regression.
- ☐ Convert the equation of a quadratic function from Vertex Form to Standard Form.
- ☐ Convert the equation of a quadratic function from Intercept Form to Standard Form.
- ☐ Convert the equation of a quadratic function from Standard Form to Intercept Form.
- ☐ Convert the equation of a quadratic function from Standard Form to Vertex Form.
- ☐ Ability to perform completing the square OR finding the vertex when converting from standard to vertex form.

6.10—311

Review 6.10

This review has material from the entirety of our study of quadratics. This material is covered from section 5.1 to 6.9.

Evaluate the following expressions with complex numbers, and leave your answer in the form $a + bi$.

$i^2 = -1$
$i^3 = -i$
$i^4 = 1$

1. $(4 + 5i) + (3 - 2i)$	2. i^{51}
$\boxed{7 + 3i}$	$4\overline{)51}$ $\;\;\;\;(i^3)^4 = \boxed{-1}$ $\;\;\;\;\;\;\;\;\frac{4\downarrow}{11}$ $\;\;\;\;\;\;\;\;\;\;\frac{8}{3}$
3. $\frac{1}{3-i}\left(\frac{3+i}{3+i}\right) = \frac{3+i}{9-i^2} = -1$ $\;\;\;\;\;\frac{3+i}{9+1} = \frac{3+i}{10} \to \boxed{\frac{3}{10} + \frac{i}{10}}$	4. $(1+i)(2-i)$ $2 + 2i - i - i^2 = -1$ $2 + i + 1$ $\boxed{3 + i}$

3. Provide the requested information and graph the function, $y = -(x+3)^2 + 4$ ← **vertex form!**

Direction of opening	
down	
Vertex (h, k)	
$(-3, 4)$	
Axis of Symmetry	
$x = h$	
$x = -3$	

x	y
2	-21

$y = -1(2+3)^2 + 4$
$y = -1(5)^2 + 4$
$y = -25 + 4 = \boxed{-21}$

6.10—312

Review 6.10

intercept form!

4. Provide the requested information and graph the function, $y = 2(x+3)(x-1)$

Direction of opening	
UP	
x – intercepts	
$(-3, 0)\ (1, 0)$	
Axis of Symmetry	
$x = -1$	
Vertex $x = -1$ $\quad (-1, 8)$	
$y = 2(-1+3)(-1-1)$	
$y = -8$	
Domain	Range
$(-\infty, \infty)$	$[-8, \infty)$

dotted,

5. Provide the requested information and graph the function, $y > 2x^2 - 8x + 3$

Direction of opening
UP
Axis of Symmetry $\quad \dfrac{-b}{2a}$
$\dfrac{-(-8)}{2(2)} = 2$
$x = 2$
Vertex $\quad x = 2$
$2(2)^2 - 8(2) + 3$
$8 - 16 + 3$
$(2, -5)$
y – intercept
$(0, 3)$

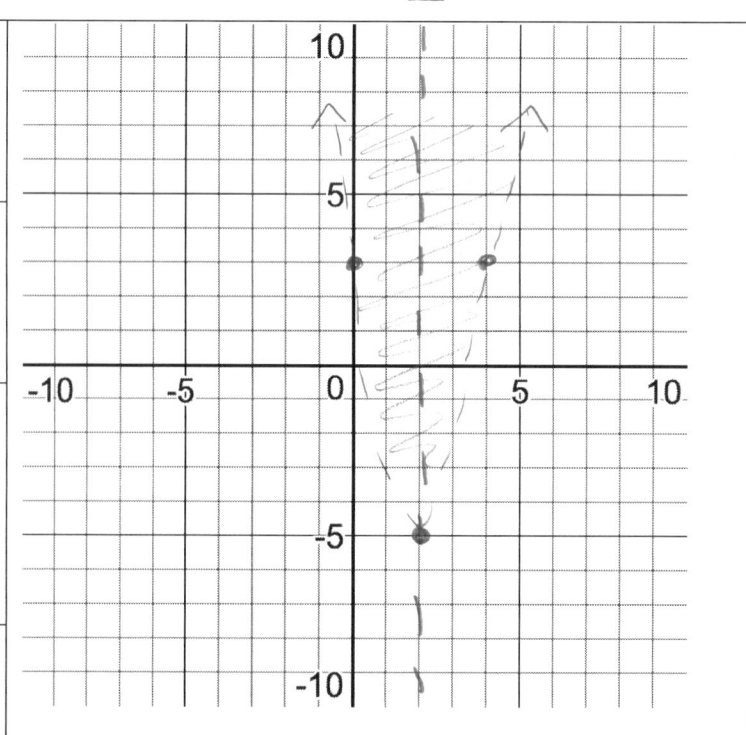

$(0,0) \rightarrow \quad 0 > 2(0)^2 - 8(0) + 3$
$\qquad\qquad\quad 0 > 3 \ \text{NO}$

Review 6.10

Solve by ANY method.

8. $x^2 - 7 - 44 = 0$ m: -44   a: -7

$(x-11)(x+4) = 0$ (-11, 4)

$\boxed{x = -4, 11}$

9. $4x^2 - 8x - 5 = 0$ m: -20   a: -8

$4(x-10)(x+2)$ -10, 2

$\boxed{x = -2, 10}$

10. $\sqrt{(x+1)^2} = \sqrt{14}$

$x + 1 = \pm\sqrt{14}$

$\boxed{x = -1 \pm \sqrt{14}}$

11. $\dfrac{2x^2 = 128}{2}$

$\sqrt{x^2} = \sqrt{64}$

$\boxed{x = \pm 8}$

12. $x^2 + 2x = -3$ m: 3   a: 2

$x^2 + 2x + 3 = 0$

$\dfrac{-b \pm \sqrt{b^2 - 4ac}}{2a}$

$\dfrac{-2 \pm \sqrt{2^2 - 4(3)}}{2} = \dfrac{-2 \pm \sqrt{-8}}{2}$

$\dfrac{-2}{2} \pm \dfrac{4}{2}i$

$\boxed{-1 \pm 2i}$

13. $4x^2 + 4x + 1 = 0$ m: 4   a: 4

$4(x+2)(x+2) = 0$ 2, 2

$\boxed{x = -2}$

6.10—313

Review 6.10

14. Convert to Standard Form.

$y = 2(x+2)^2 + 3$	$y = -3(x+1)(x-2)$
$y = 2(x+2)(x+2) + 3$	$y = -3(x^2 - 2x + x - 2)$
$y = 2(x^2 + 2x + 2x + 4) + 3$	$y = -3(x^2 - x - 2)$
$\boxed{y = 2x^2 + 8x + 11}$	$\boxed{y = -3x^2 + 3x + 6}$

15. Convert to the specified form.

$m: -28$
$a: 3$
$-4, 7$

$y = x^2 + 3x - 28$ to Intercept Form. Factor!	$y = -x^2 - 6x - 10$ to Vertex Form
$\boxed{y = (x-4)(x+7)}$	$AOS = \frac{-b}{2a} = \frac{6}{2(-1)} = \boxed{-3}$
	$x = -3$
	$y = -(-3)^2 - 6(-3) - 10$

 $y = 17 \quad (-3, 17)$ and $\boxed{y = -1(x+3)^2 + 17}$

16. Give the equation of the quadratic passing through the following points.

$(-11, 0), (7, 0) \,\&\, (0, 154)$	Vertex $(-3, 1)$ passing through $(-2, 3)$	$(1,1), (3,4), \,\&\, (4,7)$
x-int	$y = a(x-h)^2 + k$	on calculator
$y = a(x-p)(x-q)$	$y = a(x+3)^2 + 1$	
$y = a(x+11)(x-7)$	$3 = a(-2+3)^2 + 1$	
$154 = a(0+11)(0-7)$	$3 = a(1) + 1$	
$154 = -77a$	$2 = a$	
$\boxed{a = -2}$		
$\boxed{y = -2(x+11)(x-7)}$	$\boxed{y = 2(x+3)^2 + 1}$	

plug in x-int!
plug in x & y values!

17. Consider the function $y = -2x^2 + 16x - 30.$ Make sure to label/graph the AoS, Vertex, and x-intercept(s).

Parabola opens UP (DOWN)
Equation of the axis of symmetry $AOS = \frac{-b}{2a}$ $x = 4$
Vertex: $x = 4$ $-2(4)^2 + 16(4) - 30$ $y = 2$ $(4, 2)$
Domain $(-\infty, \infty)$
Range $(-\infty, 2]$
Extra Points

x	y
2	-6

$y = -2(-2)^2 + 16(-2) - 30$

Find the x-intercept(s):

$y = -2x^2 + 16x - 30$
$y = -2(x - 6)(x - 10)$
$y = -2(x - 5)(x - 3) = 0$

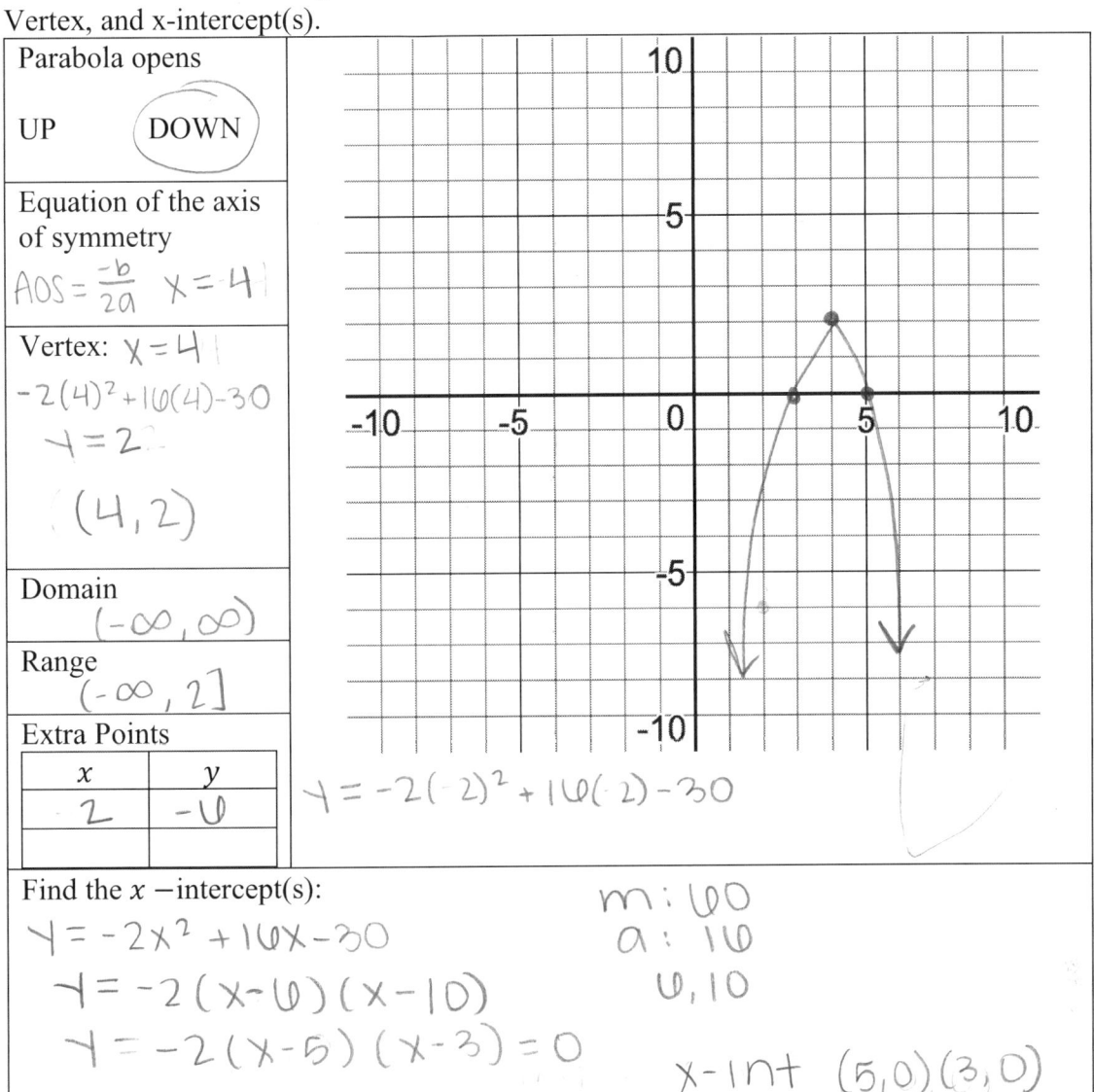

m: 60
a: 16
6, 10

x-int $(5, 0), (3, 0)$

18. Rationalize the expressions below, leaving no square roots in the denominator.

$\frac{7}{\sqrt{5}} \left(\frac{\sqrt{5}}{\sqrt{5}}\right)$

$\frac{7\sqrt{5}}{\sqrt{5} \cdot \sqrt{5}} = \frac{7\sqrt{5}}{5}$

$\frac{2}{1 - \sqrt{2}} \left(\frac{1 + \sqrt{2}}{1 + \sqrt{2}}\right)$

$\frac{2 + \sqrt{2}}{1 - \sqrt{2} + \sqrt{2} - 2} = \frac{2 + 2\sqrt{2}}{1 - 2}$

$\boxed{\frac{2 + 2\sqrt{2}}{-1}}$

19. Complete the following

Equation	$x^2 + 2x + 2 = 0$	
Value of the discriminant: $b^2 - 4ac$ $2^2 - 4(1)(2) = \boxed{-4}$		
Number of solutions	☐ 0 ☐ 1 ☑ 2 Check one	
Type of solutions (Check all that apply)	☐ Real ☑ Complex	☐ Rational ☐ Irrational

Equation	$2x^2 + 2x - 3 = 0$	
Value of the discriminant: $b^2 - 4ac$ $2^2 - 4(2)(-3) = \boxed{28}$ $\sqrt{28}$ not perfect		
Number of solutions	☐ 0 ☐ 1 ☑ 2 Check one	
Type of solutions (Check all that apply)	☑ Real ☐ Complex	☐ Rational ☑ Irrational

20. Give the equation of the parabola.

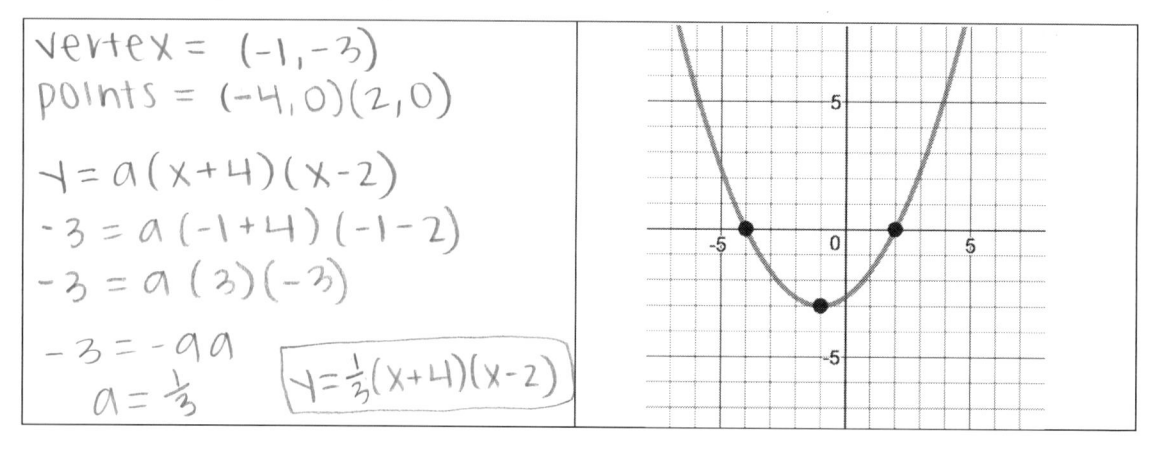

vertex = $(-1, -3)$
points = $(-4, 0)(2, 0)$

$y = a(x+4)(x-2)$
$-3 = a(-1+4)(-1-2)$
$-3 = a(3)(-3)$
$-3 = -9a$
$a = \frac{1}{3}$

$\boxed{y = \frac{1}{3}(x+4)(x-2)}$

7. QUADRATIC MODELING

In this last unit of the book, we will look at data that can be modeled with quadratic functions.

Your knowledge of quadratic functions will be used extensively. How quadratic functions are graphed, their symmetry, and how quadratic functions are solved are but a few elements from our prior unit that will carry forward into this unit.

In this unit, there are five main types of quadratic models we will be investigating.

Working with specified quadratic models.	An application problem will define a model and give you the equation. You will answer questions about the model that has been defined.
Finding quadratic models with the TI-Nspire.	Data points from a model will be given, and you will use the TI-Nspire to perform quadratic regression to find the model.
Projectile motion	Using the equations from physics that model an object thrown up or down, we will answer questions about projectile motion.
Quadratic modeling with borders	We will model application problems that involve borders, e.g. the wooden frame around a photograph.
Quadratic modeling with fencing	Models of applications that involve a fence, e.g. building a fence around a specified object.

7.1. Working with specified quadratic models

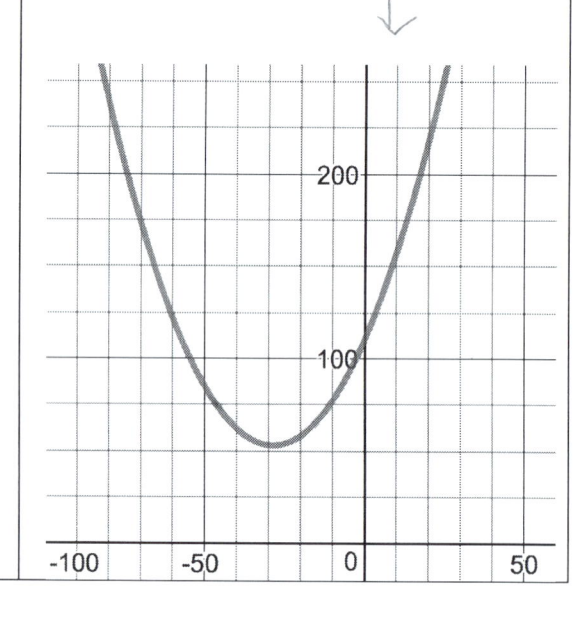

Suppose that the future population of Mathville City t years after January 1, 2015 is described (in thousands) by the quadratic model, $P(t) = 110 + 4t + 0.07t^2$

[annotation: $t =$ yrs (1000s)]

What do we know about this model? (What does looking at the equation tell us?)

y −intercept	What does this mean in context of the problem?
standard form → (0, 110)	Initial population on 1/1/15 is 110 thousand people.
Axis of Symmetry $t = \frac{-b}{2a} \quad t = \frac{-4}{2(.07)}$ $t = -28.57$	What does the model project? Growth/decline? growth

| By graphing the function $P(t) = 110 + 4t + 0.07t^2$ We see that our model is an upward opening parabola. In the context of the problem, January 1, 2015 corresponds to $t = 0$. So for all years after 2015, our model projects population growth. | *[graph of upward-opening parabola with vertex near $t = -28.57$, showing values 100, 200 on vertical axis and -100, -50, 0, 50 on horizontal axis]* |

What is the projected population of Mathville City on January 1, 2022?

Substitute $t =$ ___7___, and evaluate.

$P(7) = 110 + 4(7) + 0.07(7)^2$

$P(7) = 149.43$

Answer in the form of a sentence in context of the problem.

The population of Mathville City is projected to be ___141.43___ ___thous. people___ on January 1, 2022.

Number   units

In this same model, in how many years will the population reach 180,000?

In our model, the horizontal axis represents time, t, in years after January 1, 2015.

The vertical axis represents population in thousands.

The dotted horizontal line represents $y = 180$, or where our model would reach the population of 180,000.

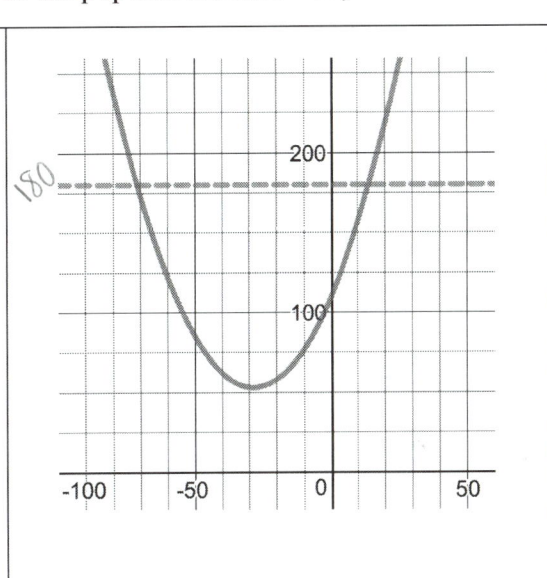

So, to solve for the time, we set

$$.07t^2 + 4t + 110 = 180$$

Subtracting 180 from both sides,

$$.07t^2 + 4t - 70 = 0$$

How do we solve such an equation?

Solve $.07t^2 + 4t - 70 = 0$

Quadratic Formula	TI-Nspire
$t = \dfrac{-b \pm \sqrt{b^2 - 4ac}}{2a}$ $a = 0.07$ $b = 4$ $c = 70$ $\boxed{b^2 - 4ac} = 35.6$ $t = \dfrac{-4 \pm \sqrt{35.6}}{2(0.07)}$ $\dfrac{-4 + \sqrt{35.6}}{0.14} \qquad \dfrac{-4 - \sqrt{35.6}}{0.14}$ $= \boxed{14.07} \qquad = \boxed{-71.189}$	1:File 1:New Document 1: Add Calculator Menu 3: Algebra 8: Polynomial Tools 3: Complex Roots of a Polynomial Enter *remember to put T in* cPolyRoots$(.07t^2 + 4t - 70, t)$ This means, find the roots, or solve $.07t^2 + 4t - 70 = 0$ for t. cPolyRoots$(-70 + 4 \cdot t + 0.07 \cdot t^2, t)$ $\{-71.1898, 14.047\}$

So, our solution has two answers. How do we interpret that in the context of the problem?

In __14.07__ years, the population of Mathville city will be 180,000.

What if the question had been "In what year will the population reach 180,000?"

The population will reach 180,000 in the year __2029__.

In addition to using the quadratic formula or the cPolyRoots command on the TI-Nspire, you can also use your calculator to solve quadratic equations graphically.

In the prior quadratic model, we were asked to find the year in which the population of Mathville reached 180,000. Mathematically, we reached the equation:

$$.07t^2 + 4t + 110 = 180$$

If we can find the rightmost intersection of $$f_1(t) = .07t^2 + 4t + 110$$ and the horizontal line $$f_2(t) = 180$$ We will have the value of t, when the population reaches 180,000.	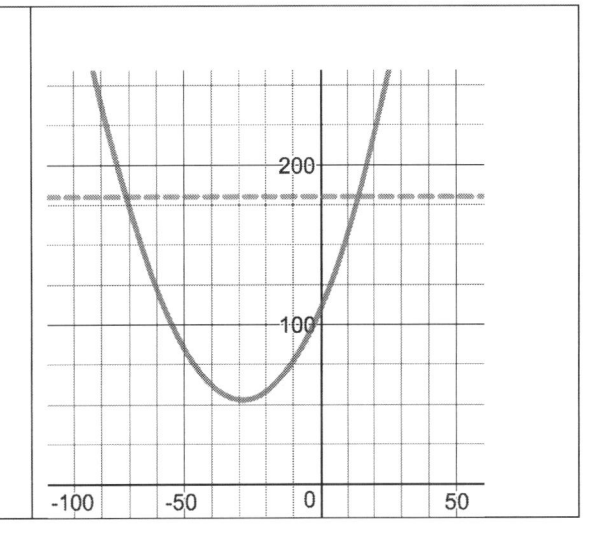

Solving a quadratic model graphically

Ctrl-i or Ctrl-Doc 2: Add Graphs Graph the quadratic	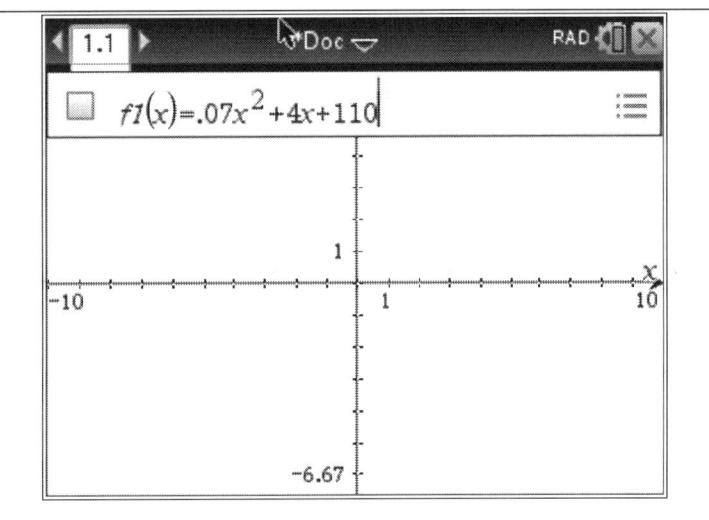
You will most likely need to adjust the window settings to see the graph. MENU 4: Window Zoom	This is not an exact science. Human intelligence of the model helps. Since our horizontal axis is time, try adjusting the x−axis to 0,50]. Since we are looking for when the y value is 180, adjust the y−axis to be [0,300].

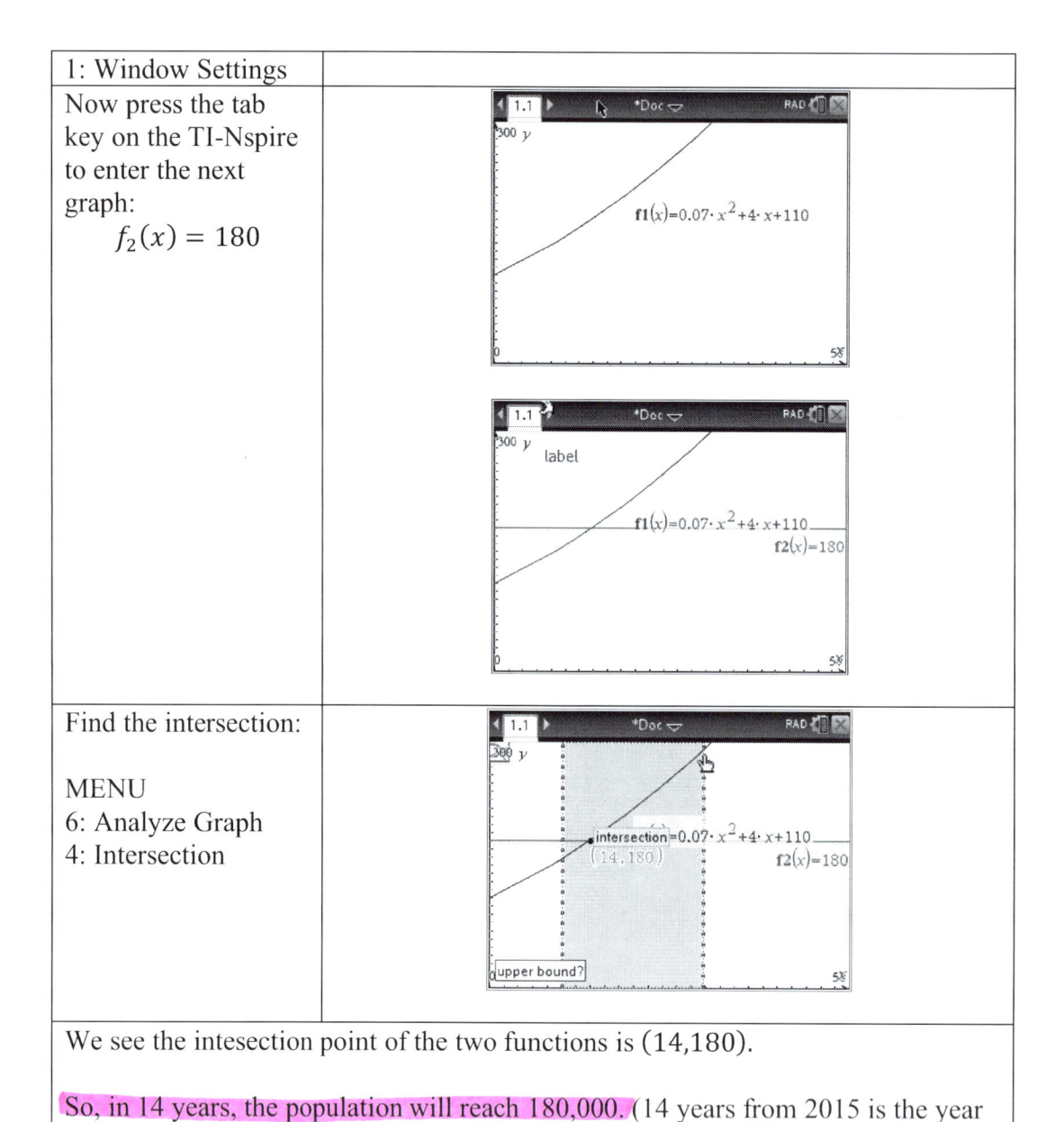

1: Window Settings	
Now press the tab key on the TI-Nspire to enter the next graph: $f_2(x) = 180$	
Find the intersection: MENU 6: Analyze Graph 4: Intersection	
We see the intesection point of the two functions is (14,180). So, in 14 years, the population will reach 180,000. (14 years from 2015 is the year 2029.)	

7.1—323

Consider the polynomial function, $I(t) = -.1t^2 + 2t$ representing the yearly income (or loss) from a real estate investment, where t is time in years. After what year does income begin to decline?

First, this is a quadratic in Standard Form. $a = -0.1$ $b = 2$ $c = 0$	What is the y-intercept, and what does it mean in context? (0, 0) Initially the income is $0.

Sketch a rough graph. By looking at the graph, what are we being asked to find?

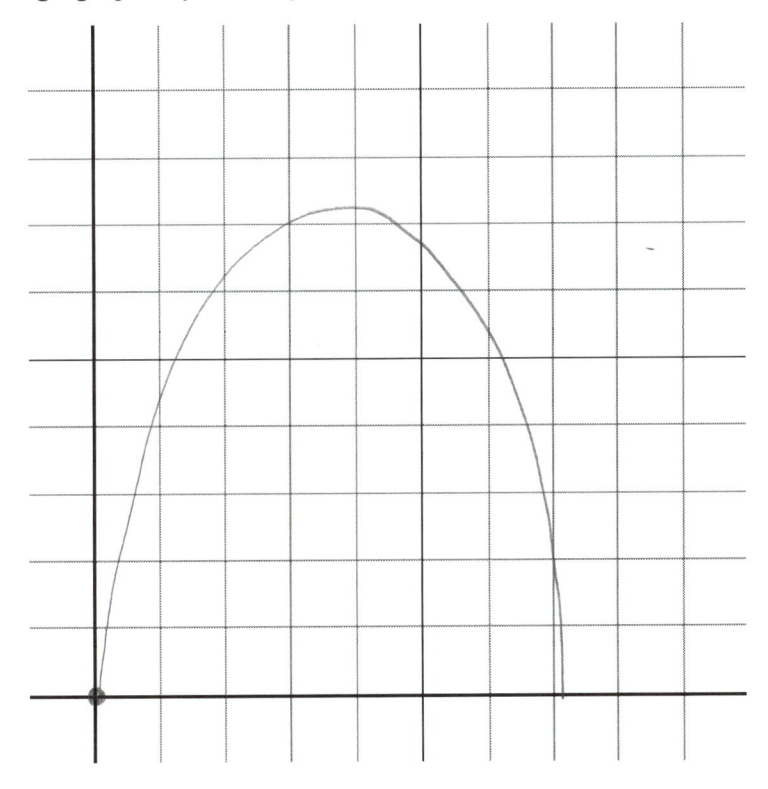

Solution (0, 10)

After 10 years, the income will begin to decline.

The number of mosquitoes $M(x)$, in millions, in a certain area depends on the June rainfall x, in inches, according to the equation

$$M(x) = 19x - x^2$$

What rainfall produces the maximum number of mosquitoes? ==What is the maximum number of mosquitos?==

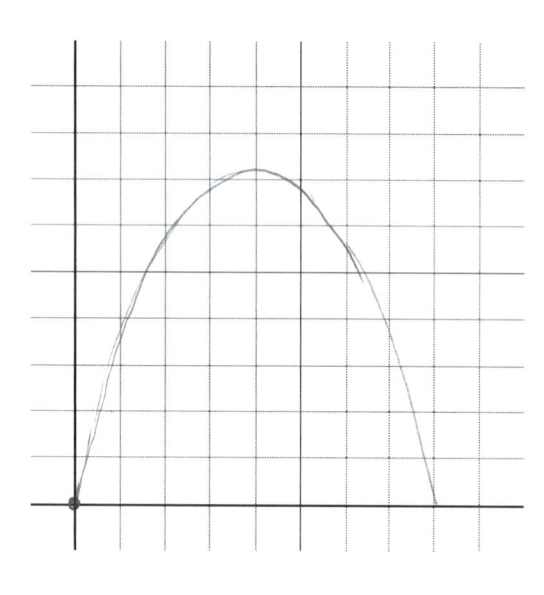

$x = \frac{-b}{2a}$

$x = \frac{-19}{2(-1)} = 9.2$

at 9.2 in reach max

$m(9.2) = 19(9.2) - 9.2^2$

$m(9.2) = 90.16$

==max pop = 90.16 mil mosquitos==

From 1970 to 1990, the prize money, P (in dollars) at the PGA Championship can be approximated using the model

$$P = 2875t^2 + 200{,}000$$

where t represents the number of years after 1970.

a. How much money did the winner of the 1970 championship win?

$t = 0$ initial = $\$200000$

b. How much did the 1980 winner receive?

$t = 10$ $P(10) = 2875(10^2) + 200000$
$P(10) = \$487500$

c. During what year did the prize money first reach $1,350,000?

$P = 1350000$

$2875t^2 + 200000 = 1350000$

plug → $2875t^2 + 200000 - 1350000 = 0$

$= (-20, 20)$

$1970 - 20 = 1950$
$1970 + 20 = 1990$

20 yrs since 1970: In 1990, the prize reaches $1.35 mil

a. Do you think this is a realistic long-term model? Why or why not?

If the prize money continues to go up, it is realistic. Short term realistic.

Homework 7.1 Working with Specified Quadratic Models

1. A football punter punts the football, and the trajectory of the ball is given by the equation $y = -0.035x^2 + 1.4x + 2$, where x is the distance horizontal distance from the kicker, and y is the vertical height of the ball, both in yards.

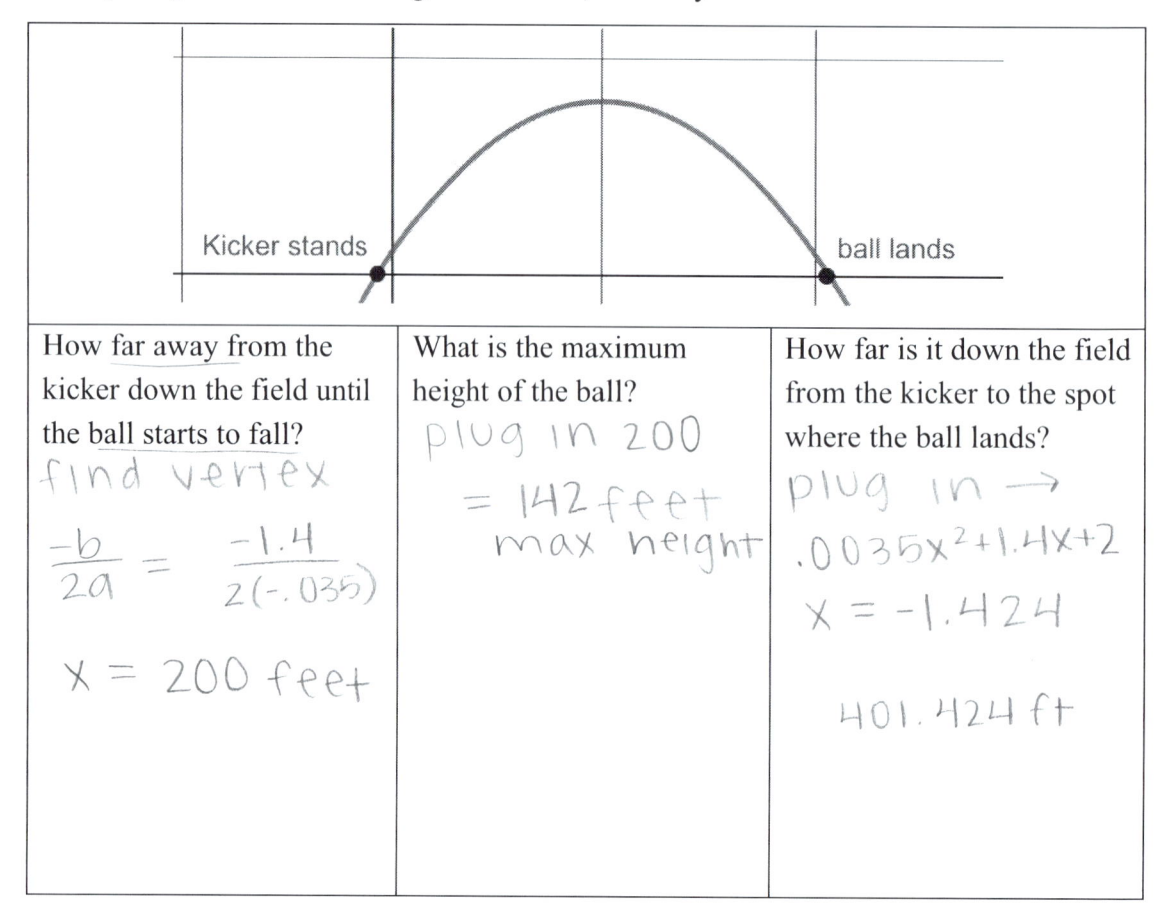

annotation: $-0.0035x^2$

How far away from the kicker down the field until the ball starts to fall?	What is the maximum height of the ball?	How far is it down the field from the kicker to the spot where the ball lands?
find vertex $\frac{-b}{2a} = \frac{-1.4}{2(-.035)}$ $x = 200$ feet	plug in 200 = 142 feet max height	plug in → $.0035x^2 + 1.4x + 2$ $x = -1.424$ 401.424 ft

2. A small business owner found that he could predict his profits, P, (in thousands of dollars) using the equation, $P = -2n^2 + 12n - 4$, where n represents the number of units, in hundreds, that he produced.

a. How many units must he produce in order to reach his maximum projected profit?

find vertex → $\frac{-b}{2a}$ $\frac{-12}{2(-2)} = 3$ 300 units

b. What is the greatest profit he should expect to earn?

$x = 3$ $P(3) = -2(3)^2 + 12(3) - 4$
$P(3) = 14$
$14000 max profit

Homework 7.1 Working with Specified Quadratic Models

3. The concentration of sugar in the bloodstream of a patient is approximated by the function: $C = -0.1t^2 + 0.1t + 5.2$, where C is the concentration of sugar in the bloodstream t hours after a sugar metabolism test.

a. What is the concentration after one hour?

$$C(1) = -0.1(1)^2 + 0.1(1) + 5.2$$
$$C(1) = 5.2$$

5.2 after an hour

b. After how many hours will the concentration reach its maximum?

find vertex →

$$\frac{-b}{2a} \qquad \frac{-0.1}{2(-0.1)} = 0.5 \text{ hours max}$$

c. What is the maximum concentration of sugar in the bloodstream?

$$x = 0.5 \rightarrow C(0.5) = -0.1(0.5)^2 + 0.1(0.5) + 5.2$$
$$C(0.5) = 5.25 \text{ max concentration}$$

d. How long will it take for the sugar concentration to drop to one unit?

solve →
$$-0.1x^2 + 0.1x + 5.2 - 4.2 = 0$$
$$-0.1x^2 + 0.1x + 1 = 0$$
$$x = -2.702, \boxed{3.702} \text{ hrs for concentration to drop 1 unit}$$

Homework 7.1 Working with Specified Quadratic Models

CHALLENGE PROBLEMS

1. You decide to go into the business of making bicycles. You know the startup costs will be a fixed amount of $350,000. Your math teacher tells you that the volume of sales will follow a demand curve and that you can expect to sell $30,000 - 100x$ bikes, where x is the sales price of the bike in dollars. You know that it costs you $55 to make each bike.

Cost Function: $C(x) = 350,000 + 55(30000 - 100x)$

Revenue Function: $R(x) = x(30000 - 100x)$

Create the profit function $P(x)$ and find the price that maximizes profits. What should the bikes be priced so as to maximize the profits? What are the maximum profits?

7.2. Finding quadratic models from data with the TI-Nspire

If you are given data points, instead of a defined quadratic function, we will use the TI-Nspire to find the quadratic function from the data points.

John realizes that a test grade is a function of the number of hours studied and knows from past experience that 1 hour of studying will result in a grade of 66; 2 hours, in a grade of 75; and 5 hours, in a grade of 90.

1	66
2	75
5	90

If the test grade, G, is a function of the hours studied, x, find the equation for G.	$y = -1x^2 + 12x + 55$
Create a Data and Statistics page Control-DOC 4: Add Lists and Spreadsheets Assign your data from the spreadsheet to the axes.	(spreadsheet with columns A: study, B: grade; values 1/66, 2/75, 5/90)
Control-DOC 5: Add Data and Statistics Menu 4: Analyze 6: Regression 4: Show Quadratic	(graph showing $y = -1 \cdot x^2 + 12 \cdot x + 55$)

Let's zoom out a bit to see more of the graph.

Menu: Window/Zoom -> Window Settings

Let's change the XMAX to 15 – you can try different values, but the goal is to see more of the parabola.

What is the minimum amount of time John must study to make a grade of 80?

We must solve:

$-x^2 + 12x + 55 = 80$

Or find the roots to

$-x^2 + 12x + 55 - 80 = 0$

$-x^2 + 12x - 25 = 0$

Using cPolyRoots

Calculator Page
Menu
3: Algebra
8: Polynomial Tools
3: Complex Roots of a Polynomial

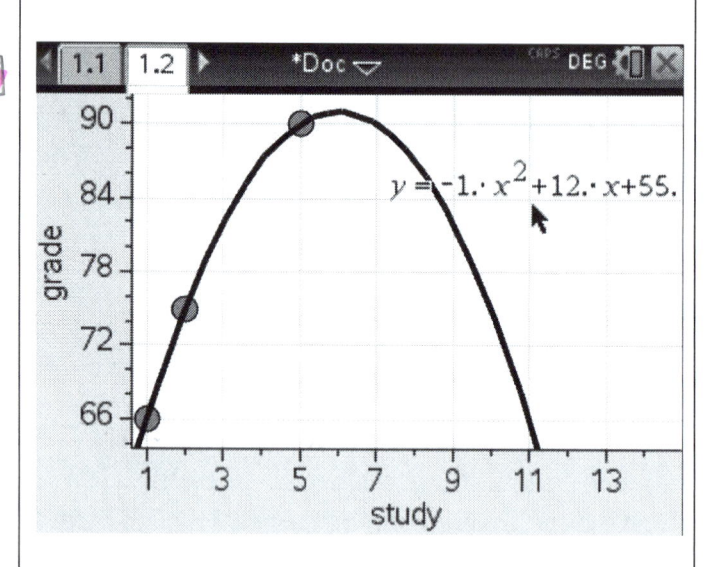

$y = -1 \cdot x^2 + 12 \cdot x + 55$

cPolyRoots$(-1 \cdot x^2 + 12 \cdot x + 55 - 80, x)$
$\{2.68338, 9.31662\}$

Remember, you can also graph the two functions on your TI-Nspire and find the intersection

$f_1(x) = -x^2 + 12x + 55$

$f_2(x) = 80$

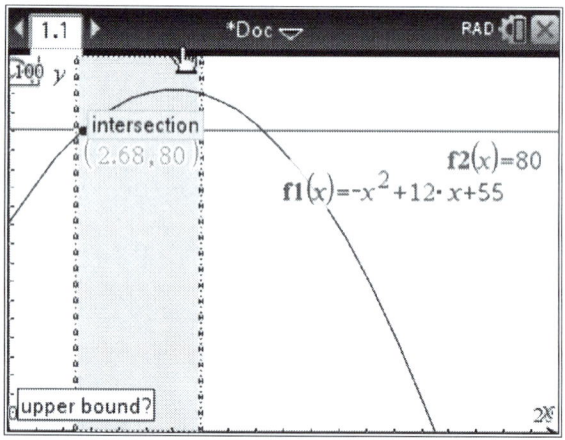

See 7.1 "*Solving a quadratic model graphically*" for more detailed instructions.

So, what is the minimum amount of time John must study to make an 80?	study 2.08 hours
What is the maximum grade that John can make on the test?	Either find the vertex from the equation or use the TI-Nspire. (See below for the TI-Nspire method.) 91 max

==Rather than finding the vertex from the original equation, you can use the TI-Nspire to find the vertex.== find vertex!

First graph the function and adjust the window accordingly. Control-DOC 1: Add Graphs	$f2(x) = -x^2 + 12 \cdot x + 55$
To find the max or min Menu 6: Analyze Graph 2: Minimum or 3: Maximum Select a point on either side of the max/min to have the calculator find the value.	maximum (6, 91) $f2(x) = -x^2 + 12 \cdot x + 55$
So, what is the maximum grade John can make? How long must he study for that grade?	91% study 6 hours

Homework 7.2 Finding Quadratic Models from the TI-Nspire

1. The number of cents per kilometer it costs to drive a car depends on how fast you drive it. At low speeds, the cost is high because the engine operates inefficiently, while at high speeds, the cost is high because the engine must overcome high wind resistance. At moderate speeds, the cost reaches a minimum. Assume that the number of cents per kilometer depends on the number of kilometers per hour (kph) the car is driven.

a. Suppose that it costs 28, 21, and 16 cents per kilometer to drive at 10, 20, and 30 kph, respectively. Write the particular equation for this situation.

$$y = 0.01x^2 - 1x + 37$$

b. How much would you spend to drive 150 kph? $150 = x$

$$y = 0.01(150)^2 - 150 + 37$$
$$y = 112$$

$\boxed{\$112}$

c. Between what two speeds must you drive to keep your cost no more than 13 cents per kilometer?

$$0.01x^2 - x + 37 = 13$$
$$0.01x^2 - x + 37 - 13 = 0$$

$(40, 60) \rightarrow$ between 40 & 60 kmph

d. Is it possible to spend only 10 cents per kilometer? Why or why not?

no, the minimum is $12

$(50, 12)$

e. The least number of cents per kilometer occurs when you get the most kilometers per liter of gas. If your tank were nearly empty, at what speed should you drive to have the best chance of making it to the gas station before being forced to walk?

$(50, 12) =$ vertex

so, to get the lowest cost, drive 50 kph

Homework 7.2 Finding Quadratic Models from the TI-Nspire

2. Suppose you are an actuary for F. Bender's Insurance Agency. Your company plans to offer a senior citizen's accident policy, and you must predict the likelihood of an accident as a function of the driver's age. From previous accident records, you find the following information:

Age in Years	20	30	40
Accidents per 100 million kilometers	440	280	200

You know that the number of accidents per 100 million kilometers driven should reach a minimum then go up again for very old drivers. Therefore, you assume a quadratic model is reasonable.

a. Find the particular equation for accidents as a function of age.

$$0.4x^2 - 36x + 1000$$

b. How many accidents per 100 mil km driven would you expect for an 80-year old driver? Plug in $x = 80$

$$0.4(80)^2 - 36(80) + 1000 \rightarrow 680$$

c. Based on your model, who is safer: a 16-year-old driver or a 70-year-old driver?

Plug in $x = 16$

$$0.4(16)^2 - 36(16) + 1000$$
$$= 526.4$$

$x = 70$

$$0.4(70)^2 - 36(70) + 1000$$
$$= 440$$

70 yr old driver is safer

Homework 7.2 Finding Quadratic Models from the TI-Nspire

CHALLENGE PROBLEMS

1. According the data site Worldometers, the USA population by year is listed below.

Year	Population in millions
2010	308
2015	319
2016	322
2017	324
2018	326

Assuming a quadratic growth model, find an equation $P(t)$, which give the population in millions for the given year. Let $t =$ the number of years AFTER 2010.

What is the estimated population in 2020?

What is the estimated population in 2050? What are your thoughts on the reliability of this number?

7.3. Projectile Motion

In this section we will discuss quadratic models for projectile motions. This involves objects that are projected up or down in the air, and the paths followed until they reach a state of rest.

ALL projectile motion problems use the equation:

$$H(t) = -\frac{1}{2}gt^2 + v_0 t + h_0$$

If you look at the equation, you see it's just a parabola.

$H(t)$ is the height

t is the time

g is the force of gravity (see below),

v_0 is the initial velocity (see below)

h_0 is the initial height.

Gravity (g)

Gravity always pulls the projectile down towards the ground. The value is constant, but it depends on the units.

If the problem is in feet, use $g = 32 \, ft/sec^2$ ☆ constant; on test

If the problem is in meters, use $g = 9.8 \, m/sec^2$

Initial Velocity

- If the object is initially moving UP, then v_0 will be **positive**. ↑ +
- If the object is initially moving DOWN, then v_0 will be **negative**. ↓ −
- If the object is initially stationary (or only moving sideways), then v_0 will be 0.

$H(t) = -\frac{1}{2}gt^2 + v_0 t + h_0$

for feet, $g = 32$ ft/sec

7.3—336

A projectile is fired vertically upward from a height of 600 feet above the ground, with an initial velocity of 803 ft/sec. v_0 — h_0	
Write a quadratic model for the height $H(t)$, t seconds after the projectile is launched.	$H(t) = -\frac{1}{2}gt^2 + v_0 t + h_0$ $\boxed{H(t) = -16t^2 + 803t + 600}$ $g = 32$ ft/sec² $v_0 = 803$ ft/sec $h_0 = 600$ ft
Draw a sketch of the scenario	(sketch showing parabola with y-axis labeled 5000 and 600, starting point at 600)
During what time interval will the projectile be more than 5000 feet above the ground? "During what time" indicates set equation = to 5000	$5000 = -16t^2 + 803t + 600$ $-16t^2 + 803t + 600 - 5000 = 0$ $t = 6.20, 43.93$ between 6.20 and 43.93 secs, it will be higher than 5000 ft.
How long will the projectile be in flight? height = 0	$-16t + 803t + 600 = 0$ $-0.73, \boxed{50.93}$ in flight for 50.93 sec

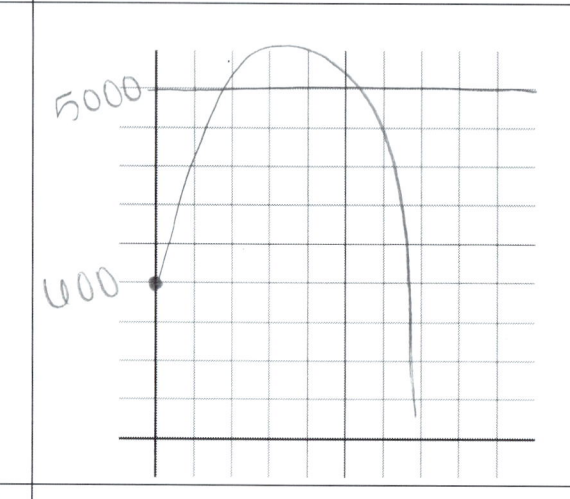

Joshua drops a rock into a well in which the water surface is 300 feet below ground level. How long does it take the rock to hit the water surface?

| Write a quadratic model for the height $H(t)$, t seconds after the projectile is launched. $g = 32$ $V = 0$ $h = 0$ | $H(t) = -\frac{1}{2}gt^2 + V_0 t + h_0$ $H(t) = -\frac{1}{2}(32)t^2 + 0t + 0$ |

Draw a sketch of the scenario

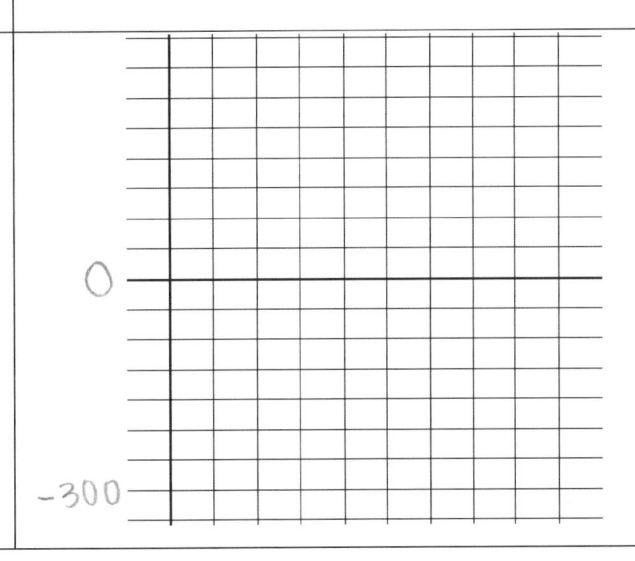

(sketch showing 0 and −300 on a grid)

Solution $H(t) = -300$

$-16t^2 = -300$

$-16t^2 + 300 = 0$

$t = \pm 4.33$

It took 4.33 sec for rock to hit water

7.3—338

A ball is thrown straight up, from the ground, with an initial velocity of 14 meters per second. Find the maximum height attained by the ball and the time it takes for the ball to return to the ground. (h ↑, v ↓)

| Write a quadratic model for the height $H(t)$, t seconds after the ball is launched.

$g = 9.8$
$v = 14$
$h = 0$ | $H(t) = -\frac{1}{2}gt^2 + V_0 t + H_0$

$H(t) = -\frac{1}{2}(9.8)t^2 + 14t + 0$

$H(t) = -4.9t^2 + 14t$ |

| Draw a sketch of the scenario | vertex (max H) — parabola opening down from 0 |

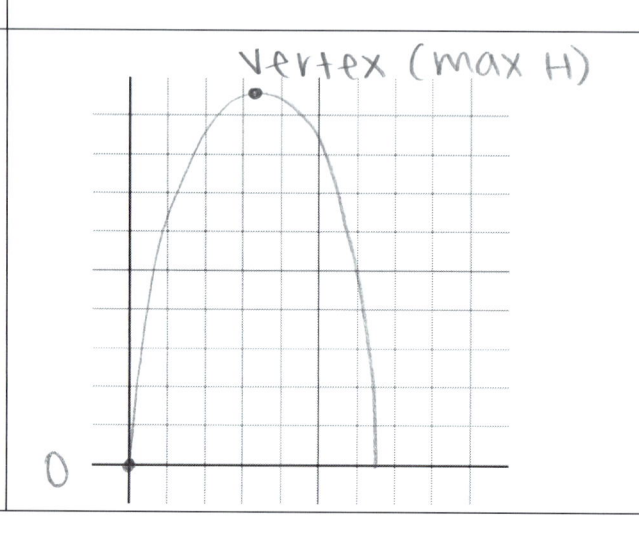

Solution

In 1.43 sec, max height is 10 m

*on the graph → always plug in "x" not any other value

$H = 0$

$-4.9t^2 + 14t = 0$

$t = 0, 2.857$ → In flight for 2.857 sec

A coin is tossed upward from a balcony 166 ft. high with an initial velocity of 16 ft./sec. During what interval of time will the coin be at a height of at least 70 feet?

Write a quadratic model for the height $H(t)$, t seconds after the coin is launched.	$H(t) = -\frac{1}{2}gt^2 + v_0 t + h_0$
$g = 32$ $v = 16$ $h = 166$	$H(t) = -16t^2 + 16t + 166$

Draw a sketch of the scenario

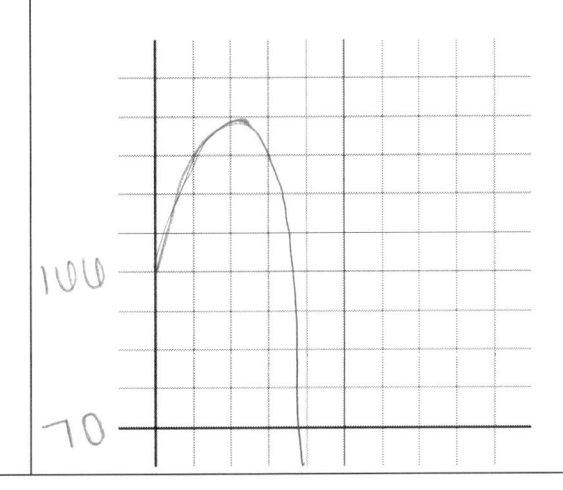

Solution

$H = 70$

$-16t^2 + 16t + 166 = 70$

$-16t^2 + 16t + 166 - 70 = 0$

$t = -2, \boxed{3}$

for 3 sec the coin is above 70 ft.

Remember Day 1 of Algebra 2? We started with this problem and you were spoon-fed some multiple choice answers.

The height of a basketball shot, $H(t)$, is given by the equation
$$H(t) = -7t^2 + 16t + 6$$
where t = time in seconds that the shot is in the air, and H = the height in feet after t seconds. Assume the rim of the basket is 10 feet tall.

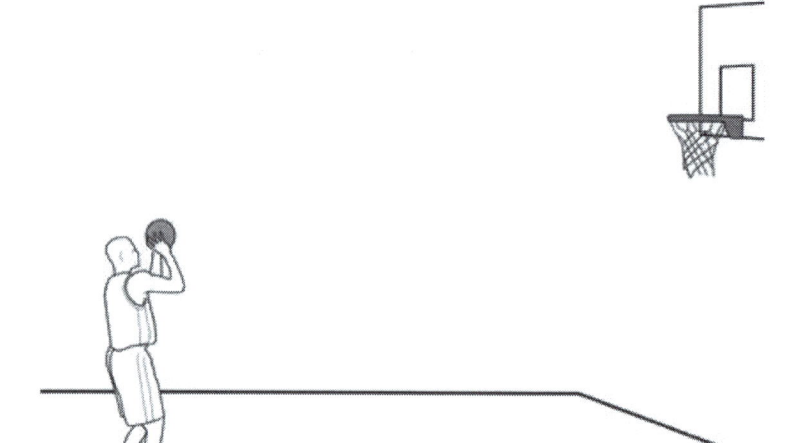

What is the initial height of the basketball? a. 5 feet b. 10 feet c. 6 feet ⓒ d. Not enough information	
What is the approximate maximum height of the basketball? e. 10 feet f. 6 feet g. 15 feet Ⓖ h. 50 feet	What is the exact maximum height? How long into the shot does it reach the max height? find vertex! $-7t^2 + 16t + 6 = 0$ 1.14, 15.1
How much time elapses from the moment the shot is taken until it enters the basket? $-7t^2 + 16t + 6 = 10$ $t = 2 \sec$ $-7t^2 + 16t + 6 - 10 = 0$	

Homework 7.3 Projectile Motion

1. Tommy Terror is in his tree house, shooting stones at the acorns directly above his head. The height in feet, h, of the stone t seconds after it leaves the sling shot can be found using the equation, $H(t) = -16t^2 + 35t + 15$

$H(t) = -\frac{1}{2}gt^2 + V_0 t + H_0$

$g = 32$
$V = 35$
$h = 15$

How high off the ground is the sling shot that Tommy is using?

What is height? $\boxed{15 \text{ ft}}$

How high is the stone 2 seconds after it leaves the sling shot?

$H(2) = -16(2)^2 + 35(2) + 15 = \boxed{21 \text{ ft}}$

What is the highest acorn that Tommy can hit using his current strategy? ★

Imagine that there is a hole in the floor of the tree house. When Tommy shoots the stone, it goes up, and then falls back down through the hole to the ground below. What will the final height of the stone be? How long will the stone be in the air before it lands on the ground?

$-16t^2 + 35t + 15 = 0$

$= \{-0.3669, 2.555\}$

The stone will be in the air for 2.6 seconds before landing.

2. A car rolls off a 120-meter cliff. $V_0 = 0 \quad h_0 = 120 \text{ m}$

Write an equation modeling the motion of the car after t seconds.

$H(t) = -\frac{1}{2}(9.8)t^2 + 120$

$H(t) = -4.9t^2 + 120$

How high will the car be after 3 seconds?

$H(3) = -4.9(3)^2 + 120 \rightarrow \boxed{75.9 \text{ m}}$

How long will it take for the car to hit the ground?

$-4.9t^2 + 120 = 0$

It will take $\boxed{4.94}$ seconds for the car to hit ground.

Homework 7.3 Projectile Motion

3. A ball is thrown upward from a height of 5.3 meters with an initial upward velocity of 21 meters per second.

Write an equation relating time and the height of the ball.

$H(t) = -\frac{1}{2}(9.8)t^2 + 21t + 5.3$

$H(t) = -4.9t^2 + 21t + 5.3$

Sketch a graph of the ball's height vs. time. Use an appropriate domain.

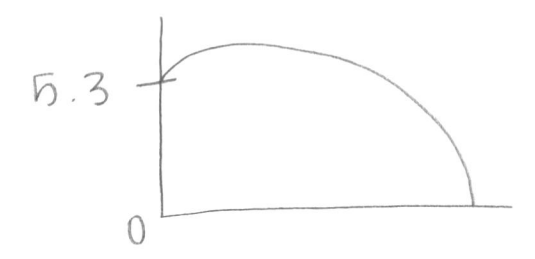

Find the height of the ball after 2 seconds.

$H(2) = -4.9(2)^2 + 21(2) + 5.3$

$= \boxed{27.7 \text{ m}}$

After how many seconds will the ball reach its maximum height?

$\frac{-b}{2a} = \frac{-21}{2(-4.9)} = \boxed{2.14 \text{ sec}}$ find vertex

How long is the ball in the air?

Until $H(t) = 0$

$-4.9t^2 + 21t + 5.3 = 0$

The ball is in air for about 4.52 secs

4. Bungee Barbie is projected with no bungee cords from the bridge between the Moss and Carlos Science building from a height of 30 feet at an initial velocity of 9 feet per second. How long will it take for Barbie to hit the ground?

$H(t) = -16t^2 + 9t + 30$

$-16t^2 + 9t + 30 = 0$

It will take about 1.68 secs for BB to hit ground

Homework 7.3 Projectile Motion

CHALLENGE PROBLEMS

A ball is launched from a height of 25 feet at an initial speed of 50 feet per second. On the way down, at 10 feet there is a spring that catches the ball and launches it again at a speed of 10 meters per second. How much time elapses until the ball comes to rest on the ground?

7.4. Quadratic Modeling with Borders

So far, we've looked at models of quadratic functions where you were given the model and asked questions about it. Additionally, we looked at how to create the model given 3 data points and most recently models of projectile motion.

In this lesson, we will look at applications that involve a border, much like a frame around a photograph.

A 4x6 inch photograph is in a wooden frame of uniform width. The area of the frame's front face is 39 square inches. How wide is the frame?

x = width

Let x represent the frame width. in inches Label the frame to the right in terms of x. Area of photo= $4 \times 6 =$ 24 in² Area of frame= 39 in² Area of frame+photo= $(6+2x)(4+2x)$	(diagram of framed photo with x labels on borders and 4, 6 labels on photo dimensions)
Write and solve the equation	Area photo + Area frame = total area $24 + 39 = (6+2x)(4+2x)$ $24 + 39 = 24 + 12x + 8x + 4x^2$ $39 = 4x^2 + 20x$ $4x^2 + 20x - 39 = 0$ $x = \left(\dfrac{-13}{2}, \dfrac{3}{2}\right)$ width of frame = $\dfrac{3}{2}$ / 1.5 in

7.4—345

In decorating your bedroom, you decide to frame your favorite math poster. The poster is 11x14 inches, and the frame that you use has an area of $150 in^2$. What is the width of the frame?

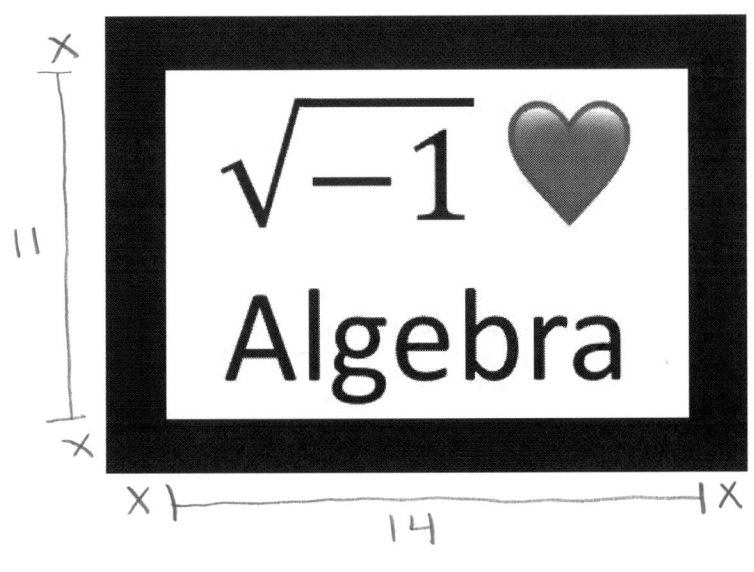

let x = width frame (in)

$11 \times 14 + 150 = (14+2x)(11+2x)$
$154 + 150 = 154 + 28x + 22x + 4x^2$
$150 = 4x^2 + 50x$
$4x^2 + 50x - 150 = 0$
$x = (-15, \frac{5}{2})$

width = $\frac{5}{2}$ in

The outer dimensions of a picture frame are 20x12 cm. The area of the picture that is exposed is 84 cm². What is the width of the frame?

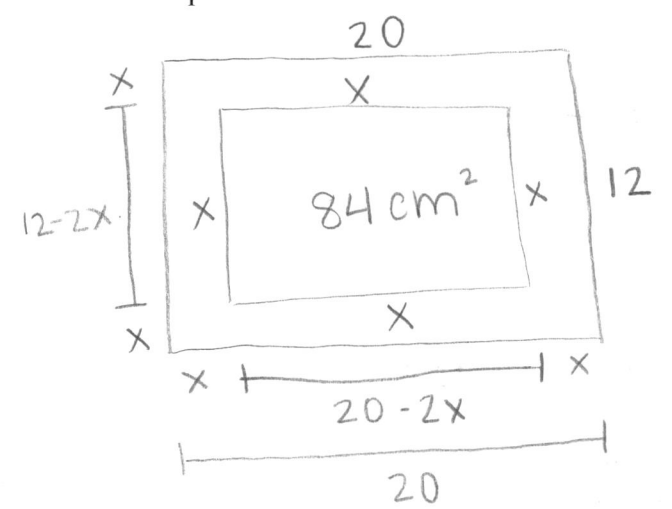

x = width of frame (cm)

$(12-2x)(20-2x) = 84$
$240 - 24x - 40x + 4x^2 = 84$
$4x^2 - 64x + 156 = 0$
$x = (3, 13)$

width = 3 cm

★ negative # answers DO NOT make sense
★ If one answer is larger than frame dimensions, does NOT work

Homework 7.4 Quadratic Modeling with Borders

1. A poster whose dimensions are 25 by 30 inches is surrounded by a border of uniform width. If the area of the border is 174 square inches, find the width of the border.

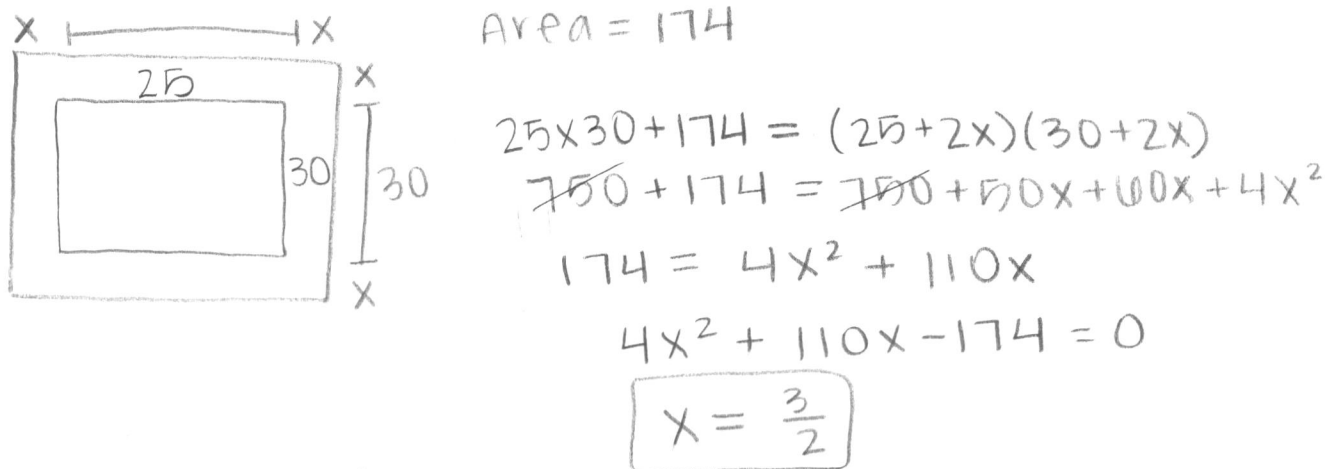

Area = 174

$$25 \times 30 + 174 = (25+2x)(30+2x)$$
$$750 + 174 = 750 + 50x + 60x + 4x^2$$
$$174 = 4x^2 + 110x$$
$$4x^2 + 110x - 174 = 0$$
$$\boxed{x = \frac{3}{2}}$$

2. On a $24 \times 18 \ cm^2$ page, the printed part is surrounded by a uniform margin. How wide is the margin if the printed part has an area of $160 \ cm^2$?

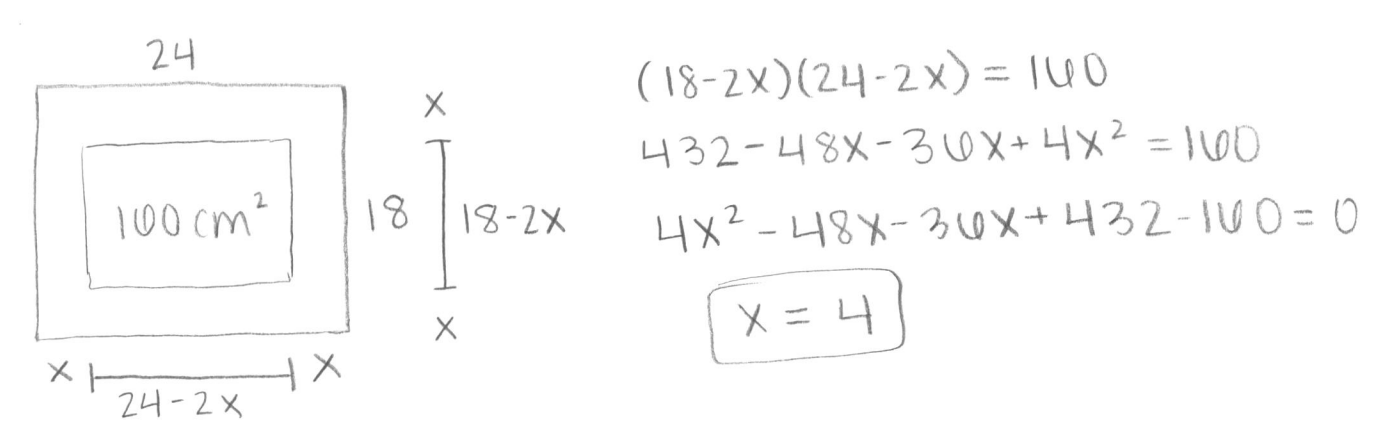

$$(18-2x)(24-2x) = 160$$
$$432 - 48x - 36x + 4x^2 = 160$$
$$4x^2 - 48x - 36x + 432 - 160 = 0$$
$$\boxed{x = 4}$$

3. You want to expand your 24 ft x 16 ft garden by planting a border of flowers. The border will have the same width around the entire garden.

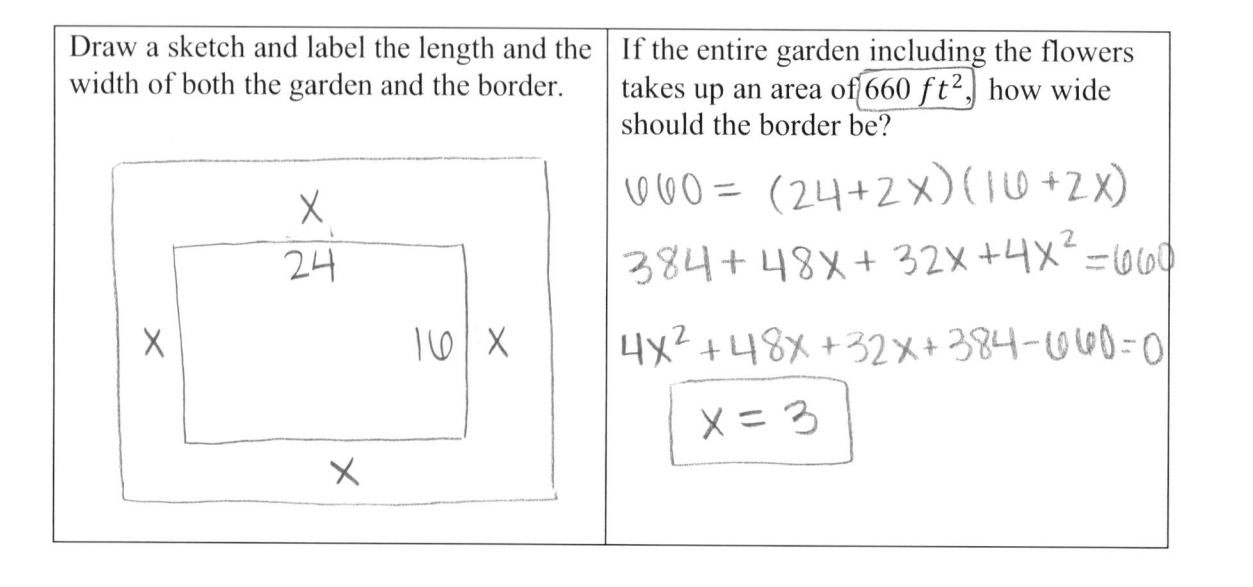

Draw a sketch and label the length and the width of both the garden and the border.	If the entire garden including the flowers takes up an area of $660 \ ft^2$, how wide should the border be?
	$660 = (24+2x)(16+2x)$ $384 + 48x + 32x + 4x^2 = 660$ $4x^2 + 48x + 32x + 384 - 660 = 0$ $\boxed{x = 3}$

Homework 7.4 Quadratic Modeling with Borders

4. A photographer is looking at frames for one of her favorite pictures. The picture is 11 in by 14 in. The frame will have the same width all the way around the picture.

a. Draw a sketch and label.

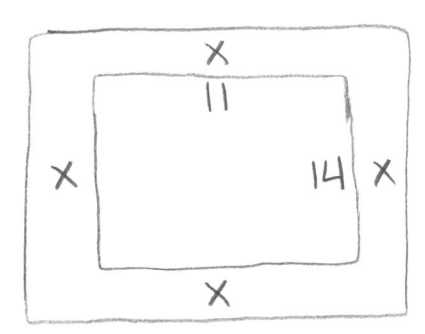

[Sketch: outer rectangle (frame) around inner rectangle (picture 11 by 14); frame width labeled x on all four sides.]

b. If she has 304 square inches of space for the framed picture on his wall, what is the width of the frame, and what are the outer dimensions of the frame?

$$(11+2x)(14+2x) = 304$$

$$154 + 22x + 28x + 4x^2 = 304$$

$$4x^2 + 22x + 28x + 154 - 304 = 0$$

$$\boxed{x = \frac{5}{2}}$$

Homework 7.4 Quadratic Modeling with Borders

CHALLENGE PROBLEMS

1. To protect your home, you have decided to create a border of broken shards of glass immediately surrounding your house, and this border surrounded by yet another rectangular border of burning coal. Your home is 100 by 80 feet. The border of glass will be 1504 square feet, and the burning coal border will be 2060 square feet. Find the width of each border.

7.5. Quadratic Modeling with fencing

Very similar to the border problems, we will visit some quadratic models with fencing. The most noticeable difference from borders, is that the fencing problems typically deal with the amount of material, or perimeter, whereas border problems deal with the width of the border.

Mr. Sherwood is going to fence in part of his yard, using the back of the house as one side of the fenced in area. If he has 500 feet of fencing to use, what dimensions would create the largest possible enclosed area?

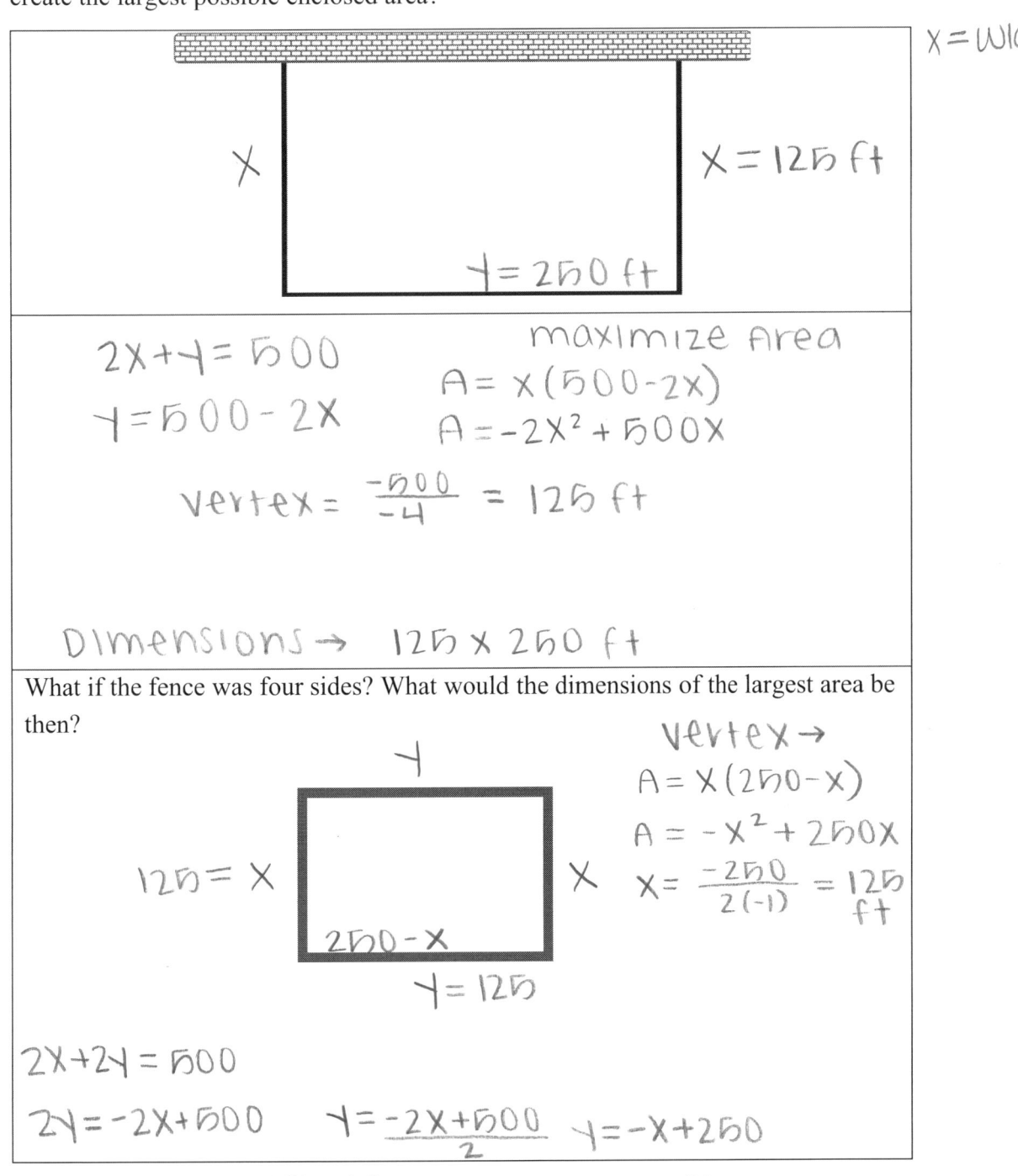

x = width (ft)

x $x = 125$ ft

$y = 250$ ft

$2x + y = 500$
$y = 500 - 2x$

maximize Area
$A = x(500 - 2x)$
$A = -2x^2 + 500x$

vertex $= \dfrac{-500}{-4} = 125$ ft

Dimensions → 125×250 ft

What if the fence was four sides? What would the dimensions of the largest area be then?

$125 = x$ y x
$250 - x$
$y = 125$

vertex →
$A = x(250 - x)$
$A = -x^2 + 250x$
$x = \dfrac{-250}{2(-1)} = 125$ ft

$2x + 2y = 500$
$2y = -2x + 500$ $y = \dfrac{-2x + 500}{2}$ $y = -x + 250$

Dimensions → 125×125 ft

A rectangular patio is surrounded on three sides by a fence (the remaining side is up against the house). If the area of the patio is $150 ft^2$, and the total length of fence is $35 ft.$, what is the length and width of the patio?

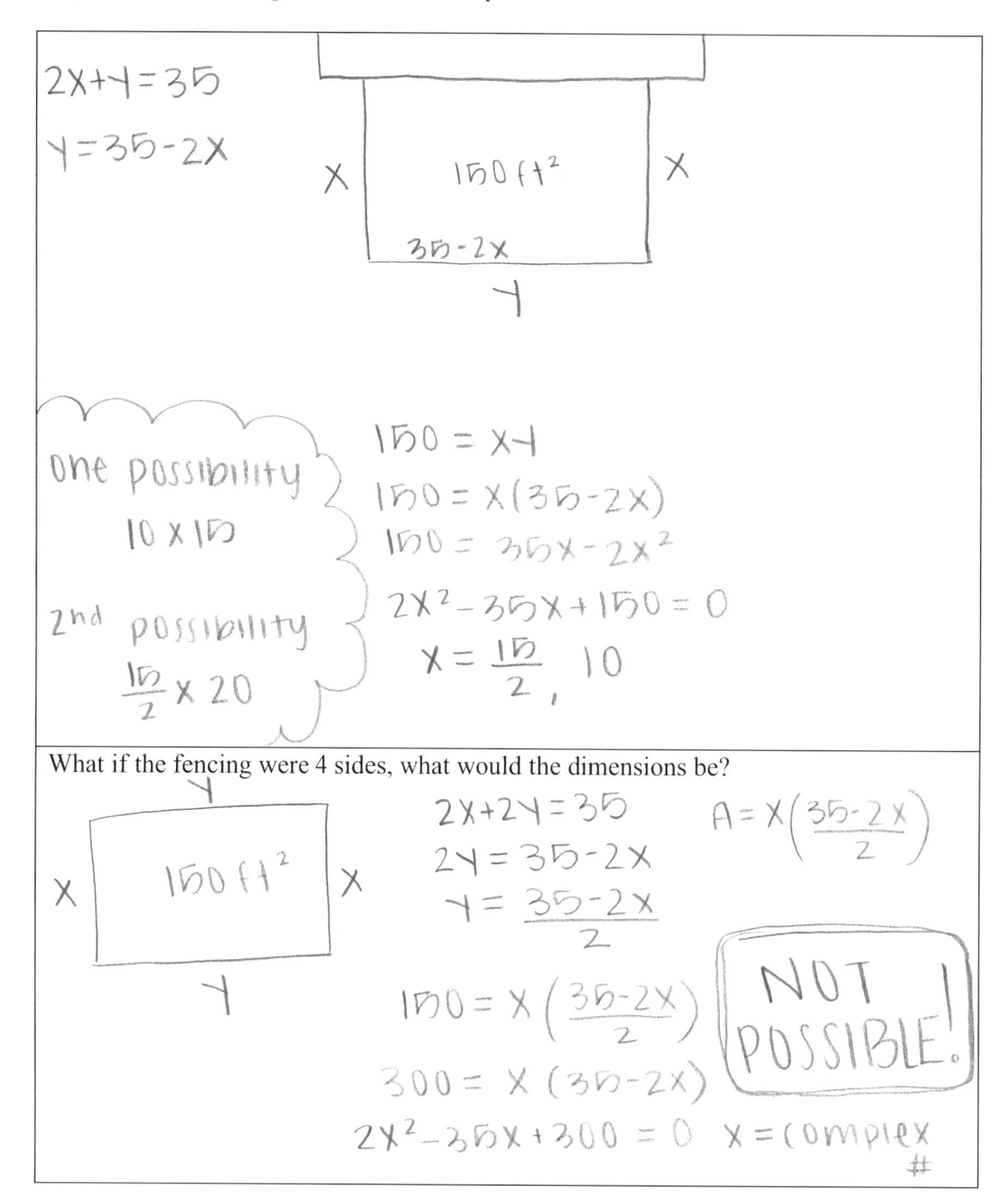

$2x + y = 35$

$y = 35 - 2x$

[Diagram: rectangle against house with sides x, x, and bottom $35-2x$ labeled y, area $150 ft^2$]

One possibility
10×15

2nd possibility
$\frac{15}{2} \times 20$

$150 = xy$

$150 = x(35 - 2x)$

$150 = 35x - 2x^2$

$2x^2 - 35x + 150 = 0$

$x = \frac{15}{2}, 10$

What if the fencing were 4 sides, what would the dimensions be?

[Diagram: rectangle with sides x, x, y, y, area $150 ft^2$]

$2x + 2y = 35$

$2y = 35 - 2x$

$y = \frac{35 - 2x}{2}$

$A = x\left(\frac{35 - 2x}{2}\right)$

$150 = x\left(\frac{35 - 2x}{2}\right)$

$300 = x(35 - 2x)$

$2x^2 - 35x + 300 = 0$ $x = $ complex

NOT POSSIBLE!

A farmer has 1500 feet of fence with which to fence a rectangular plot of land that borders a river. Find the largest **area** that can be fenced.

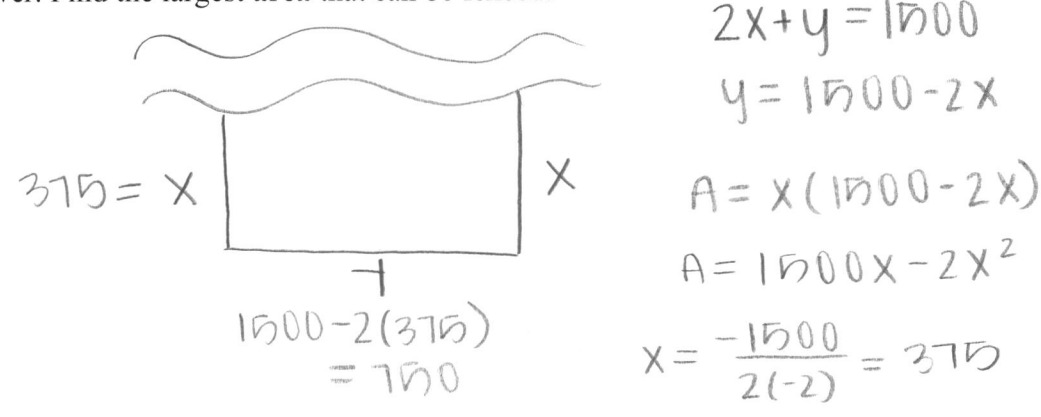

$2x + y = 1500$

$y = 1500 - 2x$

$375 = x$ ... x

$1500 - 2(375) = 750$

$A = x(1500 - 2x)$

$A = 1500x - 2x^2$

$x = \dfrac{-1500}{2(-2)} = 375$

DIMENSIONS → 375×750 ft

If 400 m of fencing encloses a rectangular field having area 8000 m², find the dimensions of the field.

$y = 200 - x$

x | 8000 m² | x

$y = 200 - x$

$2x + 2y = 400$

$y = \dfrac{400 - 2x}{2}$

$y = 200 - x$

Area = 8000

$8000 = x(200 - x)$

$8000 = 200x - x^2$

$x^2 - 200x + 8000 = 0$

$x = 55.3, \ 144.7$

$x = 55.28$

$y = 200 - 55.28$

$= 144.72$

one choice = 55.28×144.72
two choice = 144.72×55.28 } same!

Homework 7.5 Quadratic modeling with fencing

1. A rancher has 900 feet of fencing to enclose a corral that borders a river. He wants to make sure that he builds a corral that will maximize the area for his horses to roam.

Draw a picture of the situation and write an equation representing the area of the field.	What dimensions would maximize the area of the corral?
$225 = X$ [diagram: river above rectangle labeled 900 ft², sides X, bottom unlabeled] $900 - 2(225) = Y$ $= 450$	$2X + Y = 900$ $X = \dfrac{-900}{2(-2)}$ $Y = 900 - 2X$ $X = 225$ $A = X(900 - 2X)$ $A = 900X - 2X^2$ DIMENSIONS → 225×450 ft

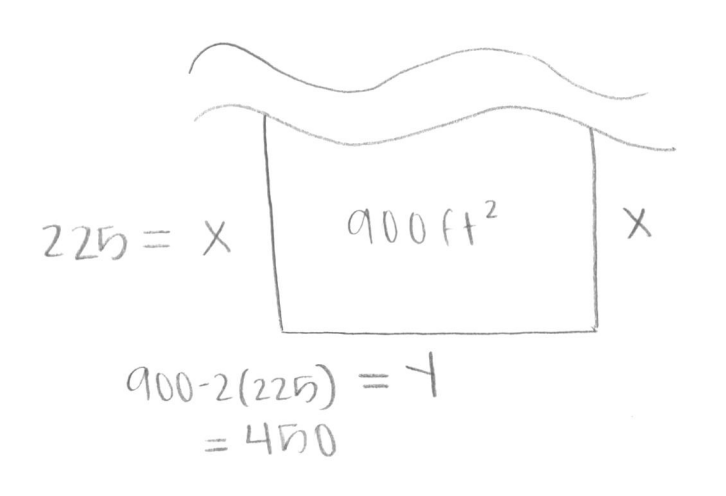

2. The perimeter of the iPhoneX is about 17 inches You have proposed an iPhoneXIV, that has a perimeter of 18 inches. What would the dimensions of the phone be if your goal is to maximize the area?

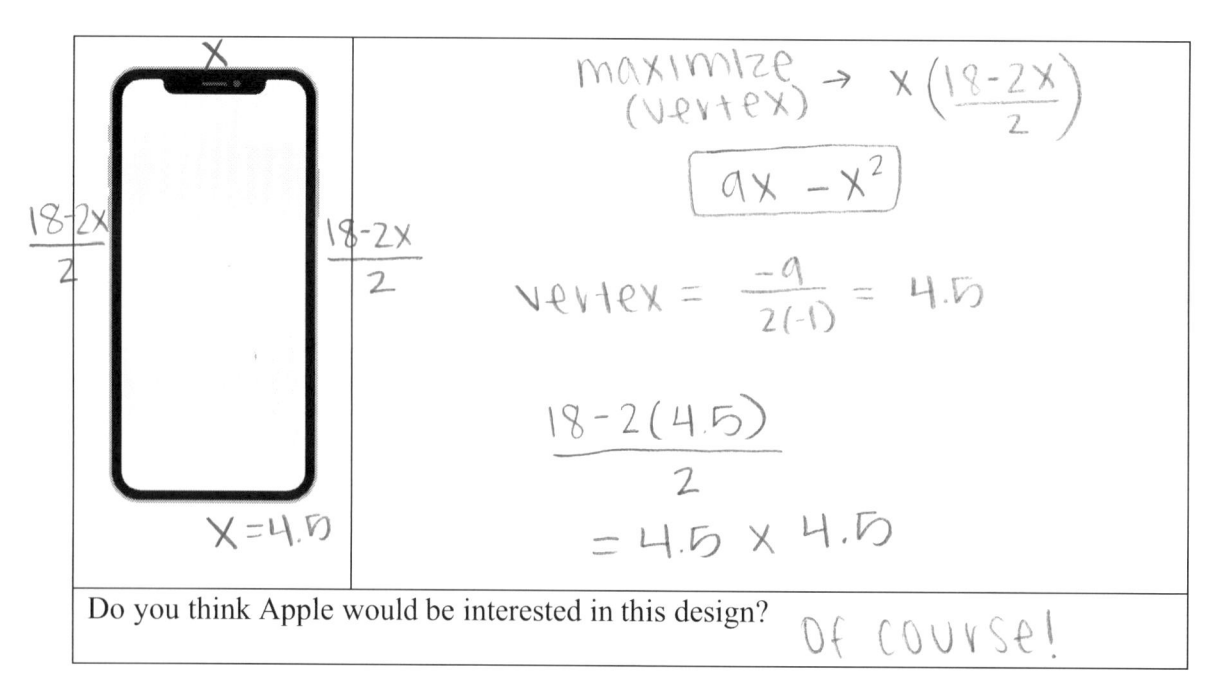

| [diagram of phone with top labeled X, sides labeled $\dfrac{18-2X}{2}$, $X = 4.5$] | maximize (vertex) → $X\left(\dfrac{18-2X}{2}\right)$

 $\boxed{9X - X^2}$

 vertex $= \dfrac{-9}{2(-1)} = 4.5$

 $\dfrac{18 - 2(4.5)}{2}$

 $= 4.5 \times 4.5$ |
| Do you think Apple would be interested in this design? || Of course! |

Homework 7.5 Quadratic modeling with fencing

3. If $500m$ of fencing encloses a rectangular field having area $10,000m^2$, find the dimensions of the field.

$2x + 2y = 500$ $\qquad A = 10000$

$y = \dfrac{500 - 2x}{2}$

$x \;\begin{array}{|c|}\hline 100000 \\ m^2 \\ \hline\end{array}\; x$

$y = \dfrac{500-2x}{2}$

$10000 = x\left(\dfrac{500-2x}{2}\right) = 10000 = x(250-x)$

$10000 = 250x - x^2 = \boxed{(50 \times 200)}$

4. A homeowner has just enough money to purchase 200 feet of fencing for her back yard. She wants to use these 200 feet to enclose the greatest possible area by building a fence with 3 sides.

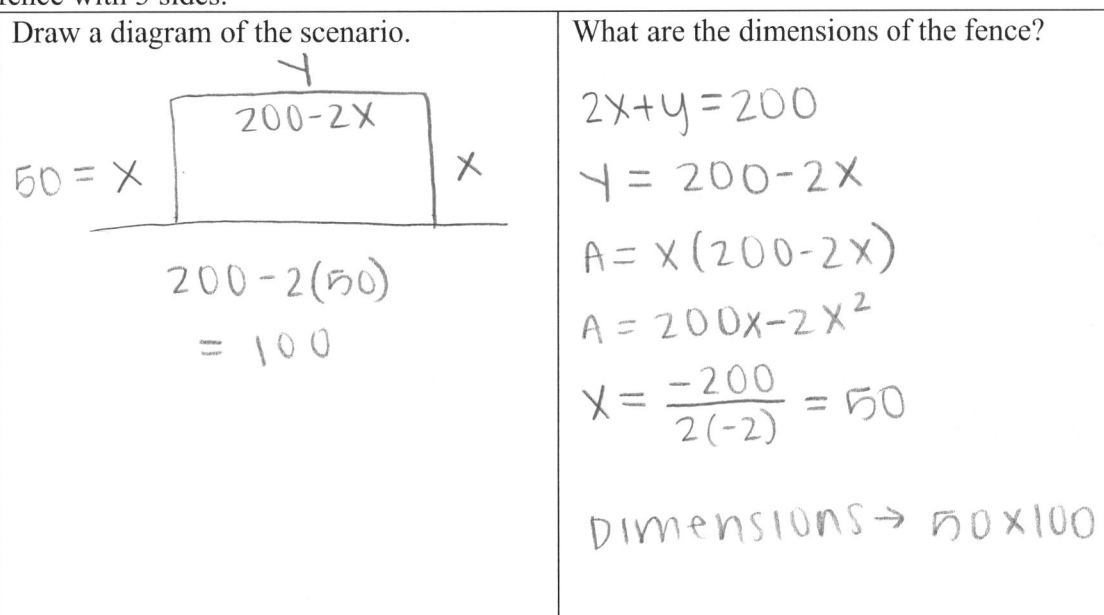

Draw a diagram of the scenario.	What are the dimensions of the fence?
$50 = x \;\begin{array}{\|c\|}\hline 200-2x \\ \hline\end{array}\; x$ $\qquad y$ $200 - 2(50)$ $= 100$	$2x + y = 200$ $y = 200 - 2x$ $A = x(200 - 2x)$ $A = 200x - 2x^2$ $x = \dfrac{-200}{2(-2)} = 50$ Dimensions → 50×100

What if the fence were 4 sides? What would the dimensions be?

$x \;\begin{array}{|c|}\hline \\ \\ \hline\end{array}\; x$ $\qquad y$ on top, $100-x$ on bottom

$2x + 2y = 200$

$y = \dfrac{200 - 2x}{2}$

$y = 100 - x$

$\dfrac{-b}{2a} \quad \dfrac{-100}{-2} = 50 \times 100 - 50$ OR $\boxed{50 \times 50}$

Homework 7.5 Quadratic modeling with fencing

CHALLENGE PROBLEMS

1. You become a parent and decide to create a fence in your back yard for your kids. Your original plan was to use the house as one side of the enclosed area, and then fence in the back yard on 3 sides. But your 4 kids will not play peacefully with each other, so you must now subdivide the backyard into 4 fenced areas. If you only have 300 feet of fencing, what is the largest area you can fence in? What are the dimensions of each kid's fenced in area?

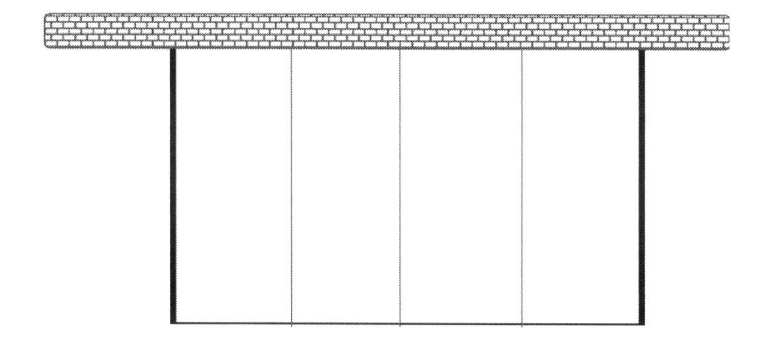

7.6. Review

Assessment Checklist. Below are the competencies one should master in preparation for an assessment on quadratic modeling.

- [ ] Understand the role of quadratics in modeling, ==the role of the vertex,== how to know if it represents ==a maximum or minimum== vertex = max/min
- [ ] Quadratic models when you are given the model or equation. Be able to answer questions based on the given model.
- [ ] Given 3 data points, use the TI-Nspire to find the exact equation of the quadratic model.
- [ ] Use the TI-Nspire to find the max/min graphically. (or vertex)
- [ ] Use the TI-Nspire to solve or find the roots of a quadratic model. This can be done graphically or by using cPolyRoots.
- [ ] Quadratic modeling with projectile motion, $H(t) = -\frac{1}{2}gt^2 + v_0 t + h_0$
- [ ] Quadratic modeling with borders.
- [ ] Quadratic modeling with fencing.
- [ ] In most all of these problems, the ability to draw the scenario accurately and represent the drawing in an equation is key.

Review 7.6 Quadratic Modeling

1. You have been given a chance to join the Woodward basketball team. The coach only asks that you answer the two questions below. Assume the path of the ball on a jump shot can be modeled by the equation $H(x) = -16x^2 + 20x + 6$, where x is the amount of time the ball is in the air in seconds, and H is the height in feet.

| What is the maximum height that the ball reaches on its way to the basket? What is the vertex? $$x = \frac{-20}{2(-16)} = 0.625$$ | The basketball rim is 10 feet tall. How long does it take for the shot to enter the basket? $$-16x^2 + 20x + 6 = 10$$ $$-16x^2 + 20x - 4 = 0$$ $$x = \tfrac{1}{4}, 1$$ takes 1 sec to enter the basket |

2. The data below shows the number of mountain bike owners in millions in the US. The data is for 2010, 2016, and 2017. Let t = the number of years after 2010. ($t = 0$ represents the year 2010).

t (years after 2010)	# of mtn bike owners in millions
0	3.962
6	2.504
7	4.62

| Find an equation B(t) that expresses the # of bike owners in millions. $$B(t) = 0.337x^2 - 2.3x + 3.96$$ | In what year was the ownership of mountain bikes at a minimum? find vertex → $$\frac{-(-2.3)}{2(0.337)} = 3.36$$ 2010 + 3.36 In 2013 there was a min bikes |

Review 7.6 Quadratic Modeling

3. Six Flags is experimenting with a new way to run Acrophobia. Instead of slowly lifting to the top and dropping you, it will start at the bottom and shoot you up in the air before free-falling back to the ground. The ride will start at a platform ==5 meters above== = h_0 ==the ground== and it will launch you upwards at a ==speed of 29.4 meters per second.== (Use $g = 9.8 \ m/sec^2$) V_0

Sketch a graph of the problem. (label the axes and denote the starting/stopping points of the ride)	What will your height be 2 seconds after the ride begins?
$H(t) = -\frac{1}{2}gt^2 + V_0 t + h_0$ $H(t) = -\frac{1}{2}(9.8)t^2 + 29.4t + 5$ $H(t) = -4.9t^2 + 29.4t + 5$	$H(t) = -4.9(2)^2 + 29.4(2) + 5$ $H(t) = 44.2 \ m$

How much time will pass before you land on the ground?

$H(t) = 0 \qquad -4.9t^2 + 29.4t + 5 = 0$

$H(t) = -0.165, \boxed{6.17 \ sec}$

4. Chris Freer would like a wooden, rectangular fence to keep the riff raff (mostly freshmen, juniors and seniors) off his front lawn. The fence will be 3 sides, with the front of the house serving as one side of the enclosed area. If you only have 84 feet of fencing, what are the dimensions of the fence providing the maximum area of seclusion?

x | 84 ft | x
y

$2x + y = 84$
$y = 84 - 2x$

$84 = xy$

$x(84 - 2x)$
$= 84x - 2x^2$

$21 \times 84 - 2(21)$
$= 42$

$\frac{-b}{2a} = \frac{-84}{2(-2)} = 21$

Dimensions → 21 × 42

Review 7.6 Quadratic Modeling

5. Woodward Academy wants to grow more of its own food for the cafeteria. Currently, the lettuce garden is 60 feet by 40 feet. By planting a border of even width around the existing garden, the school will end up with a garden **twice** the size of the existing garden.

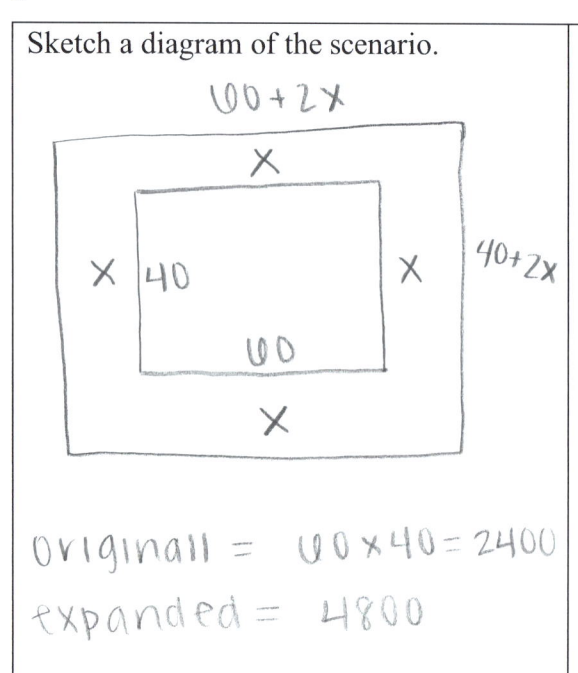

Sketch a diagram of the scenario.	Find a quadratic model for the area of the expanded garden that can be used to solve for the width of the expansion. (Leave your answer in Standard Form.)
Outer rectangle: $60+2x$ by $40+2x$; inner rectangle: 60 by 40; border width x on all sides. original = $60 \times 40 = 2400$ expanded = 4800	$x = $ width $(60+2x)(40+2x) = 4800$ $2400 + 80x + 120x + 4x^2 = 4800$ $4x^2 + 200x + 2400 - 4800$ $x = -60, 10$ **10 ft border**
What is the width of the expansion? 10 ft $x = $ width $x = 10$ ft	What are the new dimensions of the expanded garden? $60 + 2(10) = 80$ $40 + 2(10) = 60$ 80×60 ft

7.7. Algebra 2 Grades

Term 1 Grades

Assessment	Date	Assessment Grade	Current Grade in Class
Term 1 Grade:			

Term 2 Grades

Assessment	Date	Assessment Grade	Current Grade in Class
Term 2 Grade:			

Made in the USA
Columbia, SC
01 August 2019